AF905301

Non-Renewable Resource Issues

International Year of Planet Earth

Series Editors:

Eduardo F.J. de Mulder
Executive Director International Secretariat
International Year of Planet Earth

Edward Derbyshire
Goodwill Ambassador
International Year of Planet Earth

The book series is dedicated to the United Nations International Year of Planet Earth. The aim of the Year is to raise worldwide public and political awareness of the vast (but often under-used) potential of Earth sciences for improving the quality of life and safeguarding the planet. Geoscientific knowledge can save lives and protect property if threatened by natural disasters. Such knowledge is also needed to sustainably satisfy the growing need for Earth's resources by more people. Earth scientists are ready to contribute to a safer, healthier and more prosperous society. IYPE aims to develop a new generation of such experts to find new resources and to develop land more sustainably.

For further volumes:
http://www.springer.com/series/8096

Richard Sinding-Larsen
Friedrich-W. Wellmer
Editors

Non-Renewable Resource Issues

Geoscientific and Societal Challenges

Springer

Editors
Richard Sinding-Larsen
Department of Geology and Mineral Resources Engineering
Norwegian University of Science and Technology
Trondheim, Norway

Friedrich-W. Wellmer
Formerly Bundesanstalt für Geowissenschaften und Rohstoffe (BGR)
Neue Sachlichkeit 32, Hannover
Germany

ISBN 978-90-481-8678-5 ISBN 978-90-481-8679-2 (eBook)
DOI 10.1007/978-90-481-8679-2
Springer Dordrecht Heidelberg London New York

Library of Congress Control Number: 2012933152

All Rights Reserved for Chapter 12
© Springer Science+Business Media B.V. 2012
This work is subject to copyright. All rights are reserved by the Publisher, whether the whole or part of the material is concerned, specifically the rights of translation, reprinting, reuse of illustrations, recitation, broadcasting, reproduction on microfilms or in any other physical way, and transmission or information storage and retrieval, electronic adaptation, computer software, or by similar or dissimilar methodology now known or hereafter developed. Exempted from this legal reservation are brief excerpts in connection with reviews or scholarly analysis or material supplied specifically for the purpose of being entered and executed on a computer system, for exclusive use by the purchaser of the work. Duplication of this publication or parts thereof is permitted only under the provisions of the Copyright Law of the Publisher's location, in its current version, and permission for use must always be obtained from Springer. Permissions for use may be obtained through RightsLink at the Copyright Clearance Center. Violations are liable to prosecution under the respective Copyright Law.
The use of general descriptive names, registered names, trademarks, service marks, etc. in this publication does not imply, even in the absence of a specific statement, that such names are exempt from the relevant protective laws and regulations and therefore free for general use.
While the advice and information in this book are believed to be true and accurate at the date of publication, neither the authors nor the editors nor the publisher can accept any legal responsibility for any errors or omissions that may be made. The publisher makes no warranty, express or implied, with respect to the material contained herein.

Cover design: WMXDesign GmbH, Heidelberg, Germany.

Printed on acid-free paper

Springer is part of Springer Science+Business Media (www.springer.com)

Foreword

The International Year of Planet Earth (IYPE) was established as a means of raising worldwide public and political awareness of the vast, though frequently under-used, potential the earth sciences possess for improving the quality of life of the peoples of the world and safeguarding Earth's rich and diverse environments.

The International Year project was jointly initiated in 2000 by the International Union of Geological Sciences (IUGS) and the Earth Science Division of the United Nations Educational, Scientific and Cultural Organization (UNESCO). IUGS, which is a non-governmental organization, and UNESCO, an inter-governmental organization, already shared a long record of productive cooperation in the natural sciences and their application to societal problems, including the International Geoscience Programme (IGCP) now in its fourth decade.

With its main goals of raising public awareness of and enhancing research in the earth sciences on a global scale in both the developed and less-developed countries of the world, two operational programmes were demanded. In 2002 and 2003, the series editors together with Dr. Ted Nield and Dr. Henk Schalke (all four being core members of the Management Team at that time) drew up outlines of a science and an outreach programme. In 2005, following the UN proclamation of 2008 as the United Nations International Year of Planet Earth, the "year" grew into a triennium (2007–2009).

The outreach programme, targeting all levels of human society from decision makers to the general public, achieved considerable success in the hands of member states representing over 80% of the global population. The science programme concentrated on bringing together like-minded scientists from around the world to advance collaborative science in a number of areas of global concern. A strong emphasis on enhancing the role of the earth sciences in building a healthier, safer and wealthier society was adopted – as declared in the Year's logo strap-line "Earth Sciences *for* Society".

The organizational approach adopted by the science programme involved recognition of ten global themes that embrace a broad range of problems of widespread national and international concern, as follows:

- Human health: this theme involves improving understanding of the processes by which geological materials affect human health as a means identifying and reducing a range of pathological effects.

- Climate: particularly emphasizes improved detail and understanding of the nonhuman factor in climate change.
- Groundwater: considers the occurrence, quantity and quality of this vital resource for all living things against a background that includes potential political tension between competing neighbour nations.
- Ocean: aims to improve understanding of the processes and environment of the ocean floors with relevance to the history of planet Earth and the potential for improved understanding of life and resources.
- Soils: this thin "skin" on Earth's surface is the vital source of nutrients that sustain life on the world's landmasses, but this living skin is vulnerable to degradation if not used wisely. This theme emphasizes greater use of soil science information in the selection, use and ensuring sustainability of agricultural soils so as to enhance production and diminish soil loss.
- Deep Earth: in view of the fundamental importance of the deep Earth in supplying basic needs, including mitigating the impact of certain natural hazards and controlling environmental degradation, this theme concentrates on developing scientific models that assist in the reconstruction of past processes and the forecasting of future processes that take place in the solid Earth.
- Megacities: this theme is concerned with means of building safer structures and expanding urban areas, including utilization of subsurface space.
- Geohazards: aims to reduce the risks posed to human communities by both natural and human-induced hazards using current knowledge and new information derived from research.
- Resources: involves advancing our knowledge of Earth's natural resources and their sustainable extraction.
- Earth and Life: it is over 2.5 billion years since the first effects of life began to affect Earth's atmosphere, oceans and landmasses. Earth's biological "cloak," known as the biosphere, makes our planet unique but it needs to be better known and protected. This theme aims to advance understanding of the dynamic processes of the biosphere and to use that understanding to help keep this global life-support system in good health for the benefit of all living things.

The first task of the leading earth scientists appointed as theme leaders was the production of a set of theme brochures. Some 3,500 of these were published, initially in English only but later translated into Portuguese, Chinese, Hungarian, Vietnamese, Italian, Spanish, Turkish, Lithuanian, Polish, Arabic, Japanese and Greek. Most of these were published in hard copy and all are listed on the IYPE website.

It is fitting that, as the International Year's triennium terminates at the end of 2009, the more than 100 scientists who participated in the ten science themes should bring together the results of their wide ranging international deliberations in a series of state-of-the-art volumes that will stand as a legacy of the International Year of Planet Earth. The book series was a direct result of interaction between the International Year and the Springer-Verlag company, a partnership which was formalized in 2008 during the acme of the triennium.

This IYPE-Springer book series contains the latest thinking on the chosen themes by a large number of earth science professionals from around the world. The books are written at the advanced level demanded by a potential readership consisting of earth science professionals and students. Thus, the series is a legacy of the science programme, but it is also a counterweight to the earth science information in several media formats already delivered by the numerous national committees of the International Year in their pursuit of worldwide popularization under the outreach programme.

The discerning reader will recognize that the books in this series provide not only a comprehensive account of the individual themes but also share much common ground that makes the series greater than the sum of the individual volumes. It is to be hoped that the scientific perspective thus provided will enhance the reader's appreciation of the nature and scale of earth science as well as the guidance it can offer to governments, decision makers and others seeking solutions to national and global problems, thereby improving everyday life for present and future residents of planet Earth.

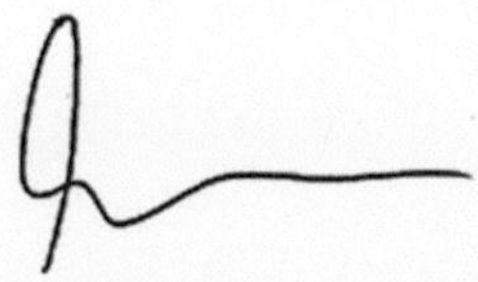

Eduardo F.J. de Mulder
Executive Director International Secretariat
International Year of Planet Earth

Edward Derbyshire
Goodwill Ambassador
International Year of Planet Earth

Series Preface

This book series is one of the many important results of the International Year of Planet Earth (IYPE), a joint initiative of UNESCO and the International Union of Geological Sciences (IUGS), launched with the aim of ensuring greater and more effective use by society of the knowledge and skills provided by the earth sciences.

It was originally intended that the IYPE would run from the beginning of 2007 until the end of 2009, with the core year of the triennium (2008) being proclaimed as a UN Year by the United Nations General Assembly. During all 3 years, a series of activities included in the IYPE's science and outreach programmes had a strong mobilizing effect around the globe, not only among earth scientists but also within the general public and, especially, among children and young people.

The outreach programme has served to enhance cooperation among earth scientists, administrators, politicians and civil society and to generate public awareness of the wide ranging importance of the geosciences for human life and prosperity. It has also helped to develop a better understanding of Planet Earth and the importance of this knowledge in building of a safer, healthier, and wealthier society.

The scientific programme, focused upon ten themes of relevance to society, has successfully raised geoscientists' awareness of the need to develop further the international coordination of their activities. The programme has also led to some important updating of the main challenges the geosciences are, and will be confronting within an agenda closely focused on societal benefit.

An important outcome of the work of the IYPE's scientific themes includes this thematic book as one of the volumes making up the IYPE-Springer Series, which was designed to provide an important element of the legacy of the International Year of Planet Earth. Many prestigious scientists, drawn from different disciplines and with a wide range of nationalities, are warmly thanked for their contributions to a series of books that epitomize the most advanced, up-to-date and useful information on evolution and life, water resources, soils, changing climate, deep earth, oceans, non-renewable resources, earth and health, natural hazards, and megacities.

This legacy opens a bridge to the future. It is published in the hope that the core message and the concerted actions of the International Year of Planet Earth throughout the triennium will continue and, ultimately, go some way toward helping to establish an improved equilibrium between human society and its home planet. As stated by the Director General of UNESCO, Koichiro Matsuura, "Our knowledge of the Earth system is our insurance policy for the future of our planet." This book series is an important step in that direction.

R. Missotten
Chief, Global Earth Observation Section
UNESCO

Alberto C. Riccardi
President
IUGS

Preface

This volume contains papers solicited by the editors to address what we believe represent important issues in the perception of geoscientific and societal challenges related to non-renewable resources. This volume attempts to document some of what has been learnt at the end of the first decade of the twenty-first century. The topics covered in this volume address policy issues; primary resources, secondary resources, socioeconomic aspects; and a last chapter which attempts to tie everything together.

The first three chapters deal with society's need for raw materials and related sustainability and policy issues. These are followed by three chapters focusing on whether depletion is likely to create significant future resource scarcities exemplified by petroleum energy and the resource status for coal, uranium and thorium. The subsequent three chapters address different resource aspects from the geological prerequisites for rare earth elements, to how metal markets can be assessed focusing on Zn, concluding with thoughts on the requirements that are needed for stretching the availability of natural resources. The two following chapters cover, respectively, the growing importance of secondary raw material sources exemplified by precious and special metals and how responsibility in mineral production can be improved through certified trading chains. The concluding chapter addresses challenges related to how raw materials efficiency can be improved through life cycle assessment.

Acknowledgements

The editors wish to express their thanks to all the authors and co-authors of these chapters that have made the effort to contribute to this United Nations International Year of Planet Earth effort within non-renewable resources. The editors are also grateful to the Norwegian University of Science and Technology for allowing time and opportunity to take part in preparing this material for publication.

Richard Sinding-Larsen
Friedrich-W. Wellmer

Contents

Contributors

Roberto F. Aguilera Centre for Research in Energy and Minerals Economics (CREME), Curtin University, Perth, Australia

Peter Buchholz Economic Geology of Mineral Resources, Federal Institute for Geosciences and Natural Resources (BGR), Stilleweg 2, Hannover, 30655, Germany

Daniel Cordier U.S. Geological Survey, Reston, VA, USA

Michel Cuney UMR G2R 7566 CNRS – CREGU, Nancy Université, BP 70239 – 54 506, Vandoeuvre Cedex, France

Luca Demicheli EuroGeoSurveys – The Geological Surveys of Europe, Brussels, Belgium

Ulrike Dorner Bundesanstalt für Geowissenschaften und Rohstoffe, Stilleweg 2, Hannover D-30655, Germany

Alain Dubreuil Natural Resources Canada, CANMET Mining and Mineral Sciences Laboratories, Ottawa, ON, Canada

Roderick G. Eggert Division of Economics and Business, Colorado School of Mines, Golden, CO, USA

Nora K. Foley U.S. Geological Survey, Reston, VA, USA

Gudrun Franken Bundesanstalt für Geowissenschaften und Rohstoffe, Stilleweg 2, Hannover D-30655, Germany

Marko Gernuks Volkswagen AG, Group Research, Environmental Affairs Product, Letter Box 1774, Wolfsburg, 38436, Germany

Simon Goldmann Bundesanstalt für Geowissenschaften und Rohstoffe, Stilleweg 2, Hannover D-30655, Germany

Bradley S. Van Gosen U.S. Geological Survey, Denver, CO, USA

Christian Hagelüken Umicore, Rodenbacher Chaussee 4, Hanau D-63457, Germany

Nawshad Haque CSIRO Process Science & Engineering, Minerals Down Under Flagship, Clayton, VIC 3169, Australia

Sigurd Heiberg Chairperson of the UNECE Committee on Sustainable Energy, Geneva, Switzerland

Corporate Strategy Analysis, Statoil ASA, Stavanger N4035, Norway

Gustavo Lagos C.C Centre for Research in Energy and Minerals Economics (CREME), Curtin University, Perth, Australia

Richard Sinding-Larsen Department of Geology and Mineral Resources Engineering, Norwegian University of Science and Technology, Trondheim, NO-7491, Norway

Maren Liedtke Economic Geology of Mineral Resources, Federal Institute for Geosciences and Natural Resources (BGR), Stilleweg 2, Hannover, 30655, Germany

Keith R. Long U.S. Geological Survey, Tucson, AZ, USA

Frank Melcher Bundesanstalt für Geowissenschaften und Rohstoffe, Stilleweg 2, Hannover D-30655, Germany

Maria Sitnikova Bundesanstalt für Geowissenschaften und Rohstoffe, Stilleweg 2, Hannover D-30655, Germany

Slavko V. Šolar Geological Survey of Slovenia, Ljubljana, Slovenia

Thomas Thielemann RWE Power AG, Stüttgenweg 2, Cologne D-50935, Germany

John E. Tilton Mining Center, Pontificia Universidad Católica de Chile, Región metropolitana Santiago, Chile

Division of Economics and Business, Colorado School of Mines, Golden, CO, USA

Jürgen Vasters Bundesanstalt für Geowissenschaften und Rohstoffe, Stilleweg 2, Hannover D-30655, Germany

Patrick Wall EuroGeoSurveys – The Geological Surveys of Europe, Brussels, Belgium

Friedrich-W. Wellmer Formerly Bundesanstalt für Geowissenschaften und Rohstoffe (BGR), Neue Sachlichkeit 32, Hannover D-30655, Germany

Mohan Yellishetty Civil & Environmental Engineering, School of Applied Science & Engineering, Monash University, Gippsland Campus, Churchill 3842, VIC, Australia

Non-renewable Resource Issues: Geoscientific and Societal Challenges: An Introduction

Richard Sinding-Larsen and Friedrich-W. Wellmer

Abstract

This chapter discusses non-renewable resource issues: the raw materials boom of the first decade of the third millennium, availability of a resource (geoscientific characteristics), society's dependence on the resource (economic, social and environmental challenges), the gradual dematerialization of advanced economies and the possibility of finding alternatives (whether the resource can be substituted or recycled). Assessing important resources against these issues determine their "criticality" and the risk for supply interruptions of e.g. rare-earths and other raw materials needed for industrial products. 'Peaks' in the production of natural resources can be driven by demand or supply. However, fundamental and important differences between a peak in the production of oil and peaks in the production of minerals have been exemplified. Concerning the future availability of natural resources the fixed stock paradigm is contrasted with the opportunity cost paradigm that takes into account the inherent dynamics of market forces and changing technology. Regardless of paradigm increasing resource efficiency through human ingenuity and creativity both in the upstream and downstream sector is shown to be vital if mankind is to successfully advance on the road to a sustainable economy based upon transparency and "good governance" that has to be mirrored on the companies' side by corporate social responsibility. The societal aspect is however, only one side of the natural resources coin, the other side being the geological and technical availability.

R. Sinding-Larsen (✉)
Department of Geology and Mineral Resources Engineering, Norwegian University of Science and Technology, Trondheim, NO-7491, Norway
e-mail: richard.sinding.larsen@geo.ntu.no

F.-W. Wellmer
Formerly Bundesanstalt für Geowissenschaften und Rohstoffe (BGR), Neue Sachlichkeit 32, Hannover D-30655, Germany
e-mail: fwellmer@t-online.de

R. Sinding-Larsen and F.-W. Wellmer (eds.), *Non-Renewable Resource Issues: Geoscientific and Societal Challenges*, International Year of Planet Earth,
DOI 10.1007/978-90-481-8679-2_1, © Springer Science+Business Media B.V. 2012

Keywords
Non-renewable resources • Primary resources • Secondary resources • Dematerialization • Supply risks • Availability • Resource efficiency • Corporate social responsibility

A New Natural Resources Boom

The first decade of the third millennium saw a raw materials boom never experienced before (Table 1). Prices skyrocketed to levels never seen earlier only to drop in the summer of 2008 with the onset of the general economic crisis but picking up in spring 2009 again (Fig. 1). Since the beginning of the millennium, we see a correlation between the development of natural resources prices and the gross domestic product (GDP) of the world (Lembke 2011) (Fig. 2). The name "decade or even century of natural resources" has been coined. The natural resources theme made headlines and cover stories in the press. In politics, the topic natural resources, nearly forgotten in the 1990s of the twentieth century by most governments of industrialized nations, suddenly received top priority. Some governments founded new raw materials agencies. What had happened?

After the general buyer's market, conditions in the last decade of the last century markets changed to a seller's market in a dramatic turnaround with the beginning of the new millennium. The main reason was the increased raw materials demand of developing populous nations like China, India, or Brazil of the so-called BRIC group. The term BRIC was invented by the financial house Goldman Sachs in 2002 for Brazil, Russia, India, and China, a country group which, at the moment, represents 40% of the world's population, but only 10% of the world's gross national product, with the chance, however, to outperform the large industrialized G8-nations by 2050.

Up to the end of the last millennium, the bulk of mineral and energy resources was used in industrialized countries that had only a very modest population growth. As Wagner and Wellmer (2009) point out, we were living in an "upside-down world": About 25% of the world's population lived in industrialized nations consuming 70–80% of the world's energy and mineral resources, coal being one of the few exceptions. Since the turn of the last millennia, however, we have witnessed the start of a new growth cycle that is driven by a surge in demand in developing nations with large populations. China has become the leading consumer in the developing world as shown in Table 2. In 2009, it had the highest consumption of every major commodity, with the exception of crude oil and natural gas. China's share in worldwide copper and aluminium consumption in 2009 was 38.8% and 36.4%, respectively, and 47.0% in steel consumption. In 2006, India joined the five largest consumer nations of steel, crude oil, and coal and replaced Germany, which is the largest consumer in the European Union.

The "super boom" from 2002 to 2008 also changed the perception of politicians and the general public in many industrialized nations concerning the importance of natural resources. Two trends were superimposed and compounded each other in a climate of a general buyer's market in the last decade of the twentieth century, i.e. a decade, during which raw materials were always available at reasonable costs for the consumer:

(a) A general trend of dematerialization
(b) A trend in some industrialized nations to scale down the manufacturing sector and instead to concentrate on the service sector, especially the financial services

The General Trend of Dematerialization

Radetzki (2008) published a table illustrating the gradual dematerialization of advanced economies (Table 3). It shows the decline in the quantitative resource needs to create a unit of economic value. The table presents the value in 2000 US$ per kilogram of a set of goods and services.

Table 1 Natural resources boom periods (as indicated by the CRB Metals Sub-index) for the last 50 years period

Period	Duration (in months)	Increase of metals sub-index
06/1963–02/1966	33	1.6 times
10/1972–03/1974	18	2.3 times
02/1978–02/1980	25	1.5 times
07/1986–04/1989	34	1.9 times
09/1993–01/1995	17	1.5 times
01/1999–12/1999	12	1.2 times
11/2002–5/2008	**67**	**4.6 times**

Modified after Wagner (2008)

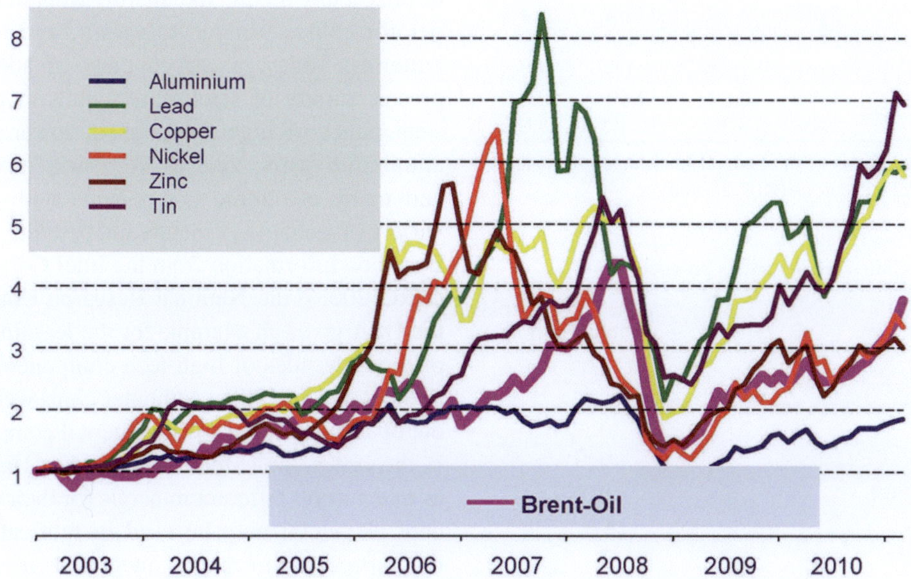

Fig. 1 Relative development of the prices of crude oil and major base metals. Prices on 01.01.2003 = 1 (Source: data bank of the Federal Institute of Geosciences and Natural Resources BGR, Hannover)

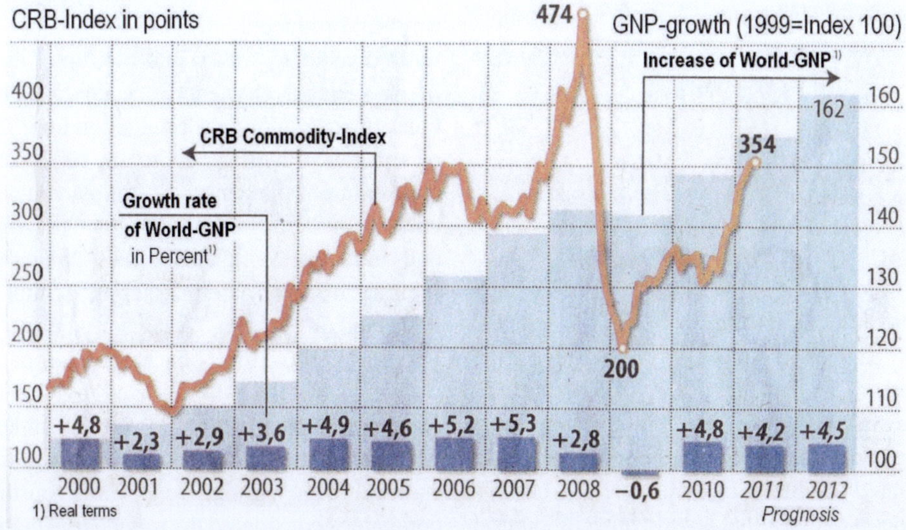

Fig. 2 Correlation of raw material prices and world GDP (Source Lembke 2011)

Table 2 Consumption of copper, aluminium, and steel by the populous developing nations Brazil, India, and China as percentage of global consumption

	1980	1990	2000	2009
Copper				
Brazil	2.6%	1.6%	2.1%	1.7%
India	0.8%	1.2%	1.5%	3.3%
China	4.2%	5.5%	12.6%	38.8%
Aluminium				
Brazil	1.9%	1.6%	1.9%	2.3%
India	1.5%	2.2%	2.3%	3.8%
China	3.5%	4.5%	13.6%	36.4%
Steel				
Brazil	2.0%	1.4%	2.0%	1.7%
India	1.6%	2.6%	3.4%	4.8%
China	5.8%	8.2%	16.4%	47.0%

Source: Data bank of the Federal Institute of Geosciences and Natural Resources BGR, Hannover

Table 3 Unit value of various products

Product	Value in US$ per kg (in 2000 US$)
Iron ore	0.02
Steam coal	0.03
Wheat	0.12
Crude oil	0.21
Standard steel	0.25
Newsprint	0.4
Supertanker	2
Motor car	15
Dishwasher	25
TV set	60
Submarine	100
Large passenger aircraft	600
Laptop computer	1,000
Mobile phone	2,000
Jet fighter	6,000
Windows 2000 software CD-ROM	20,000
Telecom satellite	40,000
Banking services	Almost ∞

Radetzki (2008)

The higher the value, the less primary material input will be needed per dollar value represented by the items. This lead many people to assume that natural resources are of lesser importance for a highly developed society creating more and more sophisticated products. In a climate of a buyer's market, people forgot that a wide range of natural resources are needed for the well-being of a nation. It starts with the construction materials to build our houses and infrastructure. It ends with a wide variety of technology metals (Hagelüken 2011) necessary for high-tech products like computers or solar panels. From a security of supply point of view, it is important to realize that the spectrum of elements necessary to supply the services mankind wants has become wider and more complex over time. A car 100 years ago consisted of only a few metals like mainly iron/steel, copper for cables, tin for soldering, and lead for batteries. Today, a car contains in addition a greater variety of steel alloy metals to produce steel plates of high strength but lower weight, aluminium parts, zinc as corrosion protection, and many electronic components with a wide variety of technology metals and precious metals. Based on information from the Intel Corporation (NRC 2008), the National Research Council of the USA gives an example for the learning curve of three decades of high-tech component development. In the 1980s, computer chips were made out of 12 minerals or their elemental components. In the 1990s, the number increased to 16. Today, as many as 60 different minerals (or their constituent elements) may be used in fabricating the high-speed, high-capacity integrated circuits that are crucial to this technology.

Unnoticed by many consumers during times of ample supply, the potential supply risks actually increased because the element spectrum necessary to produce high-quality products became more diversified and an interruption of the supply of one critical element alone could have major consequences along the production line. The onset of the raw materials boom in 2002 caused an awakening of the generally complacent consumer nations and consumer industries. Governments and industries initiated studies concerning the security of raw materials supply which will be described in section "Analysing Future Supply Risks" below. How concerned companies are showed a recent poll initiated by the German Chamber of Industry and Trade (Lembke 2011). Companies were asked where they saw the highest risks in the coming year. Three fourths of the companies considered the tremendously increased rate of price increases as the major threat (DIHK

2011). For example, in the paper industry, it was 91% of the companies interviewed; in the car industry, 83%; and in the mechanical engineering industry, still 60%.

Trend of Industrialized Countries to Reduce the Importance of Manufacturing Sector

Despite the trend of dematerialization and increasing importance of the service sector, consumption per capita of a number of mineral commodities in most developed countries has, however, remained roughly constant for the last 30 years (Menzie et al. 2001, 2005). Some countries, however, chose a strategy to reduce the importance of the manufacturing service even more than the general trend, concentrating more on the service sector, especially the financial service, which has had a pronounced effect on raw materials consumption. Most pronounced effect was observable for copper in the United Kingdom (UK) (Table 4). While in comparable nations of the European Union like France, Germany, or Italy the relative contribution of the manufacturing sector as raw material consumer to total economic output decreased between 1991 and 2008 only by about 20%, it decreased by nearly 50% in the UK (Theurer 2010). In Table 4, the reference year 2006 has been chosen to avoid the effect of the economic downturn starting 2007/2008.

Table 4 Selected metal consumption in the United Kingdom, Germany, and the USA

	1980 (in 1,000 t)	1990 (in 1,000 t)	2006 (in 1,000 t)
Copper			
United Kingdom	409	317	172
Germany	888	1,028	1,398
USA	1,868	2,150	2,096
Zinc			
United Kingdom	181	193	161
Germany	492	530	564
USA	879	992	1,153
Steel products			
United Kingdom	11,813	14,189	12,828
Germany	36,674	36,025	39,200
USA	85,936	86,720	119,600

Source: Data bank of the Federal Institute of Geosciences and Natural Resources BGR, Hannover

Analysing Future Supply Risks

For several governments of industrialized nations dependent on imports for an essential part of their natural resources needs, the market turnaround, described above in section "A New Natural Resources Boom", triggered a new way of thinking concerning raw materials. Some governments tried to analyse and understand the complexities of future raw materials supply more in detail and commissioned studies by scientific bodies and institutions. In the USA, for example, the National Research Council established the Committee on Critical Mineral Impacts on the US Economy (NRC 2008). The committee developed a methodology: a criticality matrix. On the horizontal axis, the supply risk is plotted, the likelihood of supply restrictions, on the vertical axis, the impact of restrictions on the mineral's supply, should it occur. The Committee examined 11 minerals and mineral families and concluded that the highest degree of criticality at present is exhibited by indium, manganese, niobium, the platinum group metals (PGM), and the rare earth elements (REE). The Committee assessed that the importance of their applications (in automotive catalytic converters, industrial chemical production, electronics, batteries, liquid crystal displays, or hardeners or strengtheners in steel and iron alloys), the difficulty in finding appropriate mineral substitutes for these applications, and the risk to their supply for any one of a number of reasons were high enough to place these minerals in or near the critical zone on the criticality matrix.

The German government focused more on emerging technologies and the role of high-technology elements. High-technology elements, which Hagelüken (2011) defines as the precious metals gold and silver, the platinum group metals, and the special metals, especially indium, gallium, germanium, rare earth elements, antimony, selenium, silicon, and tellurium, are of specific importance for clean technologies and high-technology

Table 5 Global demand for raw material for the analysed emerging technologies in 2006 and 2030 in relation to today's world production of each commodity (Fraunhofer ISI and IZT 2009), updated by BGR 2010 (Elsner et al. 2010)

Commodity	Indicator 2006	Indicator 2030	Selected emerging technologies
Gallium	0.18	3.97	Thin-layer photovoltaics, integrated circuits (IC), white light emitting diodes (WLED)
Indium	0.40	3.29	Displays, thin-layer photovoltaics
Scandium	Low	2.28	Solid oxide fuel cell (SOFC), aluminium alloying element
Germanium	0.28	2.20	Fibre optic cable, infrared optical technologies
Neodymium	0.40	1.66	Permanent magnets, laser technology
Platinum	Low	1.35	Fuel cells, catalysts
Tantalum	0.40	1.02	Micro-capacitors, medical technology
Silver	0.28	0.83	Radio frequency identification (RFID), lead-free soft solder
Tin	0.62	0.77	Lead-free soft solder, transparent electrodes
Cobalt	0.21	0.43	Lithium-ion batteries, synthetic fuels
Palladium	0.09	0.29	Catalysts, seawater desalination
Titanium	0.08	0.29	Seawater desalination, implants
Copper	0.09	0.24	Efficient electric motors, radio frequency identification (RFID)
Selenium	Low	0.11	Thin-layer photovoltaics, alloying element
Niobium/columbium	0.01	0.03	Micro-capacitors, ferroalloys
Ruthenium	0	0.03	Dye-sensitized solar cells, titanium alloying element
Yttrium	Low	0.01	Super conduction, laser technology
Antimony	Low	Low	Antimony tin oxide, micro-capacitors
Chromium	Low	Low	Seawater desalination, marine technologies

equipment. As Hagelüken (2011) points out, their use has experienced a tremendous growth over the last years. At least 80% of the cumulative mine production of gallium, indium, rare earth elements, silicon, and the platinum group metals, for example, has occurred over the last 30 years only.

For this reason, the German Federal Ministry of Economics and Technology commissioned a study about the future outlook of high-technology raw materials (Fraunhofer ISI and IZT 2009). The raw material needs of emerging technologies were analysed. An indicator was introduced by dividing the raw material demand of emerging technologies by today's total world production of each raw material. This indicator states which share of today's world production of a specific commodity will be required for emerging technologies in 2030. It is at the same time an indicator for the expansion needs of the mining industry (Table 5). For gallium, for example, the indicator reaches the value 4 and for neodymium 1.7. This means that in 2030, the technology-induced demand for these two commodities will be 4 and 1.7 times, respectively, higher than the total present worldwide production thereof.

Recently, the European Commission (EC 2010a) published a study "Defining critical minerals for the EU (EC 2010b)." Also, the EC like the NRC in the USA puts forward a relative concept of criticality. The EC studied 41 minerals and considered 14 of them as critical with regard to the security of supply to the European Union. These are (in alphabetical order) antimony, beryllium, cobalt, fluorspar, gallium, germanium, graphite, indium, magnesium, niobium, platinum group metals, rare earths, tantalum, and tungsten. Of greatest concern is the high concentration of the production centres in countries like China, Russia, Brazil, and Democratic Republic of Congo (DRC) (Solar et al. 2011).

This perception criticality turned into reality when the government of China which in the meantime controlled more than 95% of world's rare earths production announced to reduce its export quotas for rare earths elements by 35% in 2011. China achieved this near-monopolistic position not because rare earths deposits are merely concentrated in China but because they offered rare earths at prices no other supplier could compete with and the international customers were too complacent. In many other countries, rare earths deposits are known. Diversification of rare earths supply became a topic of high priority in government and industry circles. Long et al. (2011) give an overview of the diversification potential around the world.

Industry has to be wary of supply interruptions not only of rare earths elements but also of others of the many raw materials needed for their final products. Buchholz et al. (2011) developed a methodology for risk analysis for a big car manufacturer, illustrated here with the example of zinc. For a set of indicators (market balance, production costs, geostrategic risks, company concentration, and new supply from new projects), they analysed 50 years time series and developed benchmarks for problematic market situations, which might occur in future.

Physical Availability of Natural Resources

Although the market turbulences with a remarkable price surge for nearly every non-renewable mineral and energy commodity described above only reflects an imbalance between supply and demand of the market and not at all an exhaustion of resources, the general public suddenly started to think about possible exhaustion of resources.

The supply to cover the needs of the worldwide mineral demand which cannot be supplied from secondary resources must come from deposits which are depleted in the course of their mine life. New deposits have to be discovered to compensate for the mined production. This is the task of exploration companies or the exploration departments of natural resources companies which so far for all commodities have been successful in keeping a positive balance, i.e. the ratio of known reserves to production has been kept constant or even has been growing (Wellmer 2011). The incentive to spend part of the positive cash flow of a company on high-risk exploration activities is the possibility of high rewards once a discovered deposit is mined. Economic considerations therefore drive the exploration activities. A basic precondition, however, for exploration activities to recognize the potential of a geological environment is the presence of favourable geology and creative thinking by the exploration geologist. By way of the example of the two elements uranium and thorium (Cuney 2011), an example of geological considerations is given that illustrates the wide spectrum of mineralization accumulation possibilities. Long et al. (2011) show in a similar way the diverse geological potential for rare earth elements.

Concerning the future availability of natural resources, basically, there are two schools of thought: One line of thinking follows the fixed stock paradigm, and the other one follows the opportunity cost paradigm taking into account the inherent dynamics of market forces and technological developments towards changing and continuous demand. The continuous debate between these two schools of thinking was recently revisited when Gordon et al. (2006) claimed that the copper needed by the world's population in 2100 will exceed the estimated resources of the planet Earth. Replied that the real cost of copper to society (relative to other costs) has not increased significantly over the last 130 years and that this trend could continue through 2100, thus covering demand. Kesler (2010) demonstrated for copper that undiscovered resources are probably several orders of magnitude larger than those discovered. He postulates that human consumption could be supplied for about 5,000 years if sufficient energy and water can be supplied, if deep mining technology develops rapidly, and if access to land for exploration and exploitation can be secured.

The authors of this introductory chapter follow the arguments of the school of the opportunity cost paradigm. Recognizing that:

- Except for nutrition elements, mankind does not need raw materials as such but their functional properties.
- For finding desirable functional properties, man has at his disposal all the primary resources of the geosphere, the secondary resources of the

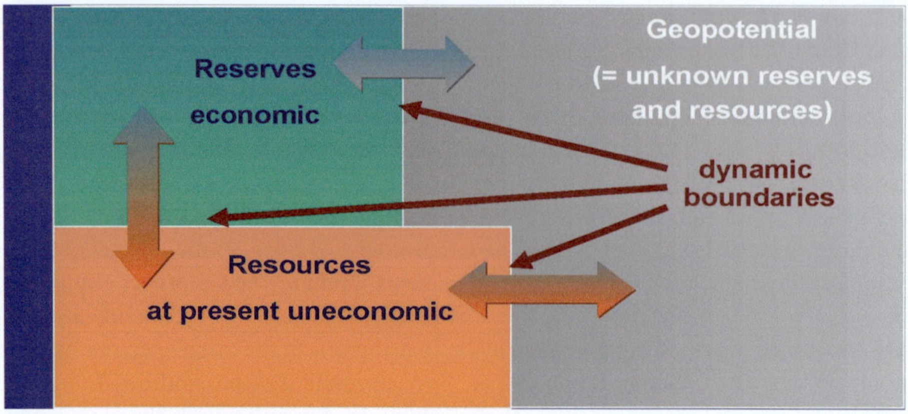

Fig. 3 Total resource box (Wellmer 2008)

technosphere, and the most important resource man's unlimited creativity (McKelvey 1973).

– In a market economy, a driver exists for finding alternative solutions for functional properties, which, besides man's curiosity for the unknown, is the price incentive triggered by scarcities; solutions for functional properties fulfilling the raw materials requirements of our society have been found in the past and will be found in future in a functioning free market economy (Wellmer 2011).

Looking into the future for analysing the availability of natural resources, three terms have to be defined:

– Reserves
– Resources
– Geopotential

These three terms are graphically displayed in Fig. 3, the "total resource box" with its share of reserves, resources, and geopotential. In general, the concept of the "total resource box" is based on the so-called McKelvey box used by the US Geological Survey (see e.g. US Geological Survey (2007) or the United Nations Framework Classification for Energy and Mineral Resources (2004)) (see also Heiberg and Wellmer 2011). It has, however, been simplified and modified to make it more understandable to politicians and non-technical people (Wellmer 2010). Reserves of a commodity are the share of the total resources that can be economically extracted with available technology under environmentally and socio-economically acceptable conditions (the "modifying factors" according to internationally accepted reserve classification systems (JORC 2004; CIM 2010). Resources are known (to various levels of certainty), but their economic viability has not been established. Looking towards the future, we have to consider a third category in addition to reserves and resources: the geopotential or potential resources. They are as yet not known, but by geological reasoning, they can be expected to be discovered by means of modern exploration technologies. Geopotential is described by Cuney (2011) for uranium and thorium, whereas Thielemann (2011) for coal deals only with reserves and resources.

Thielemann (2011) examines the known reserves and resources and concludes that for the foreseeable future, there are enough coal resources to cover our energy needs. A totally different approach is taken by Agüilera et al. (2011) concerning crude oil and natural gas. They are using the approach of estimating cumulative availability curves and examining different sources of petroleum: conventional petroleum (oil, gas, natural gas liquids) and three unconventional sources (heavy oil, oil sands, and oil shale) as a function of production costs. The higher the accepted production cost, the further into the future the constraints of availability are pushed. They conclude that large quantities of conventional and unconventional petroleum resources are available and are likely to last longer than many are predicting.

There exists another approach which will be discussed by the authors in this introductory

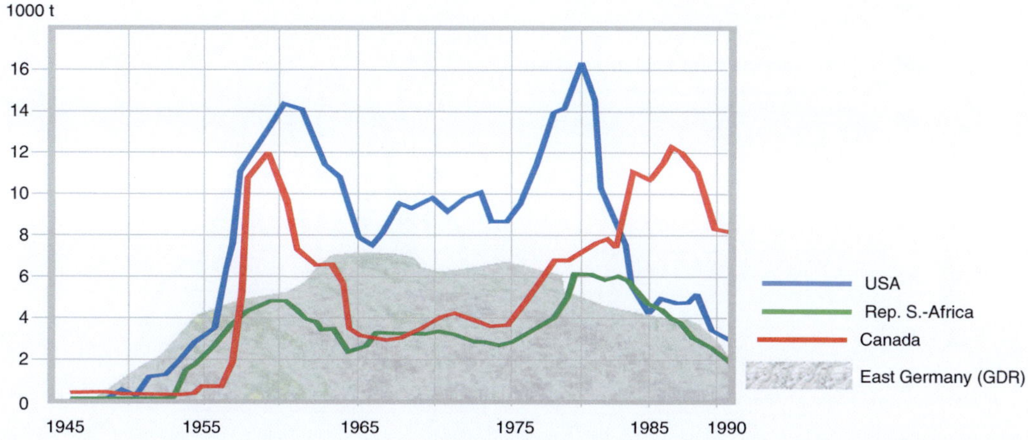

Fig. 4 Uranium production 1945–1990 in the German Democratic Republic (*GDR*), the USA, Republic of South Africa, and Canada (Source: Data bank of the Federal Institute of Geosciences and Natural Resources BGR, Hannover)

chapter. Commonly, especially in daily journals and weekly newspapers, the ratio of known reserves and production (R/P ratio) is interpreted as lifetime of reserves. This is the model, for example, of a car driving with constant speed till the last drop of fuel. It has been pointed out by many authors dealing with the availability of natural resources (e.g. Zwartendyk 1974; Wellmer 2010) that this is an inappropriate model to determine the lifetime of commodities. More suitable is a logistic bell-shaped production curve which can be observed over the lifetime of mining districts or oil or gas provinces: a normally slow rise of the production, then a plateau phase followed by a decline, more or less a mirror image of the initial start-up phase. The pertinent curve for oil has also been called the Hubbert curve, later generalized for other commodities too, especially phosphate which will be discussed below. Hubbert (1956) employed it as a tool to predict the peak of US oil production in 1971, with an error of only 1 year. The curve has a peak at which about 50% of the estimated ultimate recovery (EUR) is produced, called the depletion midpoint, also "peak oil." The discussion when and if at all "peak oil" will arrive has a high political relevance and appears also in daily newspapers or weekly magazines (see web page of the ASPO (Association for the Study of Peak Oil and Gas)).

There is, however, a pitfall in using a logistic Bell-shaped curve: It can be supply-driven or demand-driven with totally different consequences. Supply-driven means that the peak and the production plateau as well as the decline and the end of the exploitation of a mine or gas or oil field or a district is due to the depletion of exploitable reserves. Demand-driven means that due to price increases, technology changes, and/or learning effects, the consumption levels off, i.e. the growth rate of consumption decreases. It might become even negative, i.e. the consumption decreases if the consumption is reduced due to the reaction of the market because of price increases or a commodity is phased out due to, for example, environmental reasons like mercury or asbestos or for other political reasons.

For each case, an example is given:

Supply-driven Bell-shaped curve: An excellent supply-driven Hubbert curve is the curve for the uranium production of the former German Democratic Republic (GDR), a centrally planned economy.

The GDR could produce as much uranium as possible regardless of cost, up to about the mid-1970s, because it had a captive market in the former USSR. The mines were operated by a joint German-Soviet company, the Wismut Company. Only after the mid-1970s, the taking up of the

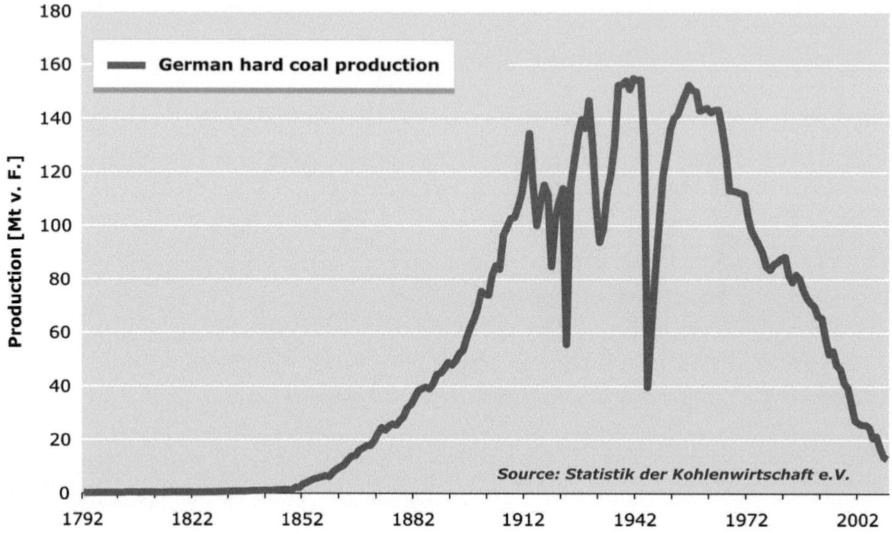

Fig. 5 German hard coal production

production by the former USSR decreased, thereby an element of demand influenced the production curve, meaning the down slope of the East German production curve in Fig. 4 (H. Richter, Dresden/Germany (formerly Wismut uranium company), Personal communication 2011). The East German uranium mines had to close after German unification in 1990. So the peak of the production curve and the plateau phase was reached long before an influence of limiting demand became a factor. Figure 4 shows also the opposite, three demand-driven curves for comparison: the uranium production curves of the USA, Canada, and South Africa, with the two peaks caused by nuclear weapon and nuclear power plant production booms.

Demand-driven bell-shaped curve: The German hard coal production which will be phased out by 2018 is the opposite case (Fig. 5). Since 1964, the German coal production was supported with gradually increasing subsidies as a political element to support the security of energy supply of Germany. German hard coal is a domestic energy source contrary to oil and gas which mainly has to be imported. So the reason for this logistic bell-shaped curve is not a lack of reserves or resources but the inability to compete on world markets. Should prices rise again, plenty of opportunities exist from a geological point of view to start hard coal production again. So after the peak during the phase of declining production, we would see an increasing reserve (resource)/production ratio—contrary to a case where a mine or a mining district has to close due to exhausted reserves.

Therefore, if for a commodity the beginning of a flattening of consumption can be observed, one always has to ask if this is caused by demand or supply. Although the topic of "peak oil" is much discussed in public and there is extrapolation of the historic production curve to determine a time for "peak oil" (e.g. BGR 2010), the historic flattening of the consumption curve is not caused by supply restrictions but by the demand, best illustrated by the "dents" in 1973 and in the years following 1979 after the oil crises with its sudden price increases (Fig. 6). The German Geological Survey BGR extrapolated and postulated a "peak oil" for conventional oil around 2025, taking into account field extensions and unconventional oil like tar sands a peak around 2035 (BGR 2010).

On the other hand, an excellent supply-driven curve for oil is the Norwegian production curve up to 2001 and the forecasted production that match remarkably the actual oil production in 2010 of 104.4 mill Sm3 (Fig. 7).

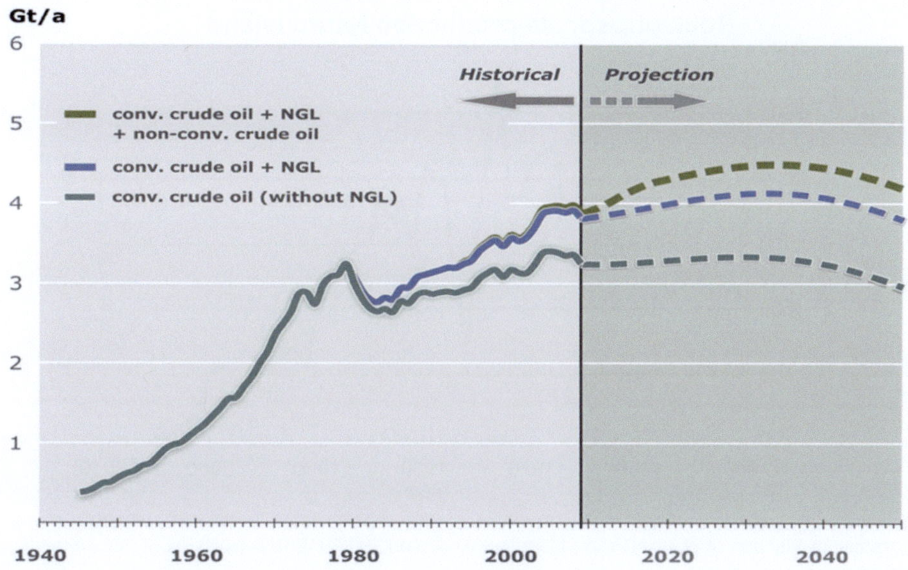

Fig. 6 Production curve of crude oil, historical development, and projection (conventional crude oil including NGL and additional field extensions and tar sands) (Source: BGR 2010)

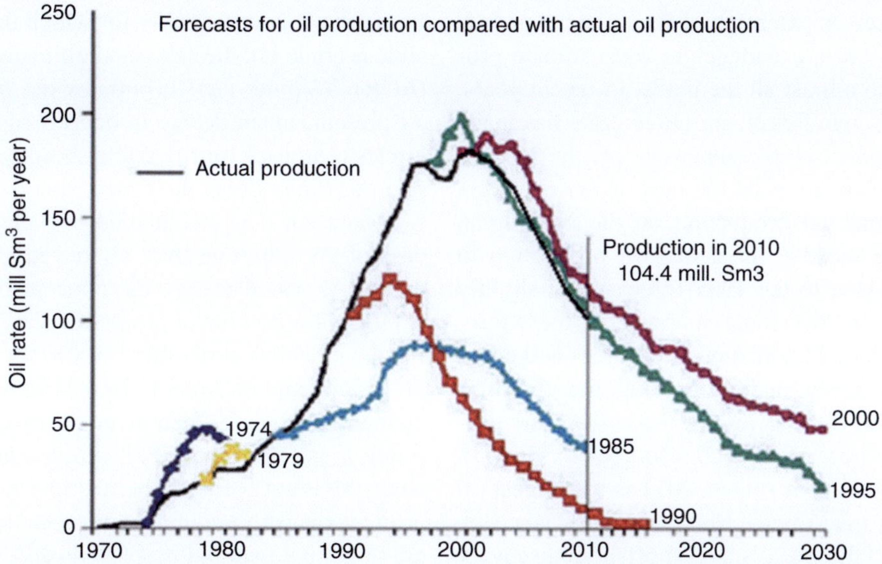

Fig. 7 Crude oil production of Norway (Source: The petroleum resources on the Norwegian continental shelf 2001). http://www.npd.no/Global/Engelsk/3%20-%20Publications/Resource%20report/Resourcereport2001.pdf

At the beginning of Norwegian oil history, actual production was considerably less than both the operators and the authorities forecasted. This continued until the mid-1980s. Between 1985 and 1995, the long-term forecasts for oil production were significantly underestimated. In this period, they did not take adequate regard for the development and implementation of new technology and improved knowledge about the reservoirs. The forecasts at that time only covered production from proven discoveries, but from 1995 onwards, they also included both undiscovered resources and possibilities for recovery over and beyond what the operators

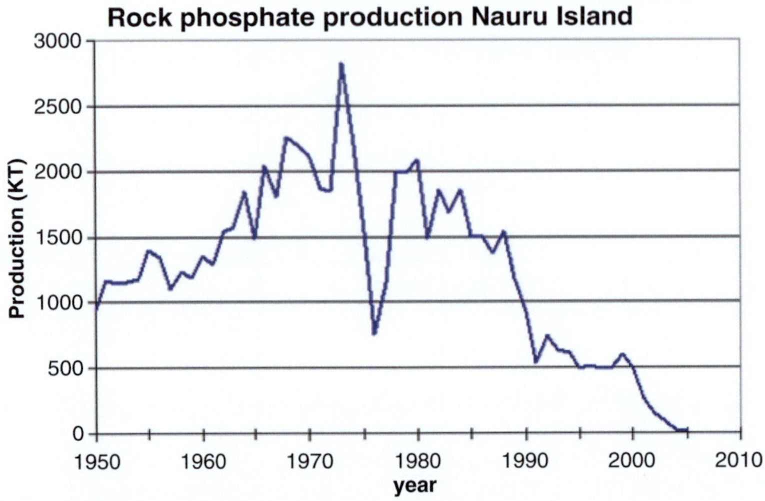

Fig. 8 Phosphate production of Nauru (From: Déry and Anderson 2007) (With permission of the authors and the Post Carbon Institute)

planned. Apart from the end of the 1980s and the beginning of the 1990s, when the large fields experienced an extension of their plateau period, there has been a tendency to overestimate production on almost all the fields. In recent years, the Norwegian Petroleum Directorate has made its own forecasts to compensate for the systematic overestimation on the part of the operators for the total oil production on the Norwegian shelf. We see that the actual oil production in 2010 is close to the 1995 forecast and slightly less than the 2000 forecast. However, these forecasts are based upon known resources and could be significantly underestimating the future if unexpected discoveries are made from the ultimate resources of the total resource box (Fig. 3). An example from August 2011 is the largest oil discovery made on the Norwegian shelf since the 1980s and the world's largest offshore discovery in 2011. It is clear that the discovery will be a positive contribution to the Norwegian Petroleum Directorate's long-term prognoses for the North Sea, and it may contribute to an increase in production possibly until 2038 and a sizeable production beyond 2050.

The concept of the Hubbert curve has also been applied to other non-fuel minerals like zinc, zircon, or especially phosphorous. It is overlooked, however, that a precondition for using a Hubbert curve approach is not fulfilled for these commodities: One must be able to at least estimate quantitatively the total resource box (Fig. 3). So far the only commodity for which this is possible is crude oil: the estimated ultimate recovery (EUR). Wellmer (2010) outlines the reasoning. At present, much debate in concerned scientific circles is aroused by a possible shortage of phosphorous (Schröder et al. 2010).

Because it is an essential nutrient element, the case of phosphorous shall be looked at in more detail. As stated above, the concept of finding solutions for functional properties does not work for the essential agricultural fertilizers nitrogen, potassium, and phosphate. The plants need these elements as such for their metabolism and substitution is not an option. For nitrogen and potassium, this is not a problem because the atmosphere is an inexhaustible source of nitrogen and oceans are full of potassium. Phosphate is different since there is no unlimited reservoir. This is the starting point to apply the Hubbert curve concept also to phosphorous (Déry and Anderson 2007; Cordell et al. 2009a, b; White and Cordell 2008).

We will start with the reasoning of the above: Is a consumption curve supply- or demand-driven? A nearly perfect example is the production curve of Nauru (Fig. 8), a Pacific island consisting mainly of guano, a phosphate ore created by bird's excrements, which has been totally mined out. Clearly, after the peak has been passed,

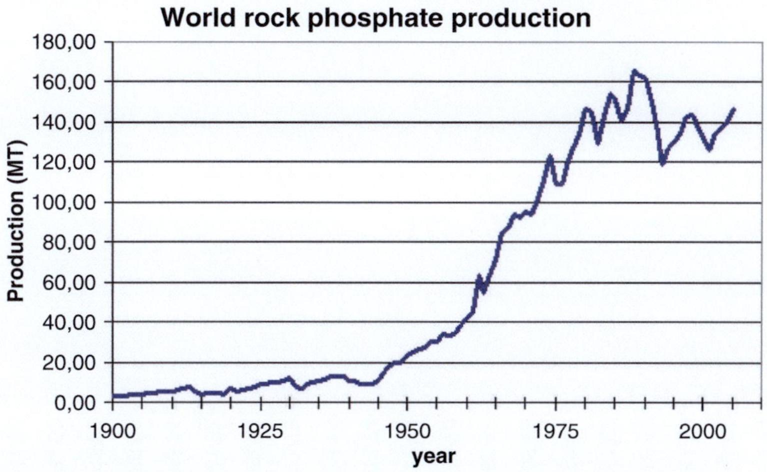

Fig. 9 Phosphorous production of the world (From: Déry and Anderson 2007) (With permission of the authors and the Post Carbon Institute)

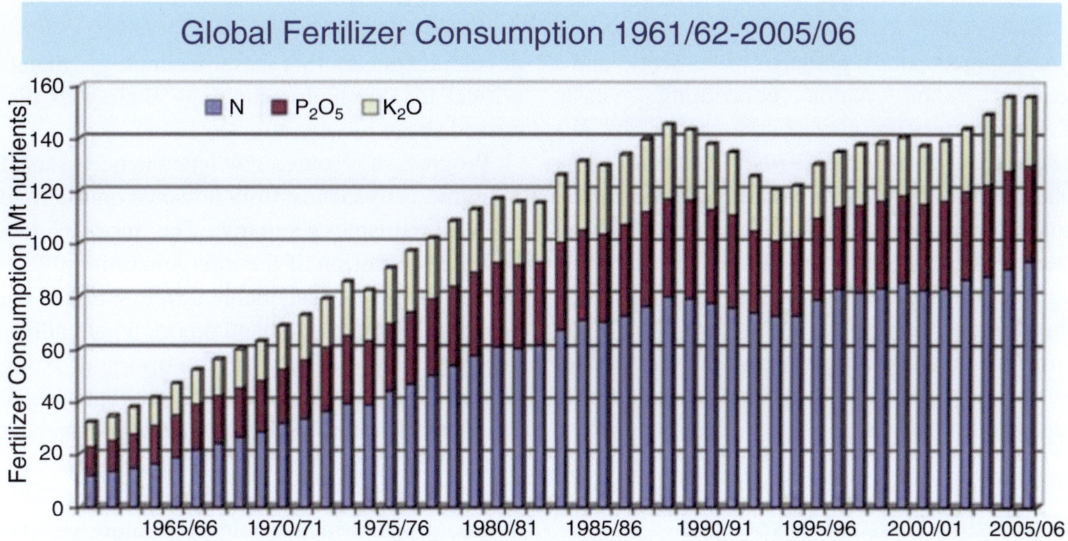

Fig. 10 Worldwide fertilizer consumption (Source: Röhling 2010; FAO 2008)

the reserve/production ratio had to decrease to nil and therefore mining had to come to an end.

Now, we will look at the phosphorous production of the world (Fig. 9) and the question arises: Is the flattening the beginning of a Hubbert curve and an indication of an imminent "peak phosphorous," or is the flattening due to a changing pattern of consumption, i.e. a demand-driven flattening?

If we study the consumption pattern of the three fertilizers K, N, and P, we see a parallel flattening of the consumption, i.e. also the two fertilizers with no constraints concerning resources (Fig. 10). Therefore, we have to conclude that the flattening is not influenced at all by resource constraints but only due to more efficient use of fertilizers, e.g. precision farming (Rickman et al. 2003).

To underline the interpretation, Fig. 11 shows the P and K use in Germany. A parallel consumption pattern can be observed—a pattern obvious in every industrialized country. Germany is the

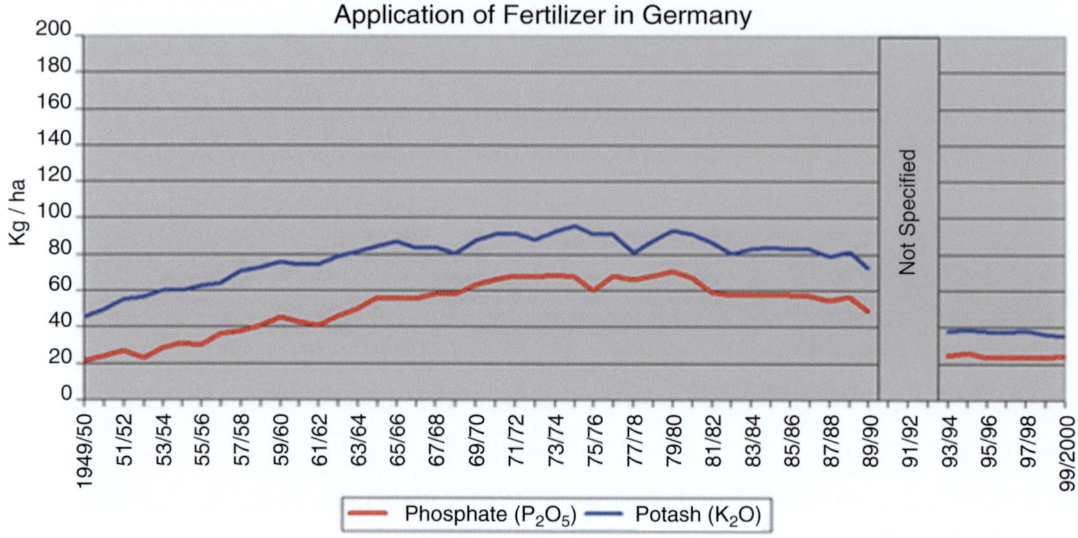

Fig. 11 Potash and phosphate consumption in Germany (Source: Bührer 2001)

fourth largest potash producer of the world and a leading export nation concerning potash. Certainly there are no resource constraints. We see a peak in the mid-1970s. But these "peak phosphorous" or "peak potash" obviously are solely demand-driven and do not permit any conclusion about resource limitation. Therefore, the authors transfer these conclusions also to the flattening of the world P production (Fig. 9) and conclude that it is solely demand-driven and does not permit any conclusion about possible future resource limitations.

Increasing Natural Resources Efficiency

Three chapters in this volume touch the subject of increasing resource efficiency. Heiberg and Wellmer (2011) examine the upstream part of natural resources development, making optimal use of information and supplying the optimum framework condition for discovering natural resources. Two chapters deal with the downstream part. Yellishetty et al. (2011) discuss issues and challenges in life cycle assessment as a tool to improve raw materials efficiency. Hagelüken (2011) examines the technosphere which has to be used to a maximum extent on the road to a sustainable development, in this case technology metals critical for our high-technology society as discussed above.

Progress in resource efficiency is necessary if mankind is to successfully advance on the road to a sustainable economy. The most widely accepted definition of the intergenerational fairness principle of sustainable development is the one given in the Brundtland report: a path allowing every future generation the option of being as well off as its predecessors (Brundtland 1987). It is intensely discussed if sustainable development concerning non-renewable resources is an oxymoron or not. Wellmer and Steinbach (2011) outline a path which would give future generations the chance to be as well off as their predecessors without running out of natural resources. They consider three key resources: (1) the geosphere or primary resources; (2) the technosphere or secondary resources, which can be recycled; and (3) human ingenuity and creativity. Mankind has two resource extremes: natural resources which are completely consumed (fossil fuels) *versus* natural resources (metals) which are wholly recyclable and can be used again:

– All concepts for a sustainable future require the availability of energy at reasonable costs acceptable to society. In the long run, this can only be provided by renewable energy. From the

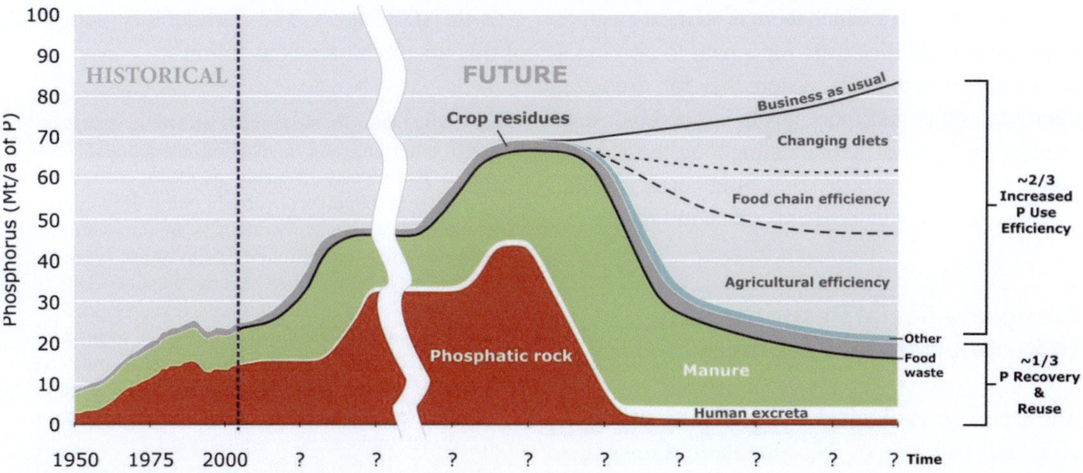

Fig. 12 A sustainable scenario for meeting long-term future phosphorus demand through phosphorus use efficiency and recovery (historical data and exit strategy after Cordell et al. (2009a, b) and Schröder et al. (2010)) (Source: Wellmer and Steinbach 2011)

present point of view, quantum jumps in technology development and increases in efficiency on a similar scale are required. This highlights the overriding role of human creativity and ingenuity as the most important resource of all (McKelvey 1973).

– Metals on the other hand survive use and are merely transferred from the geosphere to the technosphere. Metals have the best chance to become renewable resources (von Gleich 2006).

Steinbach and Wellmer (2010) developed a scenario taking the present situation of lead as an example that permits a supply equilibrium from the technosphere to develop. At the time when consumption starts to level off, chances improve of obtaining most of the material for our industrial requirements from the technosphere. There will, however, always be a need for contributions from the geosphere to offset inevitable metal losses in the technosphere. They defined four types of losses. Three loss types are unavoidable and occur (a) due to the circumstances that it is thermodynamically impossible to have a 100% perfectly closed cycle, (b) due to the different redox potential of the metals (noble and less noble metals), and (c) due to dispersal effects. The fourth loss type, however, occurs because the scrap value is too low and/or the technology is not developed far enough to enable the recovery of the metals within the bounds of thermodynamic, redox, and dispersal constraints. In market economies, recycling has to pay for itself. This type of loss is dealt within the paper of Hagelüken (2011) too, concerning precious and special metals.

Losses have to be compensated by contributions from the geosphere. But man has a choice. He does not need raw materials as such, only the intrinsic property of a material that enables it to fulfil a desirable function as mentioned in the preceding section.

Wellmer and Steinbach (2011) also deal with man's chances to find a path to sustainable use for non-metallic, non-energetic mineral resources. Except for the fertilizer elements, substitutes can always be found for all other non-metallic resources, and it is exactly for such resources that the principle for finding new solutions for functional properties by utilizing man's creativity especially applies. The concept of finding different solutions for functional properties does not work for the essential agricultural fertilizers nitrogen, potassium, and phosphate as outlined above because the plants need them as such and there is no replacement for them. But even for phosphorous, for which there is no inexhaustible reservoir, possibilities exist for a path to develop a practically closed cycle system by, for example, improving more and more fertilizing and recovery techniques. A scenario for meeting long-term future phosphorus demand based on Cordell et al. (2009a, b) and Schröder et al. (2010) is given in Fig. 12.

Corporate Social Responsibility to Exploit Natural Resources

The aspect of geological and technical availability is only one side of the same natural resources coin, the other side being societal aspects. They are condensed in the Agenda 21 of the Rio Declaration at the UN Conference on Environment and Development into the three humanitarian objectives as pillars to reach intragenerational fairness on the road to sustainable development:

1. To conserve the basic needs of life
2. To enable all people to achieve economic prosperity
3. To strive towards social justice

Instead of such a three-pillar concept, a four-pillar concept was suggested in The Mining, Minerals and Sustainable Development Project presented to the Global Mining Initiative Conference (Toronto 2002): the above three pillars of the Agenda 21 plus an additional pillar for "good governance." This should ensure that both development and use of mineral resources will optimally benefit a nation, especially so developing nations (IISD 2002). This fourth pillar of good governance has to be mirrored on the companies' side by corporate social responsibility CSR (Richards 2009).

To advance "good governance" in the natural resources industry, the Extractive Industries Transparency Initiative (EITI) was launched in 2002 at the World Summit on Sustainable Development in Johannesburg, South Africa. The 2007 G8-Summit in Heiligendamm, Germany, took up this idea. In the Heiligendamm Summit Paper, "Growth and Responsibility in the World Economy" of June 7, 2007, the second last chapter (80–87) deals with "Responsibility for raw materials: transparency and sustainable growth" (G8 2007). The introductory paragraph of this document reads as follows:

> Raw materials produced by the extractive sector are a key factor for sustainable growth in industrialised, emerging and developing economies. They are a particularly valuable asset for sustaining growth and reducing poverty in many of the poorest countries in the world. It is in our common global interest that resource wealth be used responsibly so as to help reduce poverty, prevent conflicts and improve the sustainability of resource production and supply. We firmly agree that significant and lasting progress in this area can only be achieved on the basis of transparency and good governance. Against this background, we support increased transparency with regard both to the extractive sector and the subsequent trade and financial flows. In doing so, we will work closely together with resource rich economies as well as important raw-material consuming emerging economies.

In this document, a pilot study for artisanal mining which is a significant source of some high-technology metals like tantalum and columbium/niobium was proposed. Concerning these two high-technology metals Franken et al. (2011) report on a concrete effort to establish transparency by certified trading chains in Eastern Africa in an area which is an important supplier for these metals.

Final Remarks

The editors of the International Year of the Planet Earth theme book on natural resources "Non-renewable Resource Issues: Geoscientific and Societal Challenges" attempted with this introduction to outline the present situation of mineral resources including oil and other energy resources in our societies and to place the individual papers presented in this book into a broader perspective of resources availability. On this last point, let us end with a quote from the book "On Borrowed Time?" by Tilton (2003, pp. 118–119):

> So are we living on borrowed time? Is modern civilization as we know it threatened by the depletion of oil and other mineral commodities? Are drastic public policies needed to avert disaster and to provide a secure future for generations to come? Are public policies needed simply as a precaution or as insurance against the possibility that depletion may be a problem in the future?

Modern-day prophets cry out yes, and yes again, to all of these questions. They call upon society to repent and to mend its ways, to curb its population growth, to restrain its use of mineral resources, to tame its passion for more and better things, to turn away from materialism, and to embrace a simpler life.

Standing at the other end of the spectrum are the prophet slayers. They claim the availability of mineral resources poses no problems, now and forevermore (or at least for as long as we might have an interest in the future). They contend our prophets are not prophets at all, but Chicken Littles running about crying the sky is falling.

The public is fascinated with its prophets and prophet slayers. They come with clear and uncomplicated messages, painting the world in black and white. They tell us what we need to know, what we need to think, and what we should or should not do. They are colourful, passionate, and so convinced they are right that it is hard to resist being swept up in their enthusiasm.

The real world, however, is not so simple. Rarely is it painted in black and white. Rather it is bedecked in hues of gray, and a palette of bright colours. It is full of risks, uncertainties, unknowns, and complications—all traits that make life interesting, exciting, and challenging, even if at times frustrating and troubling as well. And so it is with our fears of mineral depletion. Over the next 50 to 100 years, …mineral depletion is not likely to rank among the most pressing problems confronting society. The great beyond, however, depends on the race between the cost-increasing effects of depletion and the cost-reducing effects of new technology. The outcome will be influenced by many factors, and is simply unknown.

Acknowledgements The authors thank H. Andruleit, H.-G. Babies, P. Buchholz, B. Cramer, D. Homberg-Heumann, D. Huy, W. Neumann, S. Röhling, and S. Schmidt for support concerning data and figures.

References

Aguilera RF, Eggert RG, Lagos CCG, Tilton JE (2011) Is depletion likely to create significant scarcities of future petroleum resources? In: Sinding-Larsen R, Wellmer FW (eds) Non-renewable resource issues: geoscientific and societal challenges. Springer, Dordrecht

ASPO (Association for the Study of Peak Oil and Gas): www.peakoil.net/. Accessed 31 Dec 11

BGR (2010) (Bundesanstalt für Geowissenschaften und Rohstoffe, Federal Institute for Geosciences and Natural Resources) Kurzstudie Reserven, Ressourcen von Energierohstoffen, 2010, Hannover, 86 p

Brundtland GH (1987) Our common future: report of the world commission on environment and development. Oxford University Press, Oxford

Buchholz P, Liedtke M, Gernuks M (2011) Evaluating supply risk patterns and supply and demand trends for mineral raw materials: assessment of the zinc market. In: Sinding-Larsen R, Wellmer FW (eds) Non-renewable resource issues: geoscientific and societal challenges. Springer, Dordrecht

Bührer W (2001) 50 Jahre Düngemittelstatistik in Deutschland. Statistisches Bundesamt, Wirtschaft und Statistik 5/2001:367–371

Canadian Institute of Mining and Metallurgy (CIM) Standing Committee on Reserve Definitions (2010) CIM definition standards. http://www.cim.org/standards/MenuPage.cfm?sections=177&menu=178. Accessed 31 Dec 11

Cordell D, Drangert J-O, White S (2009a) The story of phosphorous: global food security and food for thought. Global Environ Change J. doi:10.1016/j.gloenvcha.2008.10.009

Cordell D, Schmid-Neset T, White S, Drangert JO (2009b) Preferred future phosphorus scenarios: a framework for meeting long-term phosphorus needs for global food demand. In: Mavinic D, Ashley K, Koch F (eds) Proceedings international conference on nutrient recovery from wastewater streams, Vancouver 2009. ISBN 9781843392323, IWA Publishing London 2009, pp 23–43

Cuney M (2011) Uranium and thorium: the extreme diversity of the resources of the world's energy minerals. In: Sinding-Larsen R, Wellmer FW (eds) Non-renewable resource issues: geoscientific and societal challenges. Springer, Dordrecht

Déry P, Anderson B (2007) Peak phosphorous. Www.energybulletin.net/node/33164. Accessed 31 Dec 11

DIHK (2011) Rohstoffpreise beeinträchtigen Konjunkturentwicklung. Www.dihk.de/presse/meldungen2011-02-04-rohstoffpreise-sonderauswertung. Accessed 19 Mar 11

EC (2010a) (European Commission) EU lists 14 critical mineral raw materials. EU press release 17 July 2010

EC (2010b) European Commission, Enterprise and Industry, (ECEIWG 1) Critical raw materials for the EU. Report of the Ad-hoc Working Group on defining critical raw materials, Brussels, 85 p

Elsner H, Melcher F, Schwarz-Schampera U, Buchholz P (2010) Elektronikmetalle – zukünftig steigender Bedarf bei unzureichender Versorgungslage. Commodity Top News 33, Federal Institute of Geosciences and Natural Resources (BGR), Hannover, 13 p

FAO (2008): FAOSTAT (online) Available: http://faostat.fao.org/site/575/default.aspx-ancor

Franken G, Vasters J, Dorner U, Melcher F, Sitnikova M, Goldmann S (2011) Certified trading chains in mineral production – a way to improve responsibility in mining. In: Sinding-Larsen R, Wellmer FW (eds) Non-renewable resource issues: geoscientific and societal challenges. Springer, Dordrecht

Fraunhofer ISI (Fraunhofer Institut für System- und Innovationsforschung), IZT gGmbH (Institut für Zukunftsstudien und Technologiebewertung) (2009) Rohstoff für Zukunftstechnologien.-Fraunhofer IRB Verlag, Stuttgart, 383 pp (Summary in English: Raw Materials for Emerging Technologies, http://www.bmwi.de/BMWi/Redaktion/PDF/P-R/raw-materials-for-emerging-technologies.property=pdf,bereich=bmwi,sprache=de,rwb=true.pdf)

G8 (2007) Growth and responsibility in the world economy. http://www.bmz.de/en/what_we_do/issues/goodgovernance/EITI/dokumente/index.html. Accessed 1 Jan 12

Gordon RB, Bertram M, Graedel TE (2006) Metal stocks and sustainability. Proc Natl Acad Sci USA 103/5:1209–1214, Washington, DC

Hagelüken C (2011) Secondary raw material sources for precious and special metals. In: Sinding-Larsen R, Wellmer FW (eds) Non-renewable resource issues: geoscientific and societal challenges. Springer, Dordrecht

Heiberg S, Wellmer F-W (2011) Stretching the availability of non-renewable resources. In: Sinding-Larsen R, Wellmer FW (eds) Non-renewable resource issues: geoscientific and societal challenges. Springer, Dordrecht

Hubbert MK (1956) Nuclear energy and fossil fuels. In: American petroleum institute drilling and production practice, proceedings of the spring meeting, San Antonio, TX, pp 7–25

IISD (2002) Institute for environment and development: breaking new ground – the MMSD final Report. http://www.iied.org/mmsd/finalreport/index.htm. Accessed 31 Dec 11

Joint Ore Reserve Committee (JORC) (2004) Australasian code for reporting of exploration results, mineral resources and ore reserves – the JORC code, 2004 edn. http://www.jorc.org/main.php. Accessed 31 Dec 11

Kesler SE (2010) Geological stocks and prospects for nonrenewable resources. In: Graedel T, van der Voet E (eds) Linkages of sustainability, Strüngmann forum report, vol 4. MIT Press, Cambridge, pp 109–129

Lembke J (2011) Die Lasten des Aufschwungs. Franfurter Allgemeine Zeitung 12 March 2011

Long KR, Van Gosen BS, Foley NK, Cordier D (2011) The principal rare earth elements deposits of the United States – a summary of domestic deposits and a global perspective. In: Sinding-Larsen R, Wellmer FW (eds) Non-renewable resource issues: geoscientific and societal challenges. Springer, Dordrecht

McKelvey VE (1973) Mineral resource estimate and public policy. Am Sci 60(1):32–40

Menzie WD, DeYoung JH Jr, Steblez WG (2001) Some implications of changing patterns of mineral consumption. U.S. Geological Survey Open-File Report 03–382. http://pubs.usgs.gov/of/2003/of03–382. Accessed 31 Dec 11

Menzie WD, Singer DA, DeYoung JH Jr (2005) Mineral resources and consumption in the 21st century. In: Simpson RD, Toman MA, Ayres RU (eds) Scarcity and growth revisited: natural resources and the environment in the new millennium. Resources for the Future, Washington, DC, pp 33–53

NRC (2008) National research council of the national academies: minerals, critical minerals, and the U.S. economy. The National Academies Press, Washington, DC, 245 p

Radetzki M (2008) A handbook of primary commodities in the global economy. Cambridge University Press, Cambridge, 233 p

Richards JP (ed) (2009) Mining, society and a sustainable world. Berlin, Heidelberg, New york, 506 p

Rickman D, Luvall JC, Shaw J, Mask P, Kissel D, Sullivan D (2003) Precision agriculture – changing the face of farming. Geotimes November 48:28–31

Röhling S (Federal Institute for Geosciences and Natural Resources BGR) (2010) Globale Verfügbarkeit von Dünger-und Energierohstoffen -Seminar of the AHT Niederorschel and Volksbank Dransfeld. Wernigerode 18 February

Schröder JJ, Cordell D, Smit AL, Rosemarin A (2010) Sustainable use of phosphorus, Report 357 Plant Research International, Wageningen, The Netherlands, EU Tender ENV.B.1/ETU/2009/0025, October 2010, 122 p

Solar VŠ, Demicheli L, Wall P (2011) Raw materials initiative: a contribution to the European minerals policy framework. In: Sinding-Larsen R, Wellmer FW (eds) Non-renewable resource issues: geoscientific and societal challenges. Springer, Dordrecht

Steinbach V, Wellmer FW (2010) Consumption and use of non-renewable mineral and energy raw materials from an economic geology point of view. Sustainability 2:1408–1430

Tilton JE (2003) On borrowed time? Assessing the threat of mineral depletion. Resources for the Future, John Hopkins University Press, Washington, DC, 168 pp

Theurer M (2010) Hightech statt Hedge-Fonds. Frankfurter Allgemeine Zeitung 26 Jan 2010, Frankfurt/Main, Germany

Thielemann T (2011) Coal: an energy source for future world needs. In: Sinding-Larsen R, Wellmer FW (eds) Non-renewable resource issues: geoscientific and societal challenges. Springer, Dordrecht

UN-ECE, United Nations Framework Classification for Energy and Mineral Resources (2004) http://www.unece.org/energy/se/reserves.html. Accessed 31 Dec 11

U.S. Geological Survey (2007) A resource/reserve classification for minerals. In: Mineral commodity summaries. U.S. Geological Survey, Washington, DC, 195 pp

Von Gleich A (2006) Outlines of a sustainable metals industry. In: von Gleich A, Ayres AU, Gößling-Reisemann S (eds) Sustainable metals management. Springer, Dordrecht, pp 3–39

Wagner M (2008) Rohstoffberatung der BGR für Ministerien, Politik und Industrie.-Paper Federal Institute for Geosciences and Natural Resources (BGR) Hannover 21 Jan 2008

Wagner M, Wellmer F-W (2009) A hierarchy of natural resources with respect to sustainable development – a basis for a natural resources efficiency indicator. In:

Richards JP (ed) Mining, society and a sustainable world. Berlin, Heidelberg, New York, pp 91–121

Wellmer F-W (2010) Reserves and resources of the geosphere, terms so often misunderstood. Is the life index of reserves of natural resources a guide to the future? Zeitschr Deutsche Ges Geowissenschaften 159(4): 575–590

Wellmer F-W (2011) Discovery and sustainability. In: Sinding-Larsen R, Wellmer FW (eds) Non-renewable resource issues: geoscientific and societal challenges. Springer, Dordrecht

Wellmer FW, Steinbach V (2011) Is a road to sustainable use of non-renewable mineral raw materials possible? In: Proceedings conference sustainable development in the minerals industry (SDIMI), Aachen, Germany, June 2011, pp 14–17

White S, Cordell D (2008) Peak phosphorous: the sequel to peak oil. http://phosphorousfutures.net/peak-phosphorous Accessed 31 Dec 11.

Yellishetty M, Haque N, Dubreuil A (2011) Issues and challenges in life cycle assessment in the minerals and metals sector: a chance to improve raw materials efficiency. In: Sinding-Larsen R, Wellmer FW (eds) Non-renewable resource issues: geoscientific and societal challenges. Springer, Dordrecht

Zwartendyk J (1974) The life index of mineral reserves – a statistical mirage. Can Min Metall Bull 67(750): 67–70

Raw Materials Initiative: A Contribution to the European Minerals Policy Framework

Slavko V. Šolar, Luca Demicheli, and Patrick Wall

Abstract

The European Union, which started out as the European Coal and Steel Community in 1958, put raw materials issues back at the centre of the public's and industry's attention at the end of first decade of the 21st century. This action, backed up by different policies at regional, national and EU level, was needed due to the major changes in supply.

The European Commission launched the integrative European Raw Materials Initiative in 2008, which was further updated in early 2011 by the Communication on Commodities and Raw Materials. Both communications and accompanying documents and reports reveal the proposed framework for European and Member States' minerals policy and legislative framework. An important part of minerals policy is the minerals planning policy, while on the other hand minerals policy elements should also be an important part of land use planning policy. The framework has policy considerations which are related to land access, the regulatory framework and voluntary actions, as well as being sustainability based. The recommended EU framework for national minerals policy outlines legal and information parts as two basic elements. Sustainability is represented by the balance among economic, environmental and social pillars which are maintained by public debate on potential trade-offs. The debate is far from being over, especially since new actions within the proposed European Innovation Partnership on Raw Materials are planned until 2020, as part of the Europe 2020 Strategy.

S.V. Šolar (✉)
Geological Survey of Slovenia,
Ljubljana, Slovenia
e-mail: slavko.solar@geo-zs.si

L. Demicheli • P. Wall
EuroGeoSurveys – The Geological Surveys of Europe,
Brussels, Belgium
e-mail: info@eurogeosurveys.org

R. Sinding-Larsen and F.-W. Wellmer (eds.), *Non-Renewable Resource Issues: Geoscientific and Societal Challenges*, International Year of Planet Earth,
DOI 10.1007/978-90-481-8679-2_2, © Springer Science+Business Media B.V. 2012

In this chapter the framework for minerals policy and sustainability considerations is presented and followed by a detailed description of the minerals policy framework which is part of an integrated European Commission response to ongoing public debate on access to and supply of mineral resources.

Keywords
Minerals policy • Sustainability • European union

Introduction

Mineral resources have been an important segment of humankind's well-being for a large period of our history, from the Stone Age to nowadays. They are crucial for development, to facilitate everyday life, and are also an effective economic tool of competitiveness. Human history demonstrates this through the names given to certain periods, such as the "Bronze Age," as well as historical turns or wars caused by mineral supply and demand issues. Indeed, many small- and large-scale armed conflicts from the more recent past are being put down to mineral resource-related issues.

Such a crucial place in society has demanded increased attention from governments, which can be seen in policy and management programmes or plans. These documents were either individual or part of broader strategies, in most cases harmonised within overall state policy.

Nowadays, many countries' governments are placing a lot of attention on supply and demand issues concerning resources, including minerals, in order to become or stay functional in the economy. This becomes even more important in resource-deficient periods.

Over time, there have been shifting demands on the variety of different mineral resources, depending on societal interest and the requirements of human development. Presently, the rapid advances in technology have required the substitution of resources, with previously little-used minerals now growing in demand.

During the 1990s and the beginning of the new millennium, Europe did not suffer from a lack of supply of most mineral resources, and therefore a focus on supply was not explicit. European industry and governments widely believed in fair and equal rules within the global market. But this changed as we progressed into the twenty-first century.

With the development of welfare and lifestyle, several emerging economies, such as the BRICS[1] countries, increased their demand of mineral resources. At the same time, more and more trade distortions on the global market were being observed, often caused by national protectionism in order to promote local economies. The increased global demand for raw materials resulted in higher prices for raw materials on the global trade market, whilst the prior easy access to resources became more difficult and costly.

In the view of the increasing pressure on global resources, the issue of access and supply of raw materials was discussed for the first time during the G8 Summit of Heiligendamm 2007 and resulted in a specific section in the summit's conclusions on "responsibilities for raw materials: transparency and sustainable growth" (Summit declaration 2007; Weber 2008a).

Securing reliable and undistorted access to raw materials is increasingly becoming an important factor for the EU's competitiveness and, hence, crucial to the success of the Lisbon Partnership for growth and jobs. The critical dependence of the EU on certain raw materials underlines that a shift towards a more resource efficient economy and sustainable development[2] is becoming even more pressing. It is therefore

[1] Brazil, Russia, India, China and South Africa.

[2] Sustainable development covers economic, social and environmental aspects. EC COM (2005) 658.

appropriate to develop a more coherent EU policy response to form a common approach in the international discussion on raw materials which has been addressed at the United Nations[3] and by the G8. Hence, the European Commission launched the integrative European Raw Materials Initiative[4] (RMI) in 2008 which was further updated in early 2011 by the Communication on commodities and raw materials.[5]

Mineral Resources: Policy and Sustainability Considerations

The purpose of public policy is to steer and control actions by government bodies or the public society so as to achieve desired ends or objectives. Policies can range from the very specific, i.e. a detailed course of actions or programmes of activities, to the general, i.e. an overall plan embracing identified goals, or even to the conceptual, i.e. a general expression of societal purpose. Which goals are pursued depends upon the values and interests of the people involved in policy creation. Originally, ruler made policies. However, power may devolve over time from absolute rulers to experts and elites to (more or less) democratic governments to the public. And, when a country moves along this continuum, the range of issues worthy of consideration in policy broadens to incorporate the needs and interests of the people rather than only those of the ruling classes (Solar et al. 2009).

In its most simplified form, the cycle comprises six stages: (1) identification of objectives and interests, (2) definition of policy, (3) codification of policy in laws and acts, (4) establishment of a regulatory framework, (5) monitoring and (6) review and adaptation.

The sustainable management of natural resources is not only a matter of economical and geological availability, it also relies on societal acceptance and ecologically sound performance. Policies with a wide acceptance by all parties of the process (e.g. civil society, companies, local and national government) are needed for medium- to long-term lasting strategies. Although mining has a very long history in Europe, many details which are part of such policies are not always in place. Therefore, the European Commission in close cooperation with experts from European Member States, Geological Surveys, Academia and Industry has studied the various aspects of policies in the area of mining permitting and land-use planning.[6]

Minerals Policy: In most cases, policy documents are encoded in legislation, primarily in a mining act, but quite often, the relationship between minerals policy and the mining act is not clear (even if they have a defined relationship). In some cases, a mining act would require strategy, a minerals plan, a mineral resource management plan, etc. Many of these documents have elements of minerals policy and not only operational guidelines or recommendations.

Minerals policy is a deliberate plan of action to guide decisions and achieve rational outcomes regarding the management of a nation's mineral resources. The content of the policy should reflect the goals and preferences of society, the nature of the economy and the needs of the military. A comprehensive minerals policy should include the following elements (Otto 1997):

- Policy scope (types of mineral activity, types of minerals, relationship of minerals policy to other national policies).
- Sovereignty (role of government in investment decision-making, role of State enterprises, mineral ownership, foreign participation, state equity requirement—a means to exert ownership and control, local joint venture or other equity requirements).
- Economics (taxation types, levels and distribution, export restrictions, costs, incentives, import restrictions, costs, role in economic development, employment requirements, conservation and efficiency, land use).

[3] World Investment Report (2007).

[4] COM (2008) 699 final.

[5] COM (2011) 25 final.

[6] EC report on "Exchanging Best Practice on Land Use Planning, Permitting and Geological Knowledge Sharing".

- Quality of life (social impact, environment).
- Legislative framework (applicable laws, exploration/mining rights regulatory approach, exploration and mining application priority, security of tenure).
- Regulatory agencies (role of government agencies, information availability).

They should clearly define the range of acceptable mineral activity and types of minerals that can be exploited. The objective of policy is to set the framework conditions under which an enabling economic environment could be developed that aligns the country's investments with its underlying comparative advantage, so as to improve the use of scarce capital and human resources. This economic environment includes legal, institutional and fiscal reforms that allow the nation to capture economic rent generated by mineral extraction. National minerals policies also need to provide the regulatory certainty necessary to foster investments in mineral development, including the allocation of rights to subsurface resources (Carpenter 2005). In addition, minerals policies should endeavour to ensure that mineral supply will be adequate to support the economy and the defence of the nation in question, now and in the future (Solar et al. 2009).

Minerals Planning Policy: Minerals planning policy (MPP) deals mostly with planning mineral activities, so that these activities could take place in future. MPP is often a part of a minerals policy or land-use policy and is seldom a standalone document, being a more operational, less strategic document; it can include a lot of information on minerals availability, distribution, potential and demand in paper or digital form (reports, maps, GIS).

Land-Use Planning Policy: Land-use planning is an administrative planning tool used to steer and regulate several, and moreover often controversial, needs for the same location. Mining is one component of land-use planning, as are agriculture, nature conservation, industrial and urban areas, etc.

Land Access, Regulatory Framework, Voluntary Actions

Access to land, the regulatory framework surrounding authorisation and voluntary actions are important parts of minerals policy. The recommendations of WG 2[7] are setting an effective framework for strengthening these areas.

Any land-use policy for minerals that deals with access to land must utilise a robust digital geological knowledge base, identified using a transparent methodology. Alongside information on the resource, for certain minerals of local importance, there should also be a method for estimating the long-term demand for these materials and a means by which this can be translated into a spatial plan. Recognising the contribution of recycled materials, ultimately, the aim should be to ensure fair and equal consideration of all potential uses of land including the eventual extraction of raw materials.

The most important elements of the minerals exploration and extraction application process are clarity, understanding and certainty of what needs to be provided in order to get authorisation for minerals exploration or extraction. This does not necessarily need to take the shape of a standardised application form but instead could be set out in legislation or guidance. Speeding up the minerals exploration and extraction authorisation processes may be achieved through integrating the different permits required so that they are issued by one competent authority (a one-stop shop) and with only one environmental impact assessment. However, a one-stop shop system can be difficult to implement due to the number and variety of authorisations required for minerals exploration and extraction, which often involves a number of government or advisory institutions. Parallel assessment can also speed up the minerals exploration and extraction authorisation process and should help ensure a sound

[7] Ad-hoc Working Group on Exchanging Best Practice on Land Use Planning, Permitting and Geological Knowledge Sharing—WG 2.

decision as quickly as possible. This in turn will help reduce unnecessary expenditure by all parties concerned, including industry.

Codes of practice are important instruments to achieve technical, social and environmental excellence. The use and acceptance of such codes of practice in many European countries is highly dependent on the degree to which the national legislation stipulates technical details already. Use of codes of practice, guidelines or the equivalent by industry helps to ensure protection of the environment from adverse impacts of mineral extraction. Some are set out in legislation, and an important number are improved or complemented by codes of practice promoted by the industry.

Sustainability-Based Minerals Policy

The term "sustainability" is used in a modern context to express the need for a globally balanced approach "without sudden and uncontrollable collapse" (Meadows et al. 1972) whilst ecological, economic and societal issues are fully and equally respected.

With respect to minerals, sustainable policies need to (Solar et al. 2009):

1. Facilitate the transformation of natural mineral capital into built physical, economic, environmental or social capital of equal or greater value
2. Ensure that environmental and social impacts of mining are minimised and their costs incorporated into production functions
3. Require transparency and information sharing
4. Reconsider the allocation of rights and the availability of resources across generations
5. Address benefit/risk trade-offs from the perspective of multiple stakeholders and create contingency plans that will ameliorate the effects of mineral market booms and busts (Shields and Šolar 2004)

It is also essential that a sustainable minerals policy be correlated and consistent with other governmental policies (Shields et al. 2002). A sustainable minerals policy utilises the strengths of neoclassical economics with respect to reaching market equilibrium but goes beyond that theoretical construct to incorporate the issues that are fundamental to sustainability.

Current mining policies address land ownership, access, taxation, trade and employment, etc., but not capital transformation, social impact reduction or fairness. Other issues relevant to sustainability, such as environmental protection and worker safety, may be handled in separate legislation but are seldom part of the policy set of mineral-producing countries. The issues that current minerals policies address are important and cannot be ignored; however, the scope of minerals policy will need to be broadened to incorporate such topics as the social costs of development and production, equity and transparency. On the other hand, governmental initiatives are beneficial to sustainability outcomes (Solar et al. 2009).

Minerals Policy in the Raw Materials Initiative

Minerals Policy in the Framework of the EU Raw Materials Initiative

The European Commission responded to ongoing public debate on access and supply of mineral resources through the Communication entitled "The Raw Materials Initiative—Meeting Our Critical Needs for Growth and Jobs in Europe" or in short the Raw Materials Initiative (RMI). An integrated strategy is presented which is based on three pillars:

1. Ensure access to raw materials from international markets under the same conditions as other industrial competitors
2. Set the right framework conditions within the EU in order to foster sustainable supply of raw materials from European sources
3. Boost overall resource efficiency and promote recycling to reduce the EU's consumption of primary raw materials and decrease the relative import dependence

Each of these three pillars is described in more detail. For the EU minerals policy framework, the most relevant is the second pillar which encourages the supply from EU sources. The following elements of the second pillar are exposed:

(a) Framework regulation conditions
(b) Knowledge base
(c) Networking between the national geological surveys
(d) NATURA 2000 guidelines
(e) Research projects
(f) Cohesion policy funding
(g) Skills shortage[8]
(h) Public awareness

In order to support the European Commission's work by providing experts' knowledge, two working groups that dealt with most important issues described in the RMI were formed:

- Ad-hoc Working Group on defining critical raw materials—WG 1
- Ad-hoc Working Group on Exchanging Best Practice on Land Use Planning, Permitting and Geological Knowledge Sharing—WG 2

Both Working Groups (WG), as subgroups of the Raw Materials Supply Group[9] and chaired by the European Commission, released reports in June 2010.

The Ad-hoc Working Group on defining critical raw materials (WG 1) task was to identify a list of critical raw materials at EU level. WG 1 issued an extensive report.[10] The report's main emphases and messages are described below (ECEI WG 1 2010a):

- With regard to geological availability, the WG 1 observed that, as geological scarcity is not considered as an issue for determining criticality of raw materials within the considered time horizon of the study, e.g. 10 years, global reserve figures are not reliable indicators of long-term availability.
- Changes in the geopolitical-economic framework that impact on the supply and demand of raw materials are of greater relevance. These changes relate to the growing demand for raw materials, which in turn is driven by the growth of developing economies and new emerging technologies. Moreover, many emerging economies are pursuing industrial development strategies by means of trade, taxation and investment instruments aimed at preserving their resource base for their exclusive use.
- The report analyses a prior selection of 41 minerals and metals. In line with other studies, the report puts forward a relative concept of criticality. This means that raw material is labelled "critical" when the risks of supply shortage and their impacts on the economy are higher compared with most of the other raw materials.
- Building on existing approaches, the report sets out an innovative and pragmatic approach to determining criticality. In particular:
  - It takes into account the substitutability between materials, i.e. the potential for substitution of a restricted raw material by another that does not face similar restrictions.
  - It deals with primary and secondary raw materials and recognises the recyclability of raw materials. With secondary raw materials being considered as similar to an indigenous European resource.
  - It introduces a logical way to aggregate indicators and makes use of widely recognised indexes, such as focused supply by few countries or companies and their performance.
  - It presents a transparent methodology.
- Based on a criticality methodology, calculations are made regarding the economic importance and supply risk of the 41 materials. The following is a list of critical raw materials at EU level (in alphabetical order):
  - Antimony
  - Beryllium
  - Cobalt
  - Fluorspar
  - Gallium

[8] Programmes such as Erasmus Mundus or Marie Curie are tools of the European Commission to promote and boost skills.

[9] The Raw Materials Supply Group is a group of experts mandated by Member States representatives, European Associations and Non-Government Organisations (NGO).

[10] EC report 2010: Critical raw materials to the EU economy.

  - Germanium
  - Graphite
  - Indium
  - Magnesium
  - Niobium
  - PGMs (Platinum Group Metals)/platinum, palladium, iridium, rhodium, ruthenium and osmium
  - Rare earths =/yttrium, scandium, lanthanum and the so-called lanthanides (cerium, praseodymium, neodymium, promethium, samarium, europium, gadolinium, terbium, dysprosium, holmium, erbium, thulium, ytterbium and lutetium)
  - Tantalum
  - Tungsten
- One of the most powerful forces influencing the economic importance of raw materials in the future is technological change. In many cases, their rapid diffusion can drastically increase the demand for certain raw materials.
- The recommendations are of two types: recommendations for follow-up and further support and policy-oriented recommendations to secure access to and material efficiency of critical raw materials. The WG 1 refrained from specifying detailed actions but instead indicated areas where measures should be undertaken.

Ad-hoc Working Group on Exchanging Best Practice on Land Use Planning, Permitting and Geological Knowledge Sharing (WG 2) issued a report with findings, after extensive consultation among the WG 2 members, including recommendations. These recommendations, in particular for follow-up and support, are:

1. Minerals policy, land-use planning and administrative conditions for exploration and extraction
2. Developing the knowledge base of European resources by promotion of better networking between European Geological Surveys, competent authorities and academia with a clear EU remit
3. Developing a medium- to long-term strategy for integrating subsurface components into the land services element of the Global Monitoring for Environment and Security (GMES) Land Monitoring Core Service

With regard to minerals policy and related issues, the recommendations made are the following:

- Due to the diversity of political and geological circumstances within Member States, it is not advisable to seek to impose prescriptive recommendations relating to mineral planning policy. However, analyses of practices that are in place indicate that each Member State should consider if it would be helpful to work towards adopting the recommended policy elements.
- The WG 2 recommends a National Minerals Policy to ensure that the mineral resources are provided to society in an economically viable way, harmonised with other national policies and based on sustainable development principles. This could include a commitment to provide a legal and information framework.
- Within the general outline, the Minerals Planning Policy is seen as a key component of the national minerals policy. A detailed description should be given on the ways that future minerals supply will be secured and it should demonstrate a strong link to broader land-use planning policy and regulation.
- Furthermore, a Sustainable Minerals Policy shall be based on the principles of sustainable development and incorporate economic, environmental and social requirements.
- Any land-use policy for minerals must utilise a robust digital geological knowledge base. Alongside information on the resource, for certain minerals of local importance, there should also be a method for estimating the long-term demand for these materials and a means by which this can be translated into a spatial plan whilst recognising the contribution of recycled materials. The aim of a land-use policy for minerals should be, ultimately, to ensure a fair and equal consideration of all potential uses of land including the eventual extraction of raw materials.
- A national planning framework can help to ensure that minerals are accorded due weight in the land-use planning process, and therefore, in appropriate national circumstances, it is recommended as best practice.

- The most important elements of the minerals exploration and extraction application process are clarity, understanding and certainty of what needs to be provided in order to get authorisation for minerals exploration or extraction. This does not necessarily need to take the shape of a standardised application form but instead could be set out in legislation or guidance.
- Speeding up the authorisation processes may be achieved through integrating the different permits required so that they are issued by one competent authority (a one-stop shop) and with only one environmental impact assessment or by parallel assessment. It is for individual Member States to decide which elements of best practice in authorisation to adopt, based on national circumstances.
- Codes of practice are important instruments to achieve technical, social and environmental excellence. Use of codes of practice, guidelines or the equivalent by industry helps to ensure protection of the environment from adverse impacts of mineral extraction.
- There are important issues that need to be addressed to improve the knowledge base of mineral deposits in the EU. Principal among these is the lack of harmonised EU-level data sets.

Better networking between the national Geological Surveys of Member States is the basis for cooperation between relevant institutions and the Geological Surveys, driven by the need to:

- Achieve synergies between the Geological Surveys
- Provide public data for policy making
- Facilitate investment in exploration and extraction
- Provide minerals intelligence

In order to achieve these goals, the knowledge base and networking must be structured, organised, long-term oriented and consensus-based.

GMES will provide satellite data which is needed for providing RMI-targeted information services and land-cover/land-use maps and monitoring which can benefit the RMI. Services tailored for the RMI and based on GMES data can be provided by competent national institutes or companies or, alternatively, potentially by GMES if European funding is justified.

The reports were publicly debated within extensive public consultation over summer 2010. As one result of this process, the European Commission has published a further Communication, entitled "Tackling the Challenges in Commodity Markets and on Raw Materials" (COM 2011), which was adopted in February 2011 under the European Strategy on Raw materials and a continuation of the RMI activities. This Communication deals with the Raw Materials Initiative in its fourth section, after the introduction, description and EU response to commodity markets. The RMI section is divided into subsections on (1) identifying critical materials, (2) implementing the EU trade strategy for raw materials, (3) development instruments, (4) new research, innovation and skills opportunities, (5) guidelines on the implementation of NATURA 2000 legislation and (6) increased resource efficiency and improved conditions for recycling. The next section (5th) entitled "Future Orientations of the Raw Materials Initiative" has the following subsections: (1) monitoring critical raw materials, (2) fair and sustainable supply of raw materials from global markets (pillar 1), (3) fostering sustainable supply within the EU (pillar 2), (4) boosting resource efficiency and promoting recycling (pillar 3) and (5) innovation: a cross-cutting issue.

With regard to fostering sustainable supply within the EU, the Europe 2020 Strategy is a key document because it underlines the need to promote technologies that increase investment in the EU's natural assets. Extractive industries fall under this category, but its development is hindered by a heavy regulatory framework and competition with other land uses. Many regulatory issues in this area are under the competence of Member States. The European Commission therefore acts mainly as a facilitator for the exchange of best practices. At the same time, extraction in the EU must occur in safe conditions. This is important both for the image of the sector and as a precondition for public acceptance. The Commission considers that the following practices are particularly important in promoting investment in extractive industries:

- Defining a National Minerals Policy to ensure that mineral resources are exploited in an economically viable way, harmonised with other national policies, based on sustainable development principles and including a commitment to provide an appropriate legal and information framework.

- Setting up a land-use planning policy for minerals that comprises a digital geological knowledge base, a transparent methodology for identifying mineral resources, long-term estimates for regional and local demand and identifying and safeguarding mineral resources (taking into account other land uses) including their protection from the effects of natural disasters.
- Putting in place a process to authorise minerals exploration and extraction which is clear, understandable, provides certainty and helps to streamline the administrative process (e.g. the introduction of lead times, permit applications in parallel and a one-stop shop).

The European Commission proposes to assess with the Member States, in full respect of the subsidiarity principle, the feasibility of establishing a mechanism to monitor actions by Member States in the above area, including the development of indicators (COM 2011).

Encouraged by Competitive Council Conclusion, the European Commission has started an extensive consultation process to identify possibilities, objectives, tasks and deliverables which might result in a European Innovation Partnership on Raw Materials (EIP RM). A public consultation process was supported with a Non-Paper (European Commission Background Paper 2011). In order to achieve its goals, the EIP RM focuses on three major mechanisms:

- Support to the development of innovative technologies.
- Support possible innovative regulatory action and/or cooperation with Member States.
- Promotion of standardisation and public procurement instruments.

Relevant for policy development is the second point: support possible innovative regulatory action and/or cooperation with Member States. In the Non-Paper (European Commission Background Paper 2011), one of the bottlenecks recognised is the lack of appropriate networking between Member States in relation to the implementation of the minerals policies. In particular, suitable financial engineering should be elaborated and put in place in order to mobilise public (EU and Member States) and private resources needed to achieve the goals of the present initiative, avoiding duplication or waste of resources, whilst maximising the impact.

The mechanism of how to achieve the goals is in the operational draft of the EIP RM, divided between two working packages: (WP 3) improving Europe's raw materials regulatory framework, knowledge and infrastructure base and (WP 4) improving the regulatory framework via promotion of excellence and promoting recycling through public procurement and private initiatives. WP 3 focuses more on production from primary raw materials, whilst WP 4 is targeted on secondary raw materials. Within WP 3, sustainable alternative solutions should also be related to:

- Identification and exchange of best practices in defining a minerals policy in the Member States based on principles of sustainable development and on a strict enforcement of the existing legislation, notably for what concerns the safety of mining waste facilities, the prevention of mining waste generation and the reduction of their impact on the environment.
- Identification of best practices in terms of land-use planning for minerals in the Member States and to incorporate the consideration of minerals in marine spatial plans.
- Identifying different instruments (such as a one-stop shop or parallel assessment) in order to facilitate the process for authorisation of minerals exploration and extraction in the Member States.

Basic Actions/Components of Minerals Policy

Basic components of minerals policy are well described in the document "Improving framework conditions for extracting minerals for the EU. Exchanging Best Practice on Land Use Planning, Permitting and Geological Knowledge Sharing," in short Report II of the Ad-hoc Working Group, a subgroup of the Raw Materials Supply Group chaired by the European Commission. The work on Report II has been undertaken in response to Actions 6 and 7 proposed in the Raw Materials Initiative (COM 2008), linked to the second pillar of the Initiative and in close cooperation with Member States and stakeholders. A key part of its work was a comprehensive questionnaire survey conducted in late 2009 covering:

- Exchange of best practices in land-use planning.
- Geological knowledge base and better networking.
- Integrating subsurface information in GMES.

The first two items are directly linked to minerals policy. To understand the minerals policy in a more uniform way, several definitions were proposed (ECEI WG 2 2010b):

National Minerals Policy is a clear statement of national minerals policy, setting out objectives to ensure that the mineral resources are provided to society in an economically viable way, harmonised with other national policies and based on sustainable development principles. This could include a commitment to provide a legal and information framework.

Legal framework (*regulations*) has the following tasks:

- Legal frameworks (minerals acts), covering all types of minerals to guarantee legal and planning certainty for all parties involved, and speedy and streamlined authorisation processes.
- A transparent non-distorting fiscal framework acts as a stimulus for exploration and acquisition.
- Effective safeguarding of actual and potential mineral resources through land-use planning is crucial to avoid improper land use and/or sterilisation of mineral resources.

Information framework should be:

- Reliable and comprehensive national and international statistics for trend analysis and as a decision base for authorities and the industry.
- A detailed geological knowledge base which is publicly available within the legal frameworks of Member States and includes comprehensive geological, geochemical, geophysical and general mineral data.

Minerals Planning Policy—Raw materials planning policy, as a key component of the national minerals policy, should describe in detail the ways that future minerals supply will be secured and demonstrate a strong link to broader land-use planning policy and regulation.

Sustainable Minerals Policy—An ethics-based minerals policy based on the principles of sustainable development and comprising the following three pillars:

The economic pillar:

- Providing a proper long-term economic environment for exploration and mining activities to ensure minerals supply.
- Safeguarding mineral deposits through land-use planning to secure future minerals supply, promoting research and development for resources and energy efficiency.

The environmental pillar:

- Ensuring that the negative environmental impacts of the extractive industry are controlled to acceptable levels of risk.
- Promoting sound site reclamation and aftercare practices.
- Promoting research and development, e.g. environmentally sound mining methods (cradle to grave), materials efficiency, substitution, recycling and use of Best Available Techniques (BAT).

The social pillar:

- Promoting the essential contribution of minerals in society.
- Promoting a transparency for government, authorities, industry, NGOs and the general public (from local to national) to avoid conflicts and support sound and timely decision-making.

Regulations

Following minerals policy are mining acts (laws) and their subset regulations. The regulations prescribe a set of detailed mandatory instructions on how to manage mineral resources within the whole mine cycle (prospection, exploration, mine development, operation phase, closure and remediation and after closure care). Special attention within the European Commission was put on the authorisation process.

The Working Group (WG II) stated that the most important elements of the minerals exploration and extraction application process are clarity, understanding and certainty of what needs to be provided in order to get authorisation for minerals exploration or extraction. This does not necessarily need to take the shape of a standardised application form but instead could be set out in legislation or guidance.

Representations from industry emphasised the need to speed up the minerals exploration and extraction authorisation processes. This may be achieved through integrating the different permits required so that they are issued by one competent authority (a one-stop shop) and with only one environmental impact assessment.

However, it is the view of the Working Group that a one-stop shop system can be difficult to implement due to the number and variety of authorisations required for minerals exploration and extraction, which often involves a number of government or advisory institutions.

In some cases, community consultation is undertaken in order to build social acceptance of projects. Through these consultations, projects have been able to take into account those issues that matter for the local community.

Information

The Raw Materials Initiative highlights the improvement of the EU knowledge base as a condition to enhance sustainable supply from within the EU, and this topic was debated in detail with WG II.

Data acquisition and processing activities rest with the national and regional Geological Surveys and some specialised public research institutes on the basis of the mineral potential outlined by the public data. WG II acknowledges the need for data and information at EU level on global minerals production, imports and exports, outputs and shifts in the global minerals industry. The development of the EU knowledge base on mineral resources should be based on several pillars, namely:

Intelligent mineral statistics comprising:

- Standardised and accurate statistical data on worldwide minerals production, imports and exports.
- Analysis and comments on demand, supply and underpinning factors such as technology shifts, geographic or capitalistic concentrations and geopolitical factors.

This data and information should be published on an annual (or biannual) basis to serve to analyse trends and help decision makers to better understand and monitor the EU supply and demand situation and related risks. It should be prepared in collaboration between EuroStat, interested European Commission services and national authorities already producing such data and information for their national or regional authorities.

Developing the "Europe 3D" data and knowledge basis is needed in order to assess the EU potential for deep-seated, hidden mineral deposits. Due to the size of the project, the effort it involves in data acquisition, processing and modelling and the amount of highly qualified expertise required, this should be a 20- to 30-year programme focused on known mineral-rich areas.

Public awareness is a key instrument to get broad acceptance to an idea or sector. Public events like Minerals Days, Green Week and publications and positive articles in the public press are some steps that need to be taken. Awards and scholarships for advertisement/posters promoted in particular at schools might attract young people. The working group encouraged all stakeholders and Member States to increase their visibility to the public by joining specific public events and increasing their contacts with the local press.

EU Minerals Policy Forthcoming Issues: The European Innovation Partnership

Innovation is the cornerstone of the Europe 2020 strategy for smart, sustainable and inclusive growth. The proposed European Innovation Partnership (EIP) should contribute to the mid- and long-term security of and access to the supply of raw materials essential to the competitiveness of EU industries, to increase resource efficiency in the EU and to the development of new European-based recycling activities (largely SME-based). Whilst the policy framework has been set up in the Communication on Commodities and Raw Materials (COM 2011), the Innovation Partner-ship is focused on three major mechanisms in order to meet its overall objectives:

1. Support to the development of innovative technologies
2. Support possible innovative regulatory action and/or cooperation with Member States
3. Promotion of standardisation and public procurement instruments

The second objective is directly related to minerals policy issues, those on national/Member State level as well as the minerals policy framework on the EU level. Added value for minerals policy is within special attention that should be given to promote a full implementation of existing legislation and increased exchange of best practices related to the implementation of Environmental and Minerals Policies in Member States, particularly in the field of land use and in mining waste.

The European Commission has identified five Work packages around which the main work of the Innovation Partnership could be organised:

- Work package 1—Developing new innovative technologies and solutions for sustainable raw materials supply.
- Work package 2—Developing new innovative materials by design and solutions for the substitution of critical materials.
- Work package 3—Improving Europe's raw materials regulatory framework, knowledge and infrastructure base.
- Work package 4—Improving the regulatory framework via promotion of excellence and promoting recycling through public procurement and private initiatives.
- Work package 5—International framework—horizontal approaches.

Challenges and Options

Among many challenges stated in the EU documents, the lack of appropriate networking between Member States in relation to the implementation of the minerals policies is also exposed. In particular, suitable financial engineering should be elaborated and put in place in order to mobilise public (EU and Member States) and private resources needed, avoiding duplication or waste of resources whilst maximising the impact. Other challenges/bottlenecks are related to the main one and their solution would be supportive in:

- Lack of effective structures for networking and cooperation (in particular between geological services) to develop a knowledge- and demand/user-based approach.
- Unequal application and enforcement of the European waste legislation.
- Lack of reliable projection of future supply and demand of raw materials at the EU level.
- Need for standardisation and certification schemes.
- Need for harmonised statistics on raw materials knowledge database.

Solution options are described within Work package 3—Improving Europe's raw materials regulatory framework, knowledge and infrastructure base. Sustainable alternative solutions such as those related to minerals policies look at the identification and exchange of best practices in defining a minerals policy within the Member States. This is based on principles of sustainable development and on a strict enforcement of the existing legislation, notably for what concerns the safety of mining waste facilities, the prevention of mining waste generation and the reduction of their impact on the environment; furthermore, they look at identification of best practices in terms of land-use planning for minerals in the Member States and to incorporate the consideration of minerals in marine spatial plans. It is also important to identify different instruments (such as one-stop shop or parallel assessment) in order to facilitate the process for authorisation of minerals exploration and extraction in the Member States.

Conclusions

The European Union was originally established as the European Coal and Steel Community (by six countries) in 1958 due to raw materials issues, which are presently at the centre of the public's and industry's attention. At the end of the first decade of the twenty-first century, supply of raw materials has once again become, due to different reasons (mainly demand from emerging economies), one of the most important economic topics and issues, as it was at times in the past. Over a period of more than 15 years, raw materials supply was not noticed as being problematic for European society as market-regulated supply was effective and efficient. Raw materials supply stayed in focus regardless of economic conjuncture, crisis and slow recovery.

The European Union, through the European Commission, and its Member States recently recognised that minerals supply and demand issues could be tackled through a comprehensive and integrated approach expressed within a minerals policy framework at EU level and revised Member State minerals policies.

The European Commission raised the issue by preparing several Communications that were reviewed and accepted by the European Council of Ministers and the European Parliament. The Raw Materials Initiative, as one of the most important components of EU policy at the turn of the first decade of the twenty-first century, focuses on defining the framework for minerals policy at the EU level, whilst minerals policies and regulations themselves are within the competence of Member States through the Lisbon Treaty.

Acknowledgements Special thanks are expressed to Antje Wittenberg and Paul Anciaux of DG Enterprise and Industry, European Commission, for their valuable contribution to this article, as well as to Mr. Woody Hunter of EuroGeoSurveys who substantially contributed to its review.

Disclaimer This article does not reflect the view of the European Commission or any of their officials. It is purely a private view.

References

Carpenter AS (2005) Issues that affect the competitiveness of American's mining industry. Testimony to the U.S. House Resources Sub-Committee on Energy and Mineral Resources Oversight Hearing "Improving the Competitiveness of America's Mining Industry", Washington, DC, 28 April 2005

European Commission, COM (2005) 658: On the review of the Sustainable Development Strategy – A platform for action, Brussels

European Commission, COM (2008) 699: The raw materials initiative – meeting our critical needs for growth and jobs in Europe, Brussels

European Commission, Enterprise and Industry, (ECEI WG 1) (2010a) Critical raw materials for the EU. Report of the Ad-hoc Working Group on defining critical raw materials, Brussels, 85 p

European Commission, Enterprise and Industry, (ECEI WG 2) (2010b) Improving framework conditions for extracting minerals for the EU. Exchanging Best Practice on Land Use Planning, Permitting and Geological Knowledge Sharing, Brussels, 71 p

European Commission Background Paper (2011) Public Consultation on Commission Raw Materials Initiative. DG Enterprise and Industry. URL: http://ec.europa.eu/enterprise/newsroom/cf/_getdocument.cfm?doc_id=633

European Commission, COM (2011) 25: Tackling the challenges in commodity markets and on raw materials, Brussels

G8 Summit Declaration (2007) Growth and responsibility in the world economy, Heiligendamm, Germany, 7 June 2007, pp 29–32

Meadows DH, Meadows DL, Randers J, Behrens WW (1972) The limits to growth: a report for the Club of Rome's project on the predicament of mankind, 2nd edn. Universe Books, New York

Otto J (1997) A national mineral policy as a regulatory tool. Resour Policy 23(1–2):1–7

Shields DJ, Šolar SV, Martin WE (2002) The role of values and objectives in communicating indicators of sustainability. Ecological Indicators 2 (1/2): 149–160

Shields DJ, Šolar SV (2004) Sustainable mineral resource management and indicators: case study Slovenia. Geological Survey of Slovenia. 84 p

Solar SV, Shields DJ, Miller MD (2009) Mineral policy in the era of sustainable development: historical context and future content. RMZ 56(2009):304–321

Weber L (2008a) Minerals safeguarding activities of the European Union and their impacts on national level. BHM Berg- und Hüttenmännische Monatshefte 153(8):289–295. doi:10.1007/s00501-008-0394-1

Discovery and Sustainability*

Friedrich-W. Wellmer

Abstract

Although we must rely on natural resources, especially for their functional value, our supplies are finite. A key challenge is to find solutions for sustaining development. We do not need natural resources as such, but we need their functions or physical and chemical properties. To find solutions for functions, there are three reservoirs: the combined resources of the geosphere and the technosphere and the unlimited resource of human ingenuity. With these, enough time, and the wisdom to plan ahead, we can find the necessary solutions for sustainability on Earth.

Keywords

Discovery • Ratio reserves/consumption • Snapshot of dynamic system • Geosphere • Technosphere • Human ingenuity • Hierarchy of natural resources • Learning curves • Sustainability

Article

For his well-being, man needs more than 30 billion tons of natural resources, energy and mineral resources, every year, not counting potable water which is our most important natural resource. This consumption is very unevenly spread around the globe.

Up to the end of the last millennium, the bulk of mineral and energy resources was used in industrialized countries that had only a very modest population growth. We were living in an "upside-down world": about 25% of the world's population lived in industrialized nations consuming 70–80% of the world's energy and mineral resources, coal being one of the few exceptions (Wagner and Wellmer 2009). Since the turn of the millennia, however, we have witnessed the start of a new growth cycle that is driven by a surge in demand in developing nations with large populations. China has become the leading consumer in the developing

*Expanded and modified version of article in the Official Publication of the International Year of the Planet Earth (Wellmer 2008b).

F.-W. Wellmer (✉)
Formerly Bundesanstalt für Geowissenschaften und Rohstoffe (BGR),
Neue Sachlichkeit 32, Hannover D-30655, Germany
e-mail: fwellmer@t-online.de

R. Sinding-Larsen and F.-W. Wellmer (eds.), *Non-Renewable Resource Issues: Geoscientific and Societal Challenges*, International Year of Planet Earth,
DOI 10.1007/978-90-481-8679-2_3, © Springer Science+Business Media B.V. 2012

world, as shown by Sinding-Larsen and Wellmer (2012).

In industrialized countries, the individual statistical lifetime consumption is about 1,000 t of raw materials ranging from construction material to energy resources like oil, gas, or coal, to metals like iron or copper, and to nonmetallic resources like phosphate or potash used as fertilizers. Despite talk about dematerialization, consumption per capita of a number of mineral commodities in developed countries has remained roughly constant for the last 30 years (Menzie et al. 2001, 2005). Life without raw materials is unthinkable. When we go to work by car, bus, or train, not only steel is moved but so are various other metals like zinc, chromium, nickel, or copper, as well as plastics based on oil. Even our information technology society needs a wide variety of natural resources. The more sophisticated the electronic components for information technology hardware become, the broader the variety of elements (Hagelüken 2006). Based on information from the Intel Corporation (NRC 2008), the National Research Council of the USA gives an example for the learning curve of three decades of high-tech-component development. In the 1980s, computer chips were made out of 12 minerals or their elemental components. In the 1990s, the number increased to 16. Today, as many as 60 different minerals (or their constituent elements) may be used in fabricating the high-speed, high-capacity integrated circuits that are crucial to this technology.

Deposits for natural resources first have to be discovered. It is the task of the natural resources industry to create the basis for its production activities. It wants to discover deposits which can be brought into production as fast as possible in order to recover its exploration expenditures and get a reward for exploration risks taken. Companies are neither interested to discover deposits which can be brought into production only in the far future nor to establish a world balance. The consequence is that the balance of reserves, which can be exploited economically with current technology, is always only a momentary snapshot of a dynamic system. Exploration is an ongoing process while technology is improving.

Table 1 The growing of reserves—the example of crude oil (BGR 2009)

	Production	Reserves	R/C ratio
1950	543 Mio t	11,277 Mio t	20
2008	3,900 Mio t	160,000 Mio t	41

Therefore, also the reserve/consumption ratio is a snapshot of a dynamic system. It is *not* the life index resp. lifetime of natural resources (Zwartendyk 1974). This fallacy can be illustrated for every commodity. Let us take zinc as an example. The ratio reserve/consumption fluctuated around 20 fifty years as it does today with no shortages experienced. At the same time, the zinc production was increased from 2.2 million tons in 1950 to 10.0 million tons now. Another example can be crude oil, where the reserve/consumption ratio even improved over the nearly last 60 years (Table 1).

Besides ongoing exploration, improvements in technology are important, as illustrated here again with the example of crude oil. Forty years ago, water depth from which oil was produced was less than 100 m. Indications for oil in deeper waters were considered resources or imagined geopotential, i.e. favourable geology with the potential that crude oil could be discovered. Today, oil is economically produced from water depths close to 3,000 m. In 1950, less than 5% of oil was produced offshore; in 2005, the share was 37% (Wellmer 2008a).

All such natural resources are produced from Earth, the geosphere. Since we have only one Earth to live with, it is considered that exploitation of natural nonrenewable resources and the notion of sustainability are a contradiction in itself. According to this opinion, consumption of limited resources counteracts the intergeneration challenge of sustainability, namely, that the ability of future generations to meet their own needs should not be compromised (Brundtland 1987). Although not explicitly stated, embodied in the concept of sustainability is that it be achieved over a duration of time measured in centuries or millennia, not just decades. True sustainability means "practically forever," or at least as long as humans exist (Ernst 2002).

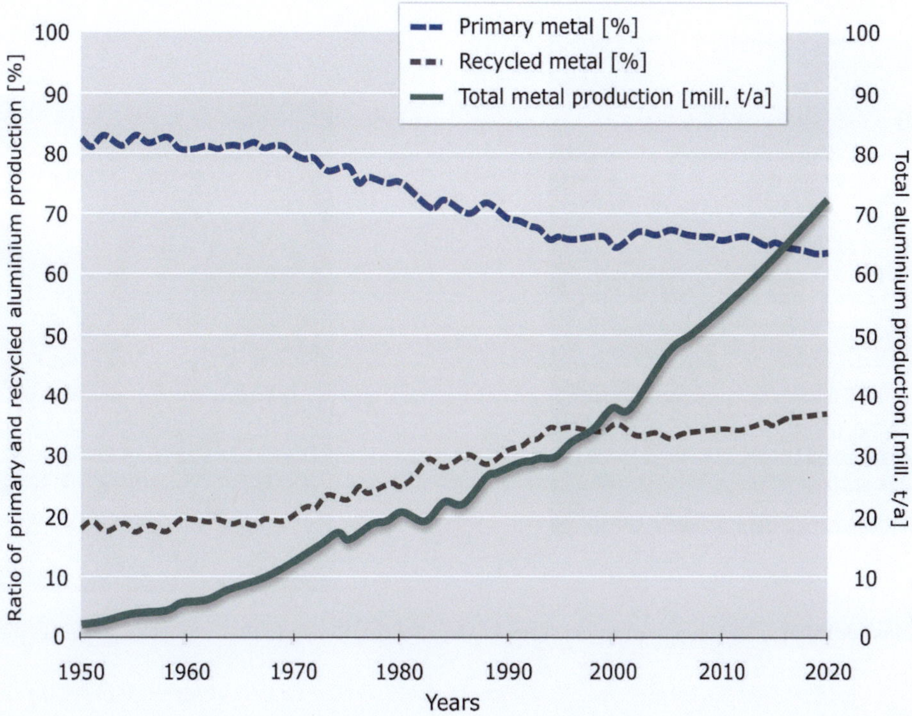

Fig. 1 Total production, and primary and secondary production of aluminium (Source: Gerber 2007)

No one will argue against intergeneration fairness. Consequently, let us analyse natural resources with regard to the challenge of intergeneration fairness. If we consider why we need natural resources, we find that we do not need the resources as such, but we need their inherent functions, or their physical and chemical properties. We do not need one tonne of copper; we need its electrical conductivity for transmitting electric power or transferring messages via electric pulses in telephone wires. Latter function can also be fulfilled by glass fibre cables, by directional antennae, or by mobile phone. Every technical solution has its own raw material profile. Until recently, silver was essential in photography, whereas modern digital cameras require totally different raw materials. In the printing industry, lead was formerly used as type metal, but it has been replaced today by offset or computer printing. These are just some examples illustrating our need for their properties and functions rather than for natural resources as such.

Moreover, establishing a cycle economy for increasing natural resources' efficiency is another important element of sustainability (Steinbach and Wellmer 2010). Cycle economy applies, for example, to water and especially to metals. Metals are not so much consumed as being used for a limited period of time. Metals survive use and are merely transferred from the geosphere to the technosphere in one form or the other. Copper, lead, or steel may be recycled many times—often with no or little reduction in quality or, in the case of alloys in particular, with somewhat lower quality after recycling. Cycle economy is the ultimate solution also for phosphorous, one of the essential plant nutrient elements. They are as important as water and cannot be replaced. Whereas for nitrogen and potassium, an unlimited reservoir exists in air and seawater, this is not so for phosphorous.

It is too short-sighted to focus on the extraction of nonrenewable resources from the geosphere and their transfer to the global industrialized

Fig. 2 Monsieur Eiffel would only need so much steel today (Source: Benitz/BGR)

civilization, the so-called technosphere. Resource efficiency can be increased by substituting primary resources of the geosphere whenever possible by secondary resources of the technosphere. It can be observed, for example, that in total aluminium consumption, the share of the recycled metal steadily increases, constituting a higher share of known resources of the technosphere in the cycle economy (Fig. 1). Substitution includes waste products. For example, gypsum from desulphurization of flue gas in power plants can replace primary gypsum. In many countries, the "3R" initiatives—reduction, reuse, and recycling—are encouraging efforts (Bleischwitz and Bringezu 2007). There are vast fields for improvements. Vehicles and many household goods are exported from industrialized to developing nations with very limited recycling systems. In these countries, recycling rates are normally high for simple but low for complex products. For example, the highest recycling rates for aluminium cans are achieved in Brazil (Rombach G, 2010, Hydro Aluminium Deutschland GmbH, Bonn, Germany, Personal communication). On the other hand, the gold recovery from electronic scrap in modern integrated smelters is 95%, whereas in "backyard recycling" operations in developing countries, the recovery is as low as 25% (Rochat et al. 2007; Hagelüken 2012).

According to the concept of sustainability as an intergeneration fairness challenge, the equation has to take into account all combined resources of the geosphere and the technosphere and the challenge to use them most thriftily. Today, the Eiffel Tower in Paris could be built with just 2,000 t of steel instead of the 8,000 t used between 1885 and 1889 (Fig. 2) (Lurgi 1992).

There is one more element in this equation to solve the intergeneration fairness postulate of sustainability in order to give future generations the means to meet their own needs: it is a "driver" to find solutions for functions and for improving natural resources' efficiency. This "driver" is man's unlimited resource of creativity and ingenuity, that of present and of future generations (McKelvey 1973).

The incentive to find new solutions for functions and improving natural resources' efficiency is on the one hand human curiosity for discovering the unknown and developing new technologies. On the other hand, it is the price incentive. If a scarcity occurs in a market economy, prices

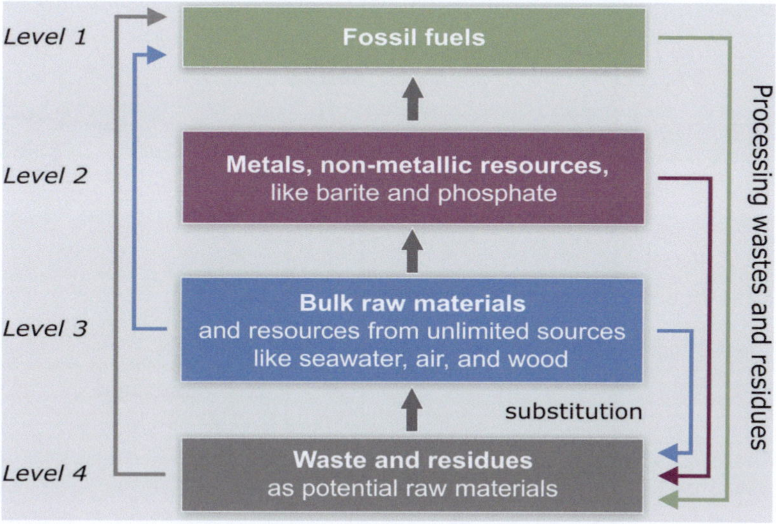

Fig. 3 Four-level hierarchy of natural resources with respect to sustainable development (From Steinbach and Wellmer 2010)

will rise. The possibility to earn high returns will induce inventiveness to find solutions. These will be on the supply-side acceleration of exploration activities to discover and to bring into production new deposits in the geosphere and in the technosphere to improve recycling rates. On the demand side, new processes will be initiated: technological changes for substitution, material savings, or inventing totally new technologies for fulfilling the required functions without using the scarce material at all as shown above in the cases of copper use, printing, or photography. In market economy systems, this scheme has worked in the past. There is no reason to believe that it will not work in future.

The need to use our ingenuity and to move up the learning curve quickly is most obvious for energy. Fossil energy sources are really consumed and cannot be recycled, but all forms of energy can be substituted—directly or indirectly via electricity—for the functions needed: heat, motion, and light. All forms of energy, therefore, have the highest potential of "substitutionability." The cycle system here is using renewable energies like sun, wind, biofuel, or geothermal power. Everywhere in the world, the potential is far larger than man's needs are. It is man's greatest challenge to make the use of fossil fuels for energy just a bridging technology as fast as possible.

For the time interval during which we have to rely on fossil fuels and nuclear power for the major portion of our energy needs, Wagner and Wellmer (2009) proposed a four-level hierarchy of natural resources with respect to sustainability as a basis for a natural resources efficiency indicator stressing the importance of efficient energy use (Fig. 3). According to this concept, sustainable development implies substituting materials at a higher level of the hierarchy, either by material from a lower level or by resources from the technosphere that replace resources from the same level in the geosphere. Energy resources (oil, natural gas, and coal, as well as uranium as fuel for nuclear power stations) occupy the highest level of the four-level hierarchy. The next hierarchy level is represented by non-energy raw materials that are derived from occurrences that developed over geological time and were formed by natural enrichment (e.g. all metal deposits and some nonmetallic deposits such as phosphate). This level also includes deposits of the technosphere which can be recycled. The third level comprises not only materials available in almost unlimited amounts on Earth, such as granite, sandstone, and clay, but also those raw materials which can be produced from air (e.g. nitrogen fertilizer) or from seawater (e.g. boron, potassium, or magnesium). Wood used for construction purposes is included in this third level

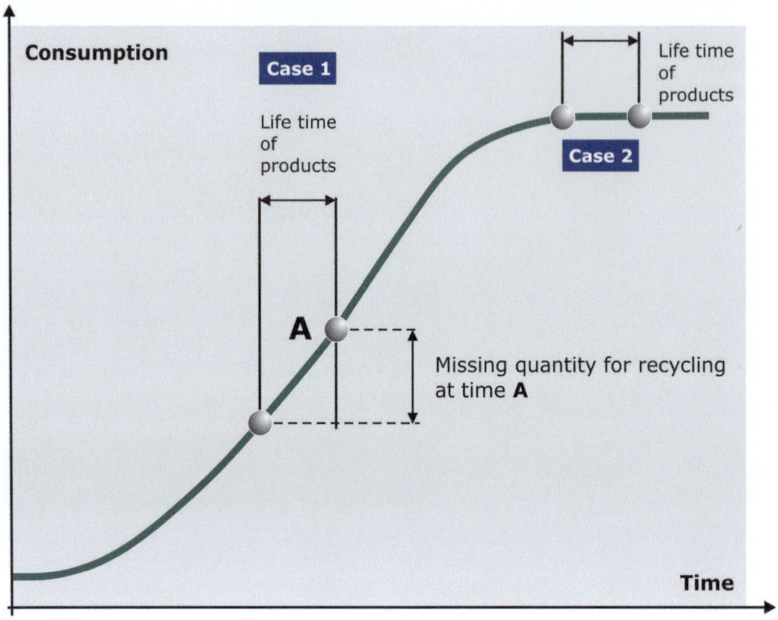

Fig. 4 Idealized growth curve of commodities illustrating constraints of possible quantities for recycling (From Steinbach and Wellmer 2010). Case 1: Times of growing consumption mean a deficit of theoretically available secondary material at a later date. Case 2: Times of constant consumption mean at a later date there is theoretically the same amount of material available as at the beginning of the life time of the products

because it is a renewable resource, although wood can become a limited resource as well if demand rises. The lowest level represents waste and residue materials from the technosphere that are potential raw materials for secondary use.

As said, the path for energy of the highest hierarchy level to reach of sustainability is to exclusively use renewable energies as soon as possible. Concerning metals as the most important resources on the next level of the hierarchy, man creates an ever increasing pool of metals in the technosphere, a pool of renewable resources, to replace limited resources of metals in enrichment deposits in the geosphere (see Fig. 1 for aluminium). Steinbach and Wellmer (2010) developed a scenario that metals practically become renewable resources when the gradient of metal consumption curve becomes zero. As Fig. 4 shows in times of growing consumption (case 1 in Fig. 4), there is always a shortfall of metals in the technosphere available for recycling and reuse. But one can foresee another future. Consumption patterns for all raw materials follow a learning curve caused by economic growth patterns of countries and country groups: after a slow start, a period of high growth, then a flattening occurs (Rostow 1971; Wagner and Wellmer 2009). This can very well be seen with zinc and is even more pronounced with steel (see Fig. 5). The flattening of steel consumption in the 90s of the last century has now been overtaken by a new growth cycle due to high growth rates, mainly in China, then India and Brazil (Sinding-Larsen and Wellmer 2012). One can envisage, however, a future world in which about 80% of the world is at the level of the industrialized nations before the recent turn of the millennia. Then, a stage is conceivable at which a maximum amount of raw material can be recovered from secondary material because the historical secondary material supply is in balance with the demand (case 2 in Fig. 4). Wellmer and Steinbach (2010) took the lead market in Germany as a test case to support this scenario. The lead market is a suitable example because it does not grow anymore in industrialized nations.

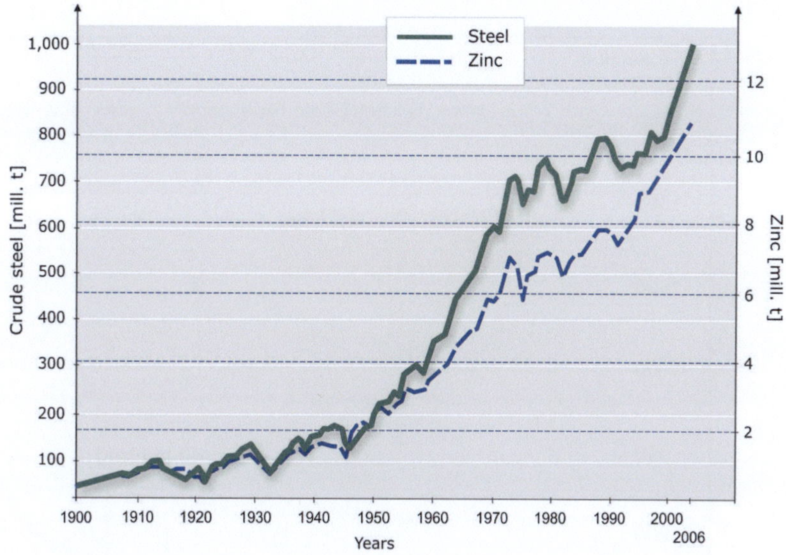

Fig. 5 Worldwide production of steel and consumption of zinc (From Steinbach and Wellmer 2010)

At this equilibrium stage, a certain contribution from the geosphere is still required to offset inevitable metal losses in the technosphere and to compensate a possible shortfall. Humans, however, have a choice. They do not need raw materials as such but rather an intrinsic property that fulfils a function.

To bridge the time interval prior to the stage when the majority of energy consumed will be renewable energy and the majority of metals are derived from the technosphere, time may be the scarcest commodity that limits the climb of the necessary learning experience. Finding solutions for substitutions, functions, and raw materials requirements through human ingenuity and creativity means providing time for progressing along the inevitable learning curves. This requires wisdom to plan ahead. The author is not of the opinion that an ingenuity gap (e.g. Homer-Dixon 2000) exists. Frequently, a period of 20 years can be observed as needed for new industrially feasible solutions in the natural resources field (Wellmer 2008a). Roodhart (2007), based on a study by the management consulting company McKinsey, compared the learning curves of the hydrocarbon exploration and production (E&P) industry with the ones in the field of consumer products, medicine, and broadband telecommunication. Whereas in these three fields, the time to reach the commercial stage varied between 8 and 15 years; in the E&P field, according to 15 case histories, the time was 22 years. As have others earlier (e.g. Mackenzie and Woodall 1983; Wellmer 2004), Roodhart (2007) sees the need of the natural resources industry to accelerate the pace of innovation. Three energy examples, one each from the fields of crude oil and natural gas, one from renewable energy, may illustrate the gradient of learning curves in the natural resources field and the need to accelerate the innovation pace:

In the German Third Reich from 1933 to 1945, there was an unlimited amount of money available for the search of crude oil to "feed" the war machine. As it turned out, the concept prevalent then that crude oil in Northern Germany, the main oil region of the country, was structurally associated with salt domes was true in some cases, but wrong for the most significant deposits. Only in later years did it become obvious that the major crude oil deposits were associated with large anticlines between the salt domes (Kockel 1997). Prior to 1945, Germany never produced more than 1 million tons/a and the peak of crude oil production in excess of 8 million tons/a was not reached until 1968, illustrating the time element necessary to step up the learning curve in exploration (Wellmer and Becker-Platen 2002).

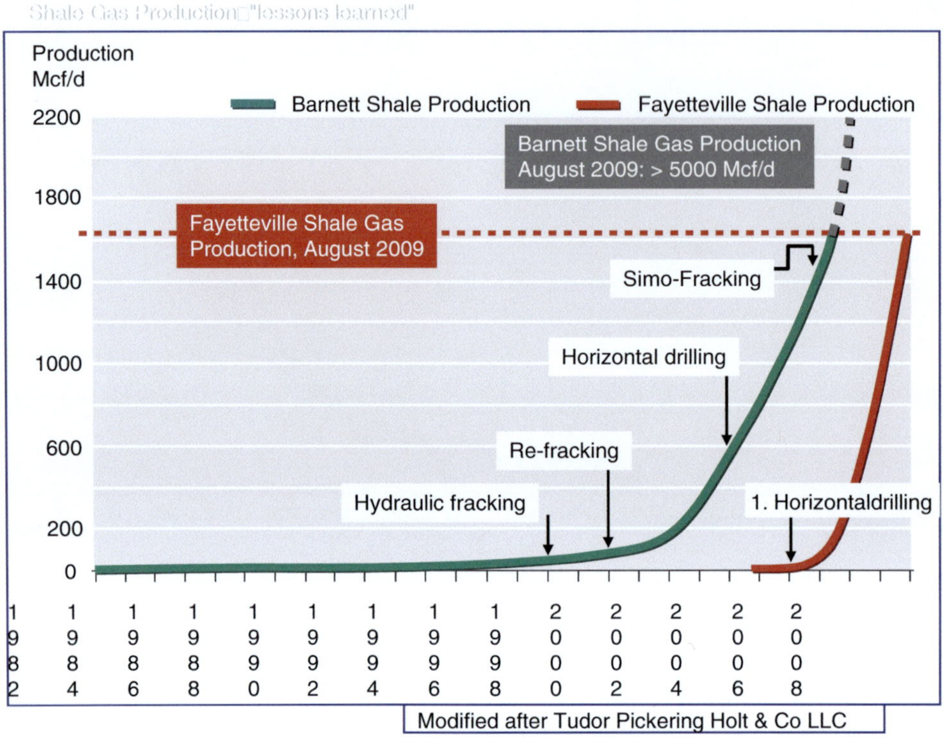

Fig. 6 Learning curve of shale gas production in the USA (From Andruleit et al. 2010a, b)

An unconventional gas resource is shale gas which is increasingly exploited in the USA. Currently about 5–6% of the domestic gas production of the US comes from Paleozoic shales rich in organic matter, i.e. the source rock is also the reservoir rock (Jenkins and Boyer II 2008). The most productive unit is the Barnett shales in Texas. It took about 20 years to develop the suitable horizontal drilling and frac techniques to generate substantial production rates (Andruleit et al. 2010a, b) (Fig. 6).

In 1983 the German Federal Ministry of Research and Technology hoped to make a quantum leap in renewable energy generation and developed a large 3-MW wind turbine, the Growian-project. constructed by a pool of large German aerospace companies. However, it never worked satisfactorily for any period of time and the project had to be terminated. This unsuccessful attempt was followed by projects of new small- to medium-sized enterprises which initially developed smaller wind turbines that worked well. It took 20 years of climbing up the learning curve before 3-MW-size units became a success (Wellmer and Kosinowski 2003). Today 5-MW wind turbines are operating. About 6.5% of electricity consumption was generated by wind power in 2009 in Germany.

To conclude, the author believes to be able to state with confidence that with human imaginativeness to find solutions, with the available resources of both the geosphere and the technosphere, and with sufficient time to go up the learning curves or wisdom to plan ahead, man has all the means to solve the conflict between use and consumption of nonrenewable natural resources and the intergeneration challenge of sustainability, thereby giving future generations the chance to be as well-off as their predecessors (Brundtland 1987) without running out of natural resources.

References

Andruleit H, Babies HG, Cramer B, Krüger M, Meßner J, Rempel H, Schlömer S, Schmidt S (2010a) Konventionell versus nicht-konventionell: Weltweite Ressourcen und Entwicklungen des "Hoffnungsträgers" Erdgas. Paper DGMK/ÖGEW-Frühjahrtstagung 2010, Celle, Germany, 12–13 Apr 2010

Andruleit H, Rempel H, Meßner J, Babies HG, Schlömer S, Schmidt S, Cramer B (2010b) Nicht-konventionelles Erdgas: Weltweite Ressourcen und Entwicklungen des "Hoffnungsträgers" unter den fossilen Energierohstoffen. Erdgas Erdöl Kohle 126(7/8):277–284

BGR (Federal Institute of Geosciences and Natural Resources, Hannover) (2009) Reserven, Ressourcen und Verfügbarkeit von Energierohstoffen. Study Hannover, Hannover, 88 p

Bleischwitz R, Bringezu S (2007) Global resource management. Policy paper 27. Development and Peace Foundation, Bonn, 12 p

Brundtland GH (1987) Our common future—report of the World Commission on Environment and Development. Oxford University Press, Oxford, 400 p

Ernst WG (2002) Global equity and sustainable resource consumption requires super-efficient extraction-conservation-recycling and ubiquitous, inexpensive energy. Int Geol Rev 44:1072–1091

Gerber J (2007) Strategy towards the red list from a business perspective. Paper ETH workshop scarce raw materials, Davos, Switzerland, 1–2 Sept 2007

Hagelüken C (2006) Mobile phone recycling—chances and challenges from a recycler's view. Paper OEWG5 (Open-ended Working Group of the Basel Convention on the Control of Transboundary Movements of Hazardous Wastes and Their Disposal), Geneva, Switzerland, 3–7 Apr 2006, MPPI-side-event (Mobile Phone Partnership Initiative) 5 Apr 2006

Hagelüken C (2012) Secondary raw material sources for precious and special metals. In: Non-renewable resource issues. Springer, Dordrecht, pp 195–212

Homer-Dixon T (2000) The ingenuity gap. Knopf, New York, 485 p

Jenkins CD, Boyer II CM (2008) Coalbed- and shale-gas reservoirs. J Petrol Technol 60(2):92–99, SPE (Soc Petr Eng) Paper 103514

Kockel F (1997) 50 Jahre Geotektonische Karte von Nordwestdeutschland 1: 100,000 – Grundlagen, Entstehung und wirtschaftlicher Nutzen. Z Angew Geol 43/1:35–41, Hannover

Lurgi AG (1992) Jahresbericht 1991, Frankfurt/Main

Mackenzie B, Woodall, R (1983) Economic productivity of base metal exploration in Australia and Canada. Paper at expert group meeting on the economics of mineral exploration, International Institute for Applied Systems Analysis, Laxenburg, Austria, 5–12 Dec 1983

McKelvey VE (1973) Mineral resource estimates and public policy. Am Sci 60(1):32–40

Menzie WD, DeYoung JH Jr, Steblez WG (2001) Some implications of changing patterns of mineral consumption. U.S. Geological Survey Open-File Report 03–382. http://pubs.isgs.gov/of/2003/of03–382

Menzie WD, Singer DA, DeYoung JH Jr (2005) Mineral resources and consumption in the 21st century. In: Simpson RD, Toman MA, Ayres RU (eds) Scarcity and growth revisited—natural resources and the environment in the new millennium. Resources for the Future, Washington, DC, pp 33–53

NRC (2008) National Research Council of the National Academies: minerals, critical minerals, and the U.S. economy. The National Academies Press, Washington, DC, 245 p

Rochat D, Hagelüken C, Keller M, Widmer R (2007) Optimal recycling for printed wiring boards in India. In: Hilty LM, Edelmann X, Ruf A (eds) Recovery of materials and energy for resource efficiency. R'07, World Congress. EMPA Materials Science & Technology, St. Gallen

Roodhart L (2007) Prepare for the future: innovate, SPE distinguished lecture series. Paper 7th European formation damage conference, 30th May 2007, Scheveningen, The Netherlands. http://www.spe.org/spe-app/spe/meetings/EFDC/2007/tech_prog.htm

Rostow WW (1971) The stages of economic growth—a non-communist manifesto, 2nd edn. Cambridge University Press, Cambridge, 253 p

Sinding-Larsen S, Wellmer F-W (2012) Non-renewable resource issues: geoscientific and societal challenges: an introduction. In: Non-renewable resource issues. Springer, Dordrecht, pp 1–19

Steinbach V, Wellmer F-W (2010) Consumption and use of non-renewable mineral and energy raw materials from an economic geology point of view. Sustainability 2(5):1408–1430. doi:10.3390/su2051408 (Special issue Sustainability and Consumption) http://www.mdpi.com/2071–1050/2/5/1408/pdf

Wagner M, Wellmer F-W (2009) A hierarchy of natural resources with respect to sustainable development – a basis for a natural resources efficiency indicator. In: Richards JP (ed) Mining, society and a sustainable world. Springer, Heidelberg/Dordrecht, etc, pp 91–121

Wellmer F-W (2004) Innovation und Lernprozesse in der Rohstoffwelt. Kali und Steinsalz 2:38–47

Wellmer F-W (2008a) Reserves and resources of the geosphere, terms so often misunderstood. Is the life index of reserves of natural resources a guide to the future? Zeitschr Deutsche Ges Geowissenschaften 159(4):575–590

Wellmer F-W (2008b) Natural resources: discoveries and sustainability. In: The international year of the planet earth, the Official Publication. Boston Hannah Intern., London, pp 64–66

Wellmer F-W, Becker-Platen JD (2002) Sustainable development and the exploitation of mineral and energy resources: a review. Int J Earth Sci (Geol Rdschau) 91:723–745, Heidelberg

Wellmer F-W, Kosinowski M (2003) Sustainable development and the use of nonrenewable resources. Geotimes 48(12):14–17

Zwartendyk J (1974) The life index of mineral reserves—a statistical mirage. Can Min Metall Bull 67(750):67–70

Is Depletion Likely to Create Significant Scarcities of Future Petroleum Resources?

Roberto F. Aguilera, Roderick G. Eggert, Gustavo Lagos C.C, and John E. Tilton

Abstract

Some energy analysts are concerned that the world will soon face a global crisis due to dwindling petroleum resources and a peak in oil production. To shed light on the subject, we have assessed the threat that depletion poses to the availability of petroleum resources by estimating cumulative availability curves for conventional petroleum (oil, gas, and natural gas liquids) and for three unconventional sources of liquids (heavy oil, oil sands, and oil shale). Our analysis extends the important study conducted by the US Geological Survey (World petroleum assessment. CD-ROM. U.S. Geological Survey, Reston, 2000) on this topic by taking account of (1) conventional petroleum resources from provinces not assessed by the Survey or other organizations, (2) future reserve growth, (3) unconventional sources of liquids, and (4) production costs.

The results indicate that large quantities of conventional and unconventional petroleum resources are available and can be produced at costs

This article copyrighted and reprinted by permission from the International Association for Energy Economics. The article first appeared in the Energy Journal (Vol. 30, No. 1). Visit the Energy Journal online at http://www.iaee.org/en/publications/journal.aspx

R.F. Aguilera (✉) • G. Lagos C.C
Centre for Research in Energy and Minerals Economics (CREME), Curtin University, Perth, Australia
e-mail: r.aguilera@curtin.edu.au; glagos@ing.puc.cl

R.G. Eggert
Division of Economics and Business, Colorado School of Mines, Golden, CO, USA
e-mail: reggert@mines.edu

J.E. Tilton
Mining Center, Pontificia Universidad Católica de Chile, Regiön metropolitana Santiago, Chile

Division of Economics and Business, Colorado School of Mines, Golden, CO, USA
e-mail: jtilton@mines.edu

R. Sinding-Larsen and F.-W. Wellmer (eds.), *Non-Renewable Resource Issues: Geoscientific and Societal Challenges*, International Year of Planet Earth, DOI 10.1007/978-90-481-8679-2_4, © Springer Science+Business Media B.V. 2012

substantially below current market prices. These findings suggest that petroleum resources are likely to last far longer than many are now predicting and that depletion need not drive market prices above the relatively high levels prevailing over the past several years.

Keywords
Availability • Cumulative availability curve • Depletion • Petroleum • Size Distribution model

Introduction

Petroleum resources are the only major mineral commodities where many experts fear resource depletion will produce significant scarcities over the next several decades. In this chapter, petroleum resources are defined by the combination of conventional and unconventional resources.

Our analysis is global in its scope, considering petroleum resources for the entire world. In addition, we assume that society can over time substitute among different types of conventional and unconventional resources.

Two concerns or questions are of particular interest. First, is depletion likely to create significant scarcities of petroleum resources over the next several decades? Among the pessimists, who believe this will be the case, are Deffeyes (2001), Campbell and Laherrere (1998), Goodstein (2004), Banks (2005), Simmons (2005), and Tertzakian (2006). On the other side are the optimists, who are less concerned. They include Adelman and Watkins (2008), Economides and Oligney (2000), Hefner (2002), Lynch (2005), Odell (2004), and Radetzki (2002). The optimists view petroleum resources as working inventories that are constantly being renewed as they are extracted. The pessimists point out that petroleum resources are fixed physical stocks that will eventually be depleted. Bentley and Boyle (2008) provide descriptions of the methodologies used by petroleum supply forecasters, including pessimists and optimists. The authors also discuss the advantages and disadvantages of the models.

The second question of interest is if society will experience a smooth or difficult transition as it moves from its current dependency on conventional petroleum to unconventional sources of liquids and other alternative energy sources. In particular, does a transition from conventional petroleum to alternative energy sources imply sharply higher energy prices with considerable economic dislocation as traditional users are forced to cut back their consumption? Here again, one finds a divide in the literature between the pessimists and the optimists.

Methodology

The estimates for conventional petroleum volumes used in this chapter are based on those provided by USGS (2000). For our purposes, the USGS study suffers from the limitation that it does not consider all of the world's petroleum provinces.[1] This is because the study adopts a 30-year time horizon and so ignores those provinces that are not expected to be producing petroleum within this period. As stated by USGS (2000), "the assessed areas were those judged to be significant on a world scale in terms of known petroleum volumes, geologic potential for new petroleum discoveries, and political or societal importance."

The resources found in the unassessed provinces are estimated in this analysis using the variable shape distribution (VSD) model. This model assumes that the number of petroleum

[1] As defined by USGS (2000), a petroleum province is "a USGS-defined area having characteristic dimensions of perhaps hundreds to thousands of squared kilometers encompassing a natural geologic entity (e.g., sedimentary basin, thrust belt, delta) or some combination of contiguous geologic entities."

provinces containing a specified volume of petroleum increases as the specified volume decreases. Additionally, it assumes that this relationship can be estimated using existing data on estimated volumes by province. Using nonlinear regression, the VSD model first estimates the lower dotted curve shown in Fig. 8 (see Appendix A) from observed data of previously assessed provinces. Then the model is extended out of sample such that (1) the cumulative number of provinces is the sum of the number of assessed and unassessed provinces, (2) the right end point of the curve does not change (all "large" provinces are assumed to be found), and (3) the curves have approximately the same shape.

Since the USGS (2000) study does not consider resources available from unconventional petroleum, these are estimated on the basis of information found in the existing literature. This information on petroleum resources is then combined with estimates of production costs to construct cumulative availability curves. The data for production costs come from the various sources cited in Appendix B and reflect only costs that producers pay. External costs, such as those associated with global warming, it is important to note, are not considered largely because there is no consensus on what these costs are.

A cumulative availability curve, which previous studies (see Tilton and Skinner 1987; Tilton 2002) have called a cumulative supply curve, shows how the total remaining volumes of a petroleum resource or other mineral commodity varies with production costs. We prefer the term cumulative availability curve since it avoids confusing this curve with the traditional supply curve. The latter shows the quantity of a good offered to the market at various prices during a specific time period, such as a month or year. The cumulative availability curve, in contrast, shows how much of a nonrenewable resource is estimated to be available and economic to extract over all time at various prices, assuming the other determinants of availability (such as technology and input costs) remain unchanged. So availability is a stock variable, whereas traditional supply is a flow variable in the sense that it can continue from one period to the next indefinitely.

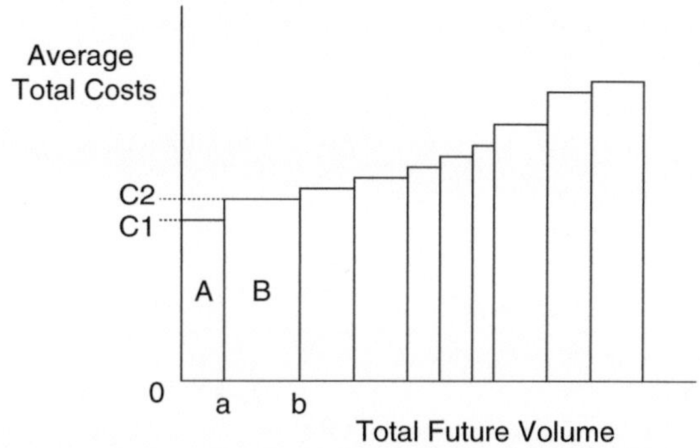

Fig. 1 Cumulative availability curve showing average total costs and total future volume

We estimate the cumulative availability curve by determining the average total costs and total future volumes for individual petroleum provinces. This information is then arranged as in Fig. 1. Column A shows the lowest-cost, highest-quality province. It can produce a total volume given by the distance 0a at a cost of $0C_1$ per unit. Column B shows the next best province. It contains the volume ab with average production costs of $0C_2$. The next column shows the third best province, and so on.

The USGS World Petroleum Assessment

The USGS (2000) study has a 30-year forecast span that covers the 1995–2025 period. It estimates undiscovered volumes of conventional oil, gas, and natural gas liquids (NGL) for provinces outside the USA. These three commodities are what the USGS refers to as conventional petroleum. The study does not assess provinces in the USA and in other parts of the world that were not expected to be producing significant volumes in the adopted time frame.

Although USGS (2000) does not assess individual provinces in the USA, it does present oil, gas, and NGL totals for the country as a whole. Table 1, which is based on Table AR-1 in USGS (2000), summarizes the results at the world level. The "world total (including the USA)" volumes

Table 1 Estimates for undiscovered conventional petroleum and reserve growth for oil, gas, and natural gas liquids (NGL) (Source: USGS 2000)

	Oil billion barrels	Gas trillion cubic feet	Gas billion barrels of oil equivalent	NGL billion barrels
World (excluding USA)				
Undiscovered conventional	649	4669	778	207
Reserve growth (conventional)	612	3305	551	42
Remaining reserves	859	4621	770	68
Cumulative production	539	898	150	7
Total	**2,659**	**13,493**	**2,249**	**324**
USA				
Undiscovered conventional	83	527	88	Combined with oil
Reserve growth (conventional)	76	355	59	Combined with oil
Remaining reserves	32	172	29	Combined with oil
Cumulative production	171	854	142	Combined with oil
Total	**362**	**1,908**	**318**	
World Total (including USA)	**3,021**	**15,401**	**2,567**	

at the bottom of the table cover 409 petroleum provinces.[2] However, as the USGS points out, the world has 937 petroleum provinces (which cover most of the earth's land area and many offshore areas down to water depths of 4,000 m). As a result, the resources in 528 provinces have not been assessed.

The USGS (2000) study assigns significant importance to reserve growth, defined as the increase in reserves of a previously discovered field through time, often due to the development of more sophisticated and efficient technologies. It forecasts potential reserve growth at the world level, using a 30-year time frame, by applying a mathematical model that incorporates the reserve growth experience of the USA (for details, see Chapter RG of USGS 2000). Its estimates of reserve growth are high for fields that are only a few years old. When fields reach approximately 15 years of age, the estimates are much lower.

[2]The provinces of the USA are individually assessed in the USGS National Oil and Gas assessment (1995) and Minerals Management Service (MMS) Outer Continental Shelf assessment (1996). For many of the 409 provinces assessed in USGS (1995), MMS (1996), and USGS (2000), volumes are either not specifically reported or reported as zero. Therefore, for reasons explained in Aguilera (2006) and Aguilera et al. (2007), those provinces could not be used in estimating the VSD model (see Appendix A).

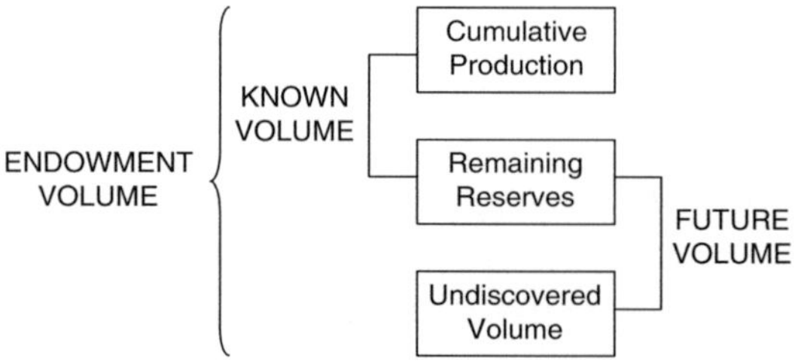

Fig. 2 The relationship between cumulative production, remaining reserves, and undiscovered volumes

At the time USGS (2000) was prepared, approximately 90% of the oil and 84% of the gas were in fields older than 15 years. Therefore, in terms of reserve growth, most of the world's discovered petroleum was "old."

To avoid confusion, we adhere to the definitions of the USGS (2000) study regarding resource terminology. Figure 2 identifies important terms used in this study and shows how they are related.

Undiscovered volumes are those resources postulated from geologic knowledge and theory to exist within a given province but outside its known fields. Undiscovered volumes, it is important to note, differ from reserve growth, which is associated with known fields, as well as from volumes from unassessed provinces. In later sections, however, we include volumes from unassessed provinces and unconventional resources, in addition to volumes from provinces assessed by the USGS, in the "future volume" category. The USGS definition of future volumes is only meant to include volumes from the provinces they have assessed.

A major objective of the USGS (2000) study is to estimate undiscovered volumes that will be added to known volumes within a 30-year time frame. To do so, it relies on various geological techniques combined with a probability assessment to account for the uncertainty. The study publicizes the mean values, which are the volumes we use in this chapter.

The USGS (2000) study has received considerable criticism. Campbell (2000), Laherrere (2000), Deffeyes (2001), and other pessimists claim that the methodologies used to estimate undiscovered volumes and reserve growth overestimate the future availability of petroleum. On the other side, the optimists argue that the study underestimates future availability by assessing only 528 provinces out of a global total of 937. This limitation the study itself recognizes. As stated in Ahlbrandt and McCabe (2002), "the USGS assessment is not exhaustive, because it does not cover all sedimentary basins of the world….The estimates are therefore conservative."

Furthermore, the US Energy Information Administration state "We believe that the USGS (2000) estimates are conservative for a variety of reasons, chief among which are that the USGS assessment did not encompass all geologically conceivable small sources of conventionally reservoired crude oil and was limited to the assessment of reserves that would be added within a 30 year time frame because, in part, '…technological changes beyond 30 years are difficult, if not impossible, to conceptualize and quantify'" (Wood et al. 2004).

There are regions that are known to contain petroleum but were not assessed in USGS (2000). For example, the USGS recently released a Circum-Arctic Resources Appraisal (USGS 2008) that estimates significant petroleum volumes in provinces that were not assessed in USGS (2000).

We critically reviewed USGS (2000) and determined that it is based on geological and statistical procedures recognized as being valid throughout the world. For this reason, we decided to use their estimates as a starting point. In a review of the USGS (2000) study by Klett et al. (2005), it was found that the estimates for new discoveries, about 10 years into the USGS (2000) study, are slightly behind schedule. On the other hand, actual reserve growth has been much larger than they anticipated (even growth of proved plus probable reserves, see Charpentier 2005). The latter is much less costly and less risky than exploration. Therefore, the fact that the USGS undiscovered estimates are not being realized as fast as expected does not necessarily mean that their estimates are overly optimistic. Rather, it means that investors have chosen to invest in development of previously discovered fields, instead of exploration. The common Hubbert-based methodology of extrapolating past production and discovery data does not consider reserves that are created through reserve growth. Other shortcomings of life cycle models (i.e., Hubbert's logistic curves) are explained in Radetzki (2002).

Conventional Petroleum Resources

This section constructs a cumulative availability curve for conventional petroleum, as well as separate cumulative availability curves for conventional liquids (i.e., oil and NGL) and for gas. It first presents the USGS (2000) estimates for conventional petroleum future volumes. It then provides future volumes for the 528 provinces that the USGS (2000) study does not cover and considers future reserve growth. It next assesses production costs and finally presents the cumulative availability curves.

Conventional Petroleum Future Volumes for Provinces Assessed by USGS (2000)

The USGS estimates of future volumes for oil and NGL are shown in Table 5 (located in Appendix C). The first column of this table indicates the region, as designated by the USGS, in which a petroleum province is located. The second column gives the name of the province. The third column identifies the specific area to which the available cost data correspond.

The fourth column provides for each province the future volumes in billions of barrels of oil equivalent (BBOE) as estimated by USGS (2000). Not every province is listed individually since production costs were not available for every province. Volumes for those provinces not specifically listed are grouped together in the rows identified as "remaining provinces." There is a "remaining provinces" row for each region. For example, those provinces in the Middle East and North Africa region, for which cost data are not available, show a combined future oil and NGL volume of 19.019 BBOE, as indicated on row 17 of Table 5. The USGS divides the world into eight regions—the former Soviet Union, the Middle East and North Africa, Asia Pacific, Europe, North America, Central and South America, sub-Saharan Africa and Antarctica, and South Asia.

Adding the individual entries in column 4 gives a total of 1,898 BBOE, as shown in row 99 of the column. This is the same figure given in USGS (2000) for world oil and NGL future volumes. However, this future volume covers only 409 provinces out of a global total of 937 provinces.

The figures shown in the remaining columns of Table 5 are discussed in the sections that follow. Table 6 is similar to Table 5 but presents data for conventional gas (see Appendix C). It indicates that world gas future volumes total 1,665 BBOE.

Conventional Petroleum Future Volumes in Unassessed Provinces

This section provides estimates of conventional petroleum future volumes for those provinces not assessed by the USGS. The volumes are estimated with the variable shape distribution (VSD) model, a size distribution model (see Appendix A) based on an estimated relationship between the cumulative number of provinces with future volumes above various sizes.

There are other models for forecasting petroleum resources, including life cycle models (e.g., Hubbert's logistic curves), rate of effort models, geologic-volumetric models, subjective probability models, discovery process models, and econometric models. However, as Adelman et al. (1983, p. 90) note in evaluating the various methods, "the concept of deposit size distribution is an essential component of models of petroleum supply designed to reflect industry behavior in a logical way."

Earlier size distribution models used to evaluate future petroleum volumes relied mainly on the log-normal and Pareto (i.e., fractal) distributions. The Pareto distribution, for instance, assumes that a log-log plot of the number of petroleum provinces versus the size of the volumes of the provinces follows a straight line with a constant negative slope. All previous methods have specified a particular form of the size distribution of nature's petroleum volumes. The VSD model is different in that it allows the data on previously assessed volumes to determine the relationship between the size and number of petroleum provinces.

We estimated the parameters of the VSD model with data from USGS (2000) and then used it to assess the total future volumes for oil and NGL and for gas for all 937 provinces of the world. The results are shown in Table 2. For oil and NGL, future volumes for all 937 provinces total 2,491 BBOE. The future volumes for 409 provinces considered by USGS (2000), as noted above, total 1,898 BBOE. The difference between these two figures—593 BBOE—is the future oil and NGL volumes estimated to exist in the unassessed provinces.

While the model provides future volumes for all 937 provinces, it does not indicate which volumes correspond to which provinces. As Table 2 shows, we have allocated the future volumes from previously unassessed USGS provinces, for both oil and NGL and for gas, among the eight regions of the world on the basis of each region's share of the future volumes in the assessed provinces. This is consistent with the idea that regions with high amounts of future volumes in previously

Table 2 Allocation by region of oil, NGL, and gas future volumes for the unassessed provinces

Future oil and NGL from VSD model for 937 provinces (BBOE)=			**2,491**	
Future oil and NGL volume from USGS (2000), including USA (BBOE)=			**1,898**	
Future oil and NGL for unassessed provinces (BBOE) =			**593**	
Region	Oil and NGL future[a] volume (BBOE)	% of total	Previously unassessed[b] oil and NGL future volume (BBOE)	Total oil and NGL future volume (BBOE)
Asia Pacific	90.047	4.7	28	118
Central and South America	177.297	9.3	55	233
Europe	69.930	3.7	22	92
Former Soviet Union	337.069	17.8	105	442
Middle East and North Africa	875.625	46.1	274	1,149
North America	222.911	11.7	70	293
South Asia	13.720	0.7	4	18
Sub-Saharan Africa and Antarctica	111.258	5.9	35	146
Total	**1,898**	**100.0**	**593**	**2,491**
Future gas from VSD model for 937 provinces (BBOE)=			**2,165**	
Future gas from USGS (2000), including USA (BBOE)=			**1,665**	
Future gas for unassessed provinces (BBOE)=			**500**	
Region	Gas future volume[a] (BBOE)	% of total	Previously unassessed[b] gas future volume (BBOE)	Total gas future volume (BBOE)
Asia Pacific	120.638	7.2	36	157
Central and South America	113.347	6.8	34	147
Europe	98.667	5.9	30	128
Former Soviet Union	542.950	32.6	163	706
Middle East and North Africa	534.251	32.1	160	695
North America	167.496	10.1	50	218
South Asia	30.442	1.8	9	40
Sub-Saharan Africa and Antarctica	56.977	3.4	17	74
Total	**1,665**	**100.0**	**500**	**2,165**

[a]Future volumes come from USGS (2000)
[b]Previously unassessed future volumes estimated by this study

assessed provinces will have unassessed provinces with generally proportional amounts of future volumes.

The largest unassessed volumes are allocated to the Middle East and North Africa, and to a lesser extent the Former Soviet Union. These regions account for almost 70% of the future oil and NGL volumes. The same procedure is followed for gas, as shown at the bottom of Table 2. The total unassessed gas future volume is 500 BBOE. The largest volumes are assigned to the Former Soviet Union, followed by the Middle East and North Africa, which together account for almost 70% of the total future gas volumes.

Future Reserve Growth

As pointed out in section "The USGS World Petroleum Assessment," the USGS defines reserve growth as the increase in reserves of previously discovered fields through time. The term, as used by the USGS, applies to cumulative past production plus remaining reserves—that is, known volumes (see Fig. 2). Here we assume that reserve growth also applies to undiscovered volumes and introduce the term "future reserve growth" to cover the expected growth in reserves associated with future volumes—that is, remaining reserves plus undiscovered volumes.

Table 3 Calculation of percentages for future reserve growth—based on data from USGS (2000) study

	Oil (Billion barrels)	Gas (BBOE)	NGL (Billion barrels)	Total petroleum (Billion barrels)
World (excluding USA)				
Undiscovered conventional	649	778	207	1,634
Reserve growth (conventional)	612	551	42	1,205
Remaining reserves	859	770	68	1,697
Cumulative production	539	150	7	696
Total	**2,659**	**2,249**	**324**	**5,232**
Known volumes	1,398	920	75	2,393
Reserve growth based on known volumes (%)[a]	**44**	**60**	**56**	**50**
USA				
Undiscovered conventional	83	88	Combined with oil	171
Reserve growth (conventional)	76	59	Combined with oil	135
Remaining reserves	32	29	Combined with oil	61
Cumulative production	171	142	Combined with oil	313
Total	**362**	**318**	Combined with oil	**680**
Known volumes	203	171	Combined with oil	374
Reserve growth based on known volumes (%)[a]	**37**	**35**	**Combined with oil**	**36**
World total				
Undiscovered conventional	732	866	207	1,805
Reserve growth (conventional)	688	610	42	1,340
Remaining reserves	891	799	68	1,758
Cumulative production	710	292	7	1,009
Total	**3,021**	**2,597**	**324**	**5,912**
Known volumes	1,601	1,091	75	2,767
Reserve growth based on known volumes (%)[a]	**43**	**56**	**56**	**48**

Note: [a]Calculated as [Reserve Growth/(Remaining Reserves + Cumulative Production)] × 100

Although the USGS claims that their undiscovered volumes are already "grown," we believe it is important to apply reserve growth to undiscovered volumes, in addition to known volumes. This is because it is likely that even undiscovered volumes will experience future growth in the same way known volumes are growing at present. To give an example, recovery factors of oil are likely to continue to increase, so that will apply to known volumes as well as presently undiscovered volumes.

We estimated future reserve growth in the following manner. First, we calculate the reserve growth associated with known volumes as shown in Table 3 (which is adapted from Table 1). For example, the "world total" section of this table indicates that known oil volumes increase by 43% due to reserve growth. This figure comes from dividing reserve growth (688 BBOE) by the summation of cumulative production plus remaining reserves (710 BBOE + 891 BBOE) and then multiplying by 100.

Second, these calculated percentages of reserve growth are then used to estimate future reserve growth for both the previously assessed and unassessed provinces on the assumption that the reserve growth percentages of future volumes will be similar to those for known volumes.

Returning to Table 5 (in Appendix C), for instance, one can see that the entries in column 5, which reflect future oil and NGL volumes plus future reserve growth, are simply the entries for future oil and NGL volumes reported in the fourth column for both assessed and unassessed provinces increased by 43%. The bottom of column 5 indicates that total future oil and NGL volumes plus future reserve growth for 937 provinces is 3,561 BBOE. This figure assumes that the estimated future reserve growth for oil future volumes, calculated at 43%, also applies to NGL.

The same procedure is followed for gas, which assumes the reserve growth for future volumes will be 56% (see Tables 3 and 6). It estimates the future volumes plus reserve growth for gas at 3,375 BBOE.

Production Costs of Conventional Future Petroleum

Column 8 of Tables 5 and 6 (in Appendix C) presents average total production costs associated with future volumes of conventional oil and NGL and of gas, respectively (in 2006 US$ per BOE). Our estimates of production costs are static and therefore do not take into account the tendency of technological advancement and other factors to reduce costs over time. Nor do they take account of cost increases. Since 2006, the year for which our estimates apply, the production costs for many petroleum products have increased. While some of these increases may be cyclical, reflecting the current boom in mining and mineral production in general, others may be permanent. In any case, rising cost over time shift our estimated curves upward and falling cost shift them downward.

Column 6 of Tables 5 and 6 shows average capital costs, while column 7 shows average operating costs. Capital costs are mainly composed of expenditures for development drilling, processing equipment, production facilities, pipelines, and abandonment. Operating costs include mainly field operating costs and transportation costs. More information on how the production costs of conventional petroleum for both assessed and unassessed provinces are estimated and assigned to specific provinces is found in Appendix B.

There is, of course, a great deal of uncertainty associated with the estimation of production costs. In addition, they can vary greatly depending on which expenditures are and are not considered. For instance, our estimated production costs include a rate of return on invested capital but do not include taxes and royalties. One can argue from the viewpoint of society as a whole that taxes and royalties are not costs since they reflect transfer payments from one group in society to another.

Furthermore, despite the considerable efforts by governments around the world to internalize over the past several decades the external costs associated with energy production, such as environmental pollution and global warming, there is still considerable disagreement over just how large the remaining external costs are. For this reason, we have not attempted to include external costs in our cost estimates. As a result, the full social costs of fossil fuels production are higher than the reported production costs. Including these costs would shift the estimated cumulative availability curves to the left, indicating that every volume would now be more expensive to produce. Moreover, various segments of the curves would shift by different amounts since external costs vary according to the type of fossil fuel. The environmental impact of producing oil sands and oil shale, for instance, is far greater than that of conventional petroleum.

Cumulative Availability Curves for Conventional Future Petroleum

The cumulative availability curve for oil and NGL in previously assessed provinces can be constructed using the data presented in Table 5 in columns 4 and 8 by graphing the future volumes that can be produced economically at various average costs of production. The result is shown as the thin solid continuous line in Fig. 3.

This line indicates that the future volume of oil and NGL volume totals 1,898 BBOE, the same as the future oil and NGL volume presented in the USGS (2000) study. The same curve in Fig. 4 shows the cumulative availability curve for conventional gas from assessed provinces and gives a future volume of 1,665 BBOE.

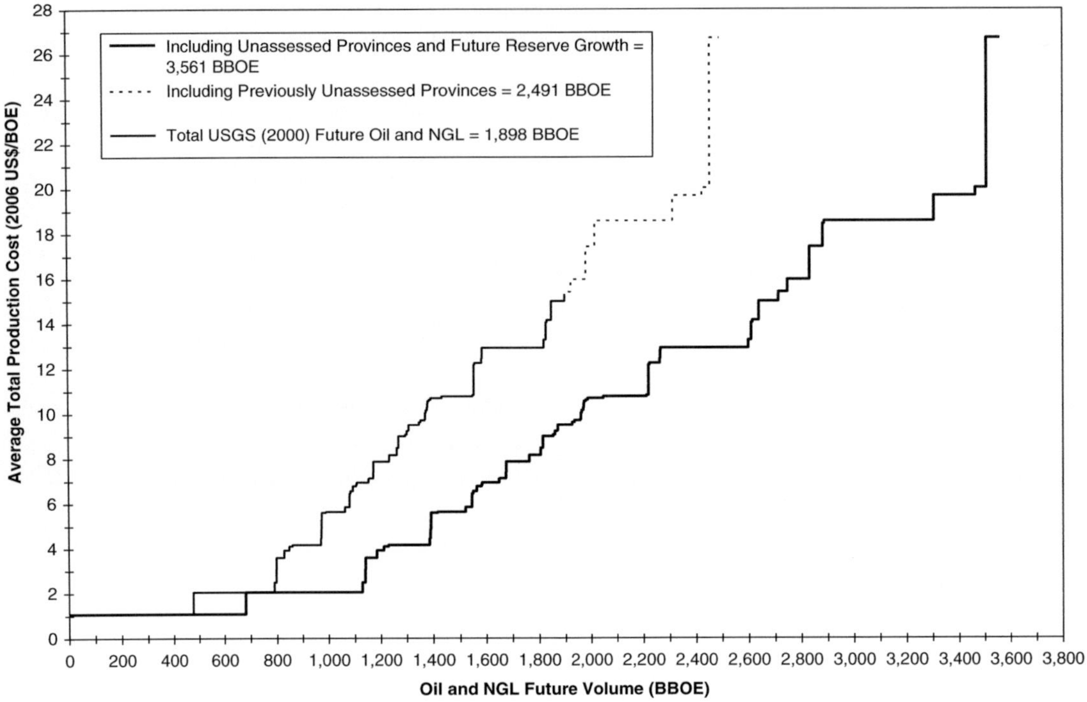

Fig. 3 Global cumulative long-run availability curves for conventional oil and NGL

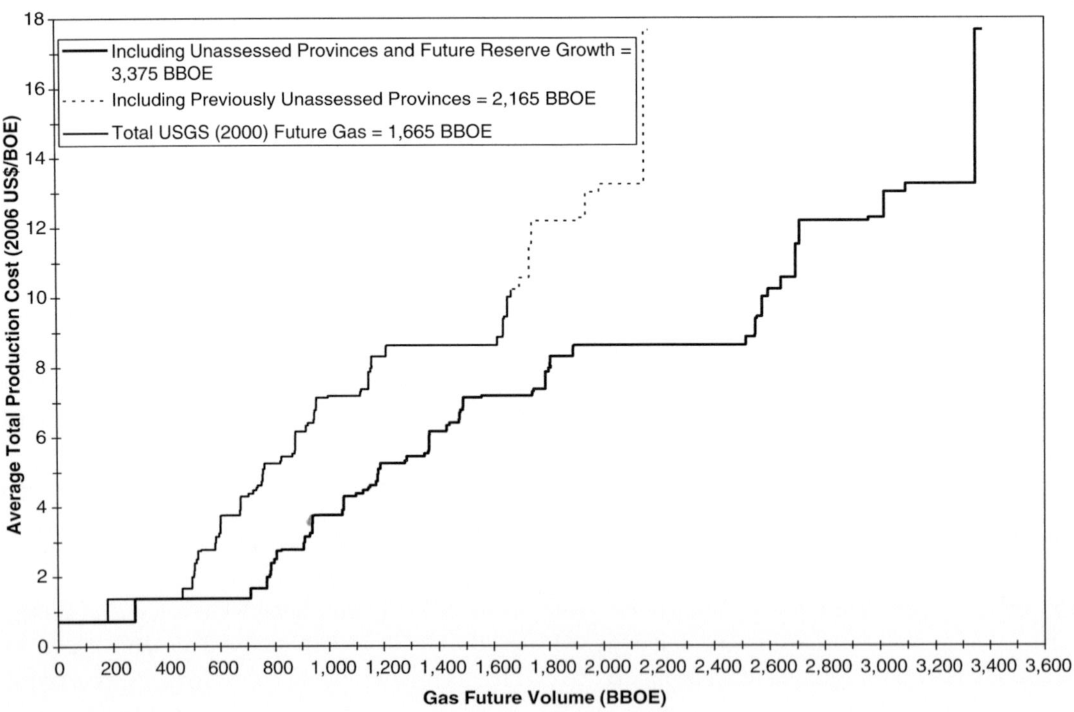

Fig. 4 Global cumulative long-run availability curves for conventional gas

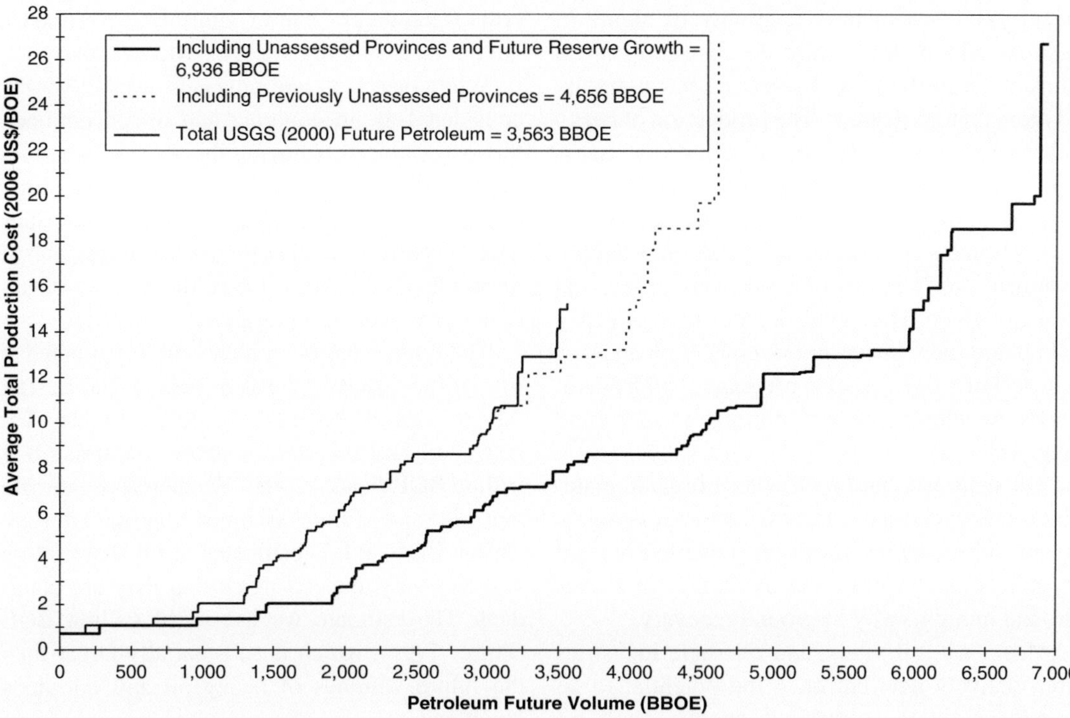

Fig. 5 Global cumulative long-run availability curves for conventional petroleum

The thin continuous line in Fig. 5 combines these two curves to produce a cumulative availability curve for conventional petroleum future volumes from assessed provinces. It shows future volumes of 3,563 BBOE, the same value given in USGS (2000).

The extensions to the thin continuous lines in Figs. 3, 4, and 5, shown as a thin broken line, cover the estimated future volumes and production costs for those provinces the USGS did not assess. In Figs. 3 and 4, these extensions start where the thin continuous curves stop since the estimated production costs for the unassessed provinces are greater than the costs of the assessed provinces. This is not the case in Fig. 5 since the average production costs for gas in some of the unassessed provinces are less than the average production costs for the oil and NGL in some of the assessed provinces. Figures 3, 4, and 5 show that combining the assessed and unassessed provinces increases the future volumes of oil and NGL to 2,491 BBOE, the future volumes of gas to 2,165 BBOE, and the future volumes of petroleum to 4,656 BBOE.

Each of these three figures also shows a thick continuous curve that takes into account estimated future reserve growth. As can be seen, these curves reflect the same costs as their thin counterparts (which do not consider future reserve growth), but each block or step is now wider, due to the larger expected future volumes arising from expected future reserve growth. Incorporating future reserve growth increases future volumes for oil and NGL to 3,561 BBOE (Fig. 3), for gas to 3,375 BBOE (Fig. 4), and for petroleum to 6,936 BBOE (Fig. 5).

Three Unconventional Petroleum Resources

Heavy oil, oil sands, and oil shale are the three main unconventional sources of liquids considered to be potential alternatives to conventional oil. Heavy oil is a dense and viscous oil that is chemically characterized by its content of asphaltenes. Although the definition may vary, heavy oil tends to have a gravity of about 15 degrees API or less, while the conventional oil considered in the previous section has a gravity of more than 15 degrees. The production of heavy oil usually requires measures to reduce its viscosity, such as the injection of super-heated steam and the addition of diluents into the reservoir.

Oil sands, also called tar sands and natural bitumen, are mixtures of sand, water, clay, and crude bitumen. They share the attributes of heavy oil but are more dense and viscous, with an average API of 8 degrees. One can extract oil from oil sands by either open-pit mining or in-ground recovery.

Oil shale is a hard rock whose organic material is chiefly kerogen. The extraction of kerogen, which subsequently has to be converted into oil by pyrolysis, is carried out by surface or underground mining, or by in-ground recovery.

Heavy oil, oil sands, and oil shale industries are relatively new but have the potential to be important future sources of energy. High oil prices, along with continued technological improvements, are increasing the viability of these resources significantly. Still, there are challenges associated with unconventional resource production. First, the "energy returned on energy invested" ratios (EROEI) are much lower than for conventional petroleum. Second, unconventional sources are usually more expensive and so less economical mainly because they are harder to extract and refine. Third, the environmental problems are more serious. Heavy oil, oil sands, and oil shale production causes much higher amounts of greenhouse gas emissions than conventional petroleum. Other factors that could limit production are the potential lack of availability of inputs such as water and natural gas. However, the role of advancing technology will be critical in reducing the need for these inputs, while mitigating environmental damage.

A wide range of estimates can be found in the existing literature for the future volumes and production costs associated with unconventional resources. This section relies upon estimates that both seem reasonable to us and are put forth by reliable sources. As with conventional petroleum, production costs cover both operating and capital costs (including a rate of return on investment), but not taxes and royalties, or external costs.

While heavy oil and oil sands are still widely considered as uneconomic and unconventional, this perception is changing thanks to cost-reducing technological improvements as well as higher energy prices. For example, Canadian authorities have recently now classified a large amount of the country's oil sands (those under active development) as proved oil reserves.

The future volume of heavy oil is estimated at four trillion BOE (Chevron Texaco 2009). The Alberta Energy Research Institute estimates that heavy oil and oil sands together comprise 8–9 trillion BOE (Isaacs 2005). We therefore estimate that oil sands future volumes approximate five trillion BOE. While estimates of oil shale future volumes vary, experts agree that they are abundant. The estimate we use is 14 trillion BOE (Porter 1995), which is substantially larger than the future volumes of heavy oil and oil sands combined.

Although the costs of heavy oil, oil sands, and oil shale presumably vary across different areas, the analysis here presumes a single average production cost for each resource. This is because the world's unconventional resources—at least those that we are currently aware of—are not as widely spread as conventional petroleum. Heavy oil is found mostly in Venezuela's Orinoco Belt, oil sands in northern Alberta, Canada, and oil shale in the western USA.

Heavy oil is typically more expensive to produce than conventional oil, though production costs have been falling. Based on Mommer (2004), we estimate the average production costs of heavy oil at around US $15 per BOE.

Many major oil corporations are investing heavily in the oil sands industry, and as this occurs, production costs are falling. Based on Suncor (2006) and the National Energy Board of Canada (2006), we estimate that average production costs are approximately US $25 per BOE.

Although there is currently very little production from oil shale, large oil corporations are investing considerably in this resource as well. Shell Oil believes it has the technology to produce oil shale commercially at a cost of US $25 per barrel (AOC Petroleum Support Services for US DOE 2004), while the US Department of Energy's Office of Petroleum Reserves (2006) believes production is economic at a cost of US $35 per barrel. We use this latter estimate.

These estimated production costs for heavy oil, oil sands, and oil shale, it is interesting to note, are substantially below current market prices. Indeed, heavy oil and oil sands are even less than the cost of producing conventional oil in certain areas.

Conventional and Unconventional Resources

This section first presents a combined cumulative availability curve that takes into account both conventional petroleum as well as heavy oil, oil sands, and oil shale. It then considers the life expectancies of these resources, that is, how long they would last under different scenarios for future consumption growth.

The Combined Cumulative Availability Curve

Figure 6 shows the cumulative availability curve for conventional petroleum, heavy oil, oil sands, and oil shale. It is derived from the cost and volume data from sections "Conventional Petroleum Resources" and "Three Unconventional Petroleum Resources." The curve has three relatively wide blocks, each representing an unconventional source of liquid. The combined future volumes of conventional petroleum, heavy oil, oil sands, and oil shale total 29.9 trillion BOE.

Although the production mechanisms between conventional and unconventional resources differ

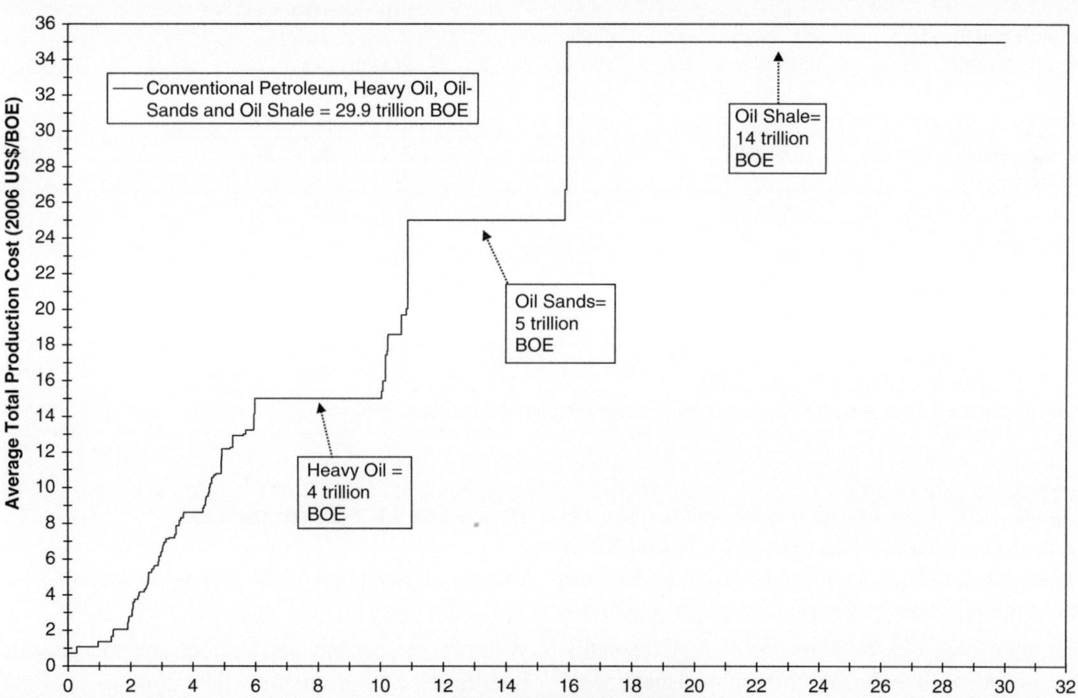

Fig. 6 Global cumulative long-run availability curve for conventional petroleum and unconventional sources of liquids including heavy oil, oil sands, and oil shale

Table 4 Life expectancies

1	2	3	4			5
Conventional resources	Future volumes (BOE)	2006–2008[a] Average annual production (BOE)	Life expectancy in[b] years, at various growth rates in production			Average annual[c] growth in production, 1978–2008 (%)
			0%	**2%**	**5%**	
Oil and NGL, from USGS (2000)	1.898E+12		**64**	**41**	**29**	0.89
Oil and NGL, including unassessed provinces	2.491E+12	2.98E+10	**84**	**49**	**33**	
Oil and NGL, including unassessed provinces and future reserved growth	3.561E+12		**120**	**61**	**39**	
Gas, from USGS (2000)	1.665E+12		**96**	**53**	**35**	2.81
Gas, including unassessed provinces	2.165E+12	1.74E+10	**124**	**62**	**40**	
Gas, including unassessed provinces and future reserve growth	3.375E+12		**194**	**79**	**47**	
Petroleum, from USGS (2000)	3.563E+12		**75**	**46**	**31**	1.48
Petroleum, including unassessed provinces	4.656E+12	4.72E+10	**99**	**54**	**39**	
Petroleum, including unassessed provinces and future reserve growth	6.936E+12		**147**	**69**	**42**	
Conventional and unconventional and resources	Cumulative future volumes (BOE)	Cumulative average annual production (BOE)	Life expectancy in years, at various growth rates in production			
			0%	**2%**	**5%**	
Conventional petroleum and heavy oil	1.09E+13	4.72E+10	**231**	**86**	**51**	–
Conventional petroleum, heavy oil and oil sands	1.59E+13	4.72E+10	**337**	**102**	**58**	–
Conventional petroleum, heavy oil, oil sands and oil shale	2.99E+13	4.72E+10	**633**	**131**	**70**	–

[a] Average annual production comes from British Petroleum (2009)
[b] Life expectancies estimated by this study
[c] Average annual growth in production calculated from British Petroleum (2009)

significantly, it is common to combine the resources in the same availability curve (e.g., see Farrell 2008). This consolidation is appropriate, first of all, because the different production mechanisms are reflected in the reported production costs. More fundamentally, we and others include both conventional and unconventional resources on the same curve because they substitute for one another in actual fuel use.

Resource Life Expectancies

The life expectancies for any particular energy resource depends on three factors—its future volumes, its current production, and the growth over time of its production. The upper portion of Table 4 reports life expectancies for conventional oil and NGL, gas, and petroleum.

The rows indicate how many years the future volumes for these resources would last assuming production grows in the future at 0%, 2%, or 5% a year.[3] In the case of conventional petroleum, for example, the table indicates that with production growth of 2% a year (which is above the average annual growth in production over the past several decades), future volumes from assessed provinces assuming no future reserve growth would last for 46 years. Adding in future volumes from unassessed provinces increases this figure to 54 years and considering future reserve growth pushes it to 69 years.

The lower portion of Table 4 expands the analysis to include unconventional sources of liquids. It shows that the life expectancy of 69 years increases to 86, 102, and 131 years when future volumes from heavy oil, then oil sands, and finally oil shale are taken into account. If we consider all three unconventional resources but a future growth rate of 0%, the life expectancy of 131 years increases to 633 years. Alternatively, Table 4 shows that assuming 5% future production growth reduces the life expectancy to 70 years. It should be noted that all the estimated life expectancies in Table 4 are based on the assumptions underlying the availability estimates.

Conclusions

Two important implications flow from the findings of the previous two sections. First, the quantity of available conventional petroleum is greater than often assumed, given the tendency to overlook unassessed provinces and future reserve growth. Furthermore, volumes of unconventional resources are even more abundant than conventional resources. As a result, conventional and unconventional resources combined are likely to last far longer than many now expect.

Second, a huge break does not appear between the average total production costs of conventional and unconventional petroleum resources. Instead, the costs of producing certain unconventional resources are actually lower than the costs of certain conventional resources. In addition, all the sources of conventional and unconventional petroleum resources considered in this analysis, according to our estimates, can be produced at costs far below the current prices. As a result, even if conventional petroleum peaks, as some predict will happen soon, this does not necessarily mean a "hard landing" with sharply increasing prices and declining consumption. Rather, society may rely increasingly on unconventional resources, including heavy oil, oil sands, and oil shale, as well as nonfossil sources of energy.[4] Should nonfossil fuels become more important, some of the higher cost fossil fuels including conventional petroleum resources may never need to be exploited.

These implications suggest that expectations of a sharp rise in real oil prices over the next decade or two may be misplaced. Indeed, a case can be made that the high prices observed over the past years for conventional oil are not sustainable given the available conventional and unconventional energy resources that can be exploited at costs below current prices.

It is important to note that our cumulative availability curves are solely for several fossil fuels. They do not include coal, nuclear power, or renewable energy sources. Nor do they consider geopolitical factors that may limit access to energy resources in certain geographic areas for short or long periods of time. In addition, the life expectancies shown in Table 4 implicitly assume

[3] Of course, production would not grow exponentially until all future volumes were exhausted and then abruptly stop. Instead, we would expect the production profile to follow a logistic curve or perhaps a series of declining plateaus.

[4] Holland (2008) provides further economic reasons explaining why the transition to substitute resources (e.g., renewables) is likely to be smooth. In particular, he points out that transitions to alternatives should occur when oil is depleted such that its price rises to the production cost of the alternatives.

that the future volumes for an energy resource represent the fixed stock of that resource available over all time to society. Particularly for heavy oil, oil sands, and oil shale, this is unlikely to be the case. As Tilton (2002) has argued, the fixed stock paradigm for assessing the future availability of resources can be misleading. A more appropriate approach is the opportunity cost paradigm, which sees the long-run availability of fossil fuels as depending largely on a race between the cost-increasing effects of depletion and the cost-reducing effects of new technology and substitutions among energy sources.

In light of these limitations, the cumulative availability curves presented and the conclusions regarding the future availability of petroleum resources should be treated with some caution and considered as simply first approximations. Changes in the underlying assumptions would alter the findings.

What history clearly indicates, however, is that it is not appropriate to assume no new resources will be forthcoming from unassessed provinces, reserve growth, and unconventional resources—assumptions needed to conclude that the world will very soon face serious petroleum depletion problems. Over time, of course, depletion may force the world to reduce its reliance on conventional petroleum. This, however, does not imply a difficult transition period with sharply higher prices and lower consumption of energy since unconventional sources of petroleum—and presumably other alternative sources of energy—are available in substantial volumes at costs below current market prices.

Appendix A: The Variable Shape Distribution (VSD) Model

Although geological methods are the most dependable means of estimating undiscovered resources, size distribution models have proven useful as well. The advantage of the latter is that they are faster and easier to employ as they require far less time and resources. In addition, the models are "applicable to different types of data including accumulations, fields, and reservoirs, over varying regions from plays, to petroleum systems to petroleum provinces, to the world" (Barton 1995).

Previous size distribution models used to estimate the volumes of energy resources have assumed that the relationship between a given volume size and the number of areas that contain or exceed that volume follows either the log-normal or the Pareto distributions.[5] The earliest efforts used the log-normal distribution (Kaufman 1962). Subsequently, however, researchers at the USGS (Drew 1997) observed that the log-normal distribution provides overly pessimistic results and concluded that the Pareto distribution was superior. The difference between the two distributions can be seen in Fig. 7, where they are shown as density distributions. While it is now generally acknowledged that the log-normal distribution underestimates petroleum resources, for various reasons, the Pareto distribution may overestimate them. Aguilera (2006) shows the sensitivity of results to these methods.

Historically, all the methods used to forecast petroleum volumes have been "based on an assumed form of the size-frequency distribution of the natural population of oil and gas accumulations" (Barton 1995). The variable shape distribution (VSD) model is different in that it allows the data to determine the size distribution relationship rather than specifying this relationship ex ante. Specifically, we start by observing the curvature given by the USGS (2000) data points on a log-log plot (see Fig. 8). We then estimate the VSD model whose predicted values for the relationship between size and number of petroleum provinces mostly closely match the actual data. As with all size distribution models, the original sample used to estimate the parameters

[5] The Pareto distribution is also known as the power law, Bradford, hyperbolic, fractal, scaling, Zipf (when the slope is 1.0), log-geometric, and J-shape distribution.

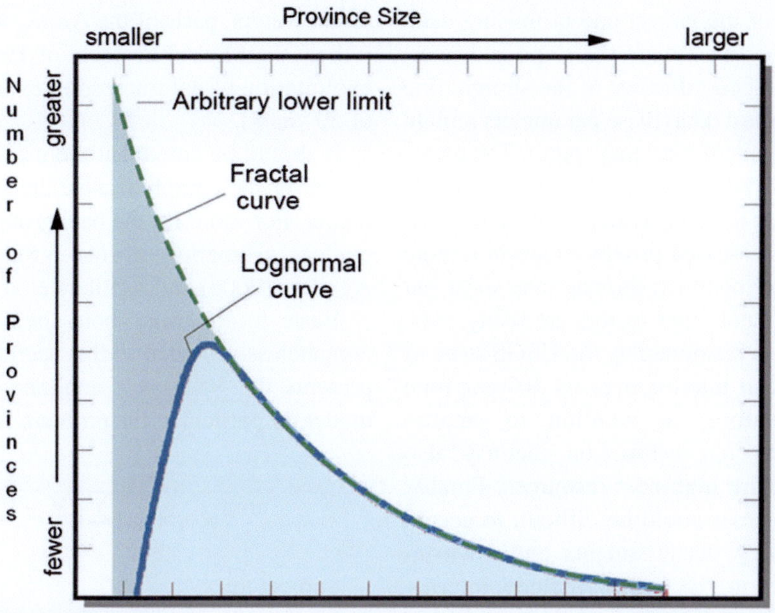

Fig. 7 Density distributions of number of provinces versus province size (Adapted from Barton 1995)

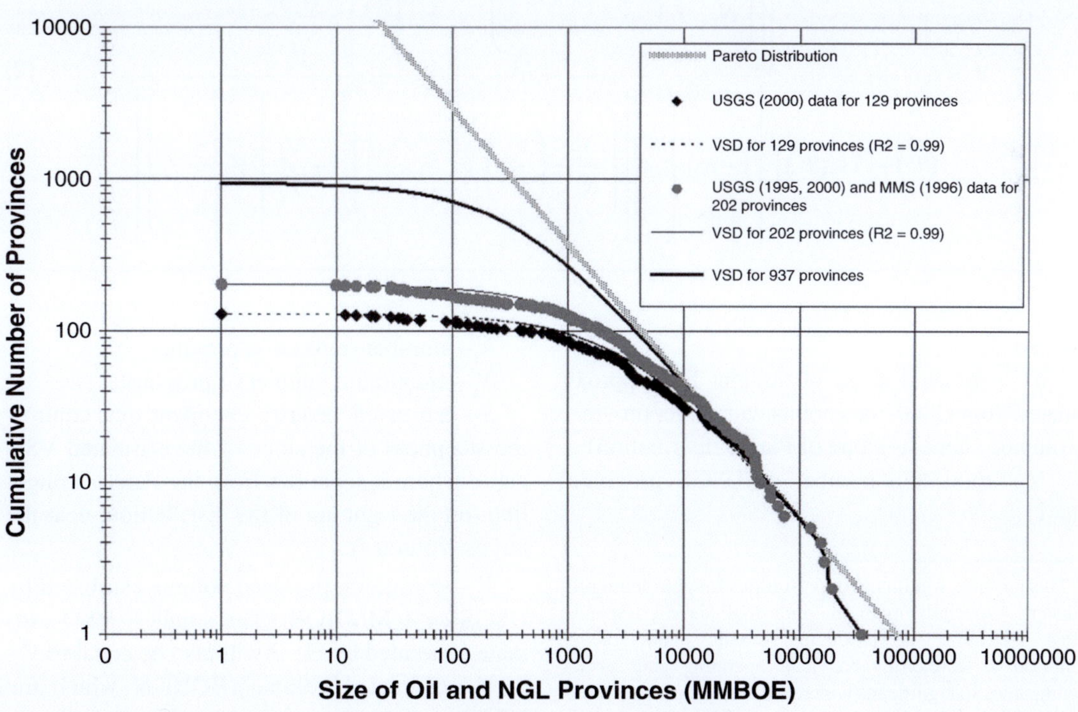

Fig. 8 Endowment of oil and NGL—number of provinces versus province size

contains most of the largest and promising data. This allows one to estimate the slope and intercepts, on log-log coordinates, of the straight line given by the largest data (these parameters remain constant during the forecasting stage). The previously unassessed data will then generally contain smaller volumes than the assessed.[6] Furthermore, many of the unassessed provinces are in remote areas where petroleum resources may exist but due to location and other factors are likely to be high-cost and so presumed by the USGS to be of little commercial interest over its 30-year time horizon. Therefore, in addition to smaller expected sizes, our estimation method also assumes these are high-cost resources. Finally, some of these areas could be difficult to access due to boundary disagreements and environmental protection. Examples include international waters, parts of the Arctic, and Antarctica (where the Madrid Protocol of 1991 forbids the exploitation of natural resources for a minimum of 50 years).

It should be noted that the present study does not cover the comprehensive details of the VSD model. For more on the background and mathematical estimation of the VSD method, see Aguilera (2006) and Aguilera et al. (2007).

Basic information about the VSD model is nonetheless provided in this section. Equation 1 presents the VSD as a nonlinear least squares model. In particular, the problem is:

$$\min_{\{V_x, a_p, V_s, \psi, S\}} \sum_{i=1}^{n} \left(V_i - \hat{V}_i \right)^2 \tag{1}$$

Subject to:

$$\hat{V}_i = \frac{\left\{ \left[\left(\frac{1}{N_i} - \left(\frac{V_m}{V_x} \right)^{\left(\frac{\log N_x - \log N_m}{\log V_x - \log V_m} \right)} \right)^{\frac{1}{a_p}} + \frac{V_m}{V_x} \right] \cdot V_x \right\} \times (\Psi)}{(\Psi) + [1 - (\Psi)] \cdot \left[1 - \exp\left(- \left\{ \left[\left(\frac{1}{N_t} - \left(\frac{V_m}{V_x} \right)^{\left(\frac{\log N_x - \log N_m}{\log V_x - \log V_m} \right)} \right)^{\frac{1}{a_p}} + \frac{V_m}{V_x} \right] \cdot V_x \right\} \div V_s \right) \right]^S} \tag{2}$$

where

a_p—estimated slope of straight line approximated from USGS data points with larger province volumes (same as slope of Pareto distribution)

N_m—minimum number of USGS provinces (= 1)

N_i—number (rank) of a province

N_x—maximum number of provinces

S—estimated severity exponent that controls the steepness of the slope of the estimated VSD curve where it separates from the Pareto straight line (on the right tail of the distribution, near the largest volumes)

V_m—minimum province volume estimated by USGS (= 1 MMBOE). The smallest VSD estimate generated by Eq. 2 will also be equal to V_m

V_s—estimated volume (BOE) at which the USGS data begins to deviate from the Pareto

[6] This presents significant opportunities for petroleum producers, especially medium- and small-sized companies that specialize in producing smaller endowments. Innovative technology has lead operating and service companies to significantly increase recoveries from small endowments (Aguilera and Aguilera 2008).

straight line (on the right tail of the distribution, near the largest volumes)

$\hat{V}_i$—estimated volume of a province (BOE)

V_x—estimated maximum volume (BOE) based on Pareto straight line (at $N_m = 1$)

Ψ—estimated separation ratio that controls the amount of separation between the Pareto straight line and the estimated VSD curve (on the right tail of the distribution, near the largest volumes).

As an illustration, USGS (2000) provides estimates of the oil and NGL endowment for 129 provinces, excluding the provinces of the USA. We then use nonlinear regression to estimate the parameters of the VSD (see Eq. 2) that provides the best fit of the USGS (2000) data. As Fig. 8 shows, the curve generated by the estimated VSD model provides a very good fit of the actual data.

The estimated VSD model is next employed to assess the size distribution relationship among all assessed provinces, those in the USA, as well as those elsewhere. The parameter values estimated with the previous sample are used. Volumes of the oil and NGL endowment for provinces in the USA are available from the USGS National Oil and Gas assessment (1995) and Minerals Management Service Outer Continental Shelf assessment (1996). When combined with the provinces of the rest of the world, a total of 202 assessed provinces are available.[7] Figure 8 shows that the estimated VSD curve again provides a very good fit of the actual data. In addition, the volumes for individual provinces generated by the estimated VSD model compare very well with the values estimated by geological methods (see Aguilera 2006; Aguilera et al. 2007).

Finally, the VSD model (i.e., Eq. 2) is used to estimate the size distribution relationship for all of the 937 provinces of the world, including those not previously assessed. Again, the model is run using the estimated parameter values from the first sample. The size distribution relationship is shown in Fig. 8 and allows us to estimate the endowment volumes of oil and NGL in the previously unassessed provinces, assuming most of the larger provinces have already been assessed. The good fit for the previous two samples (including a high R^2 and comparable volumes) provides some confidence that the estimated volumes for 937 provinces are reasonable. Aguilera (2006) further validates the VSD model using the cases of known, future, and endowment volumes oil, gas, NGL, and petroleum.

Appendix B: Production Costs for Conventional Petroleum

The estimated production costs for conventional petroleum shown in columns 6, 7, and 8 of Tables 5 and 6 (in Appendix C) are based on data from Bloomberg (2005), Center for Global Energy Studies (2003), Compañía Española de Petróleo Sociedad Anónima—CEPSA (Personal Communication with H. Quiroga, Emailed March 16, 2005), Ecopetrol SA (2003), Energy Information Administration—EIA (2006), ESP Oil (ESP Oil Co – Maracaibo, Venezuela, Personal communication with D. Quispe, Emailed March 17, 2005), Global Policy Forum (2002), Goldman Sachs (2005), Institute for the Analysis of Global Security (2003), Oil and Natural Gas Commission of India (2004), Pakistan Oil Fields Limited (2005), Rosetta Exploration Inc. (Rosetta Exploration Inc – Calgary, Canada, Personal communication with P. Pedersen, Emailed March 21, 2005), Wood Mackenzie (2004–2006), and personal communication with contacts in the energy industry.[8] For simplicity, the procedures followed are illustrated for oil and NGL, though gas production costs are estimated in the same way.

Column 8 of Table 5 (see Appendix C) gives the average total cost (ATC) of oil and NGL production by province in US$ per BOE, which is

[7] In total, 409 provinces are assessed by USGS (1995, 2000) and MMS (1996), but point estimates greater than zero are only given for 202 provinces.

[8] For further information about costs of each specific area of Tables 5 and 6, contact Roberto F. Aguilera; e-mail: r.aguilera@curtin.edu.au.

simply the sum of average capital costs (CC) and average operating costs (OC).

The majority of these costs were available in 2006 US dollars. Costs from earlier years were converted to 2006 US dollars by assuming an average inflation rate of 2.5% annually. The costs do not include taxation and royalties.

The average capital cost per BOE (column 6 of Table 5) for each province is calculated from [9, 10]:

$$CC = \frac{UCC}{URR} \tag{3}$$

where *UCC* is the ultimate capital cost in a province in US dollars and *URR* is the ultimately recoverable resources for the province in BOE.[11]

The average operating cost per BOE (column 7 of Table 5) for each province is calculated from:

$$OC = \frac{OE}{VP} \tag{4}$$

where *OE* is operating expense during 1 year of production in US dollars and *VP* is the volume produced in that year in BOE.

The production of oil, gas, and NGL occurs in combination throughout some of the assessed areas. To get an average total cost for the main product in an area (*ATCM*, in US$ per BOE), we use:

$$ATCM = ATC \times \frac{PMP}{TPC} \tag{5}$$

where *PMP* is the production of the main product during 1 year in BOE and *TPC* is the total production of the combined commodities in that year in BOE.

To get the average total cost of a secondary product (*ATCS*, in US$ per BOE) that is not the main product in an area, the cost of the main product is calculated first using Eq. 5. We then use:

$$ATCS = ATCM \times ROG \tag{6}$$

where *ROG* is the constant fraction representing the relative total production cost of oil to gas, or vice versa. It is assumed that oil costs approximately 1.5 times as much as gas to produce.

Cumulative availability curves are estimated by attaching production costs to the volumes of USGS petroleum *provinces*. Since most available data on average production costs apply to countries, fields, or areas operated by a corporation, these costs must be assigned to particular petroleum *provinces* in order to construct the desired availability curves. To do so, we match the area for which costs are available with provinces using geographic maps, on the assumption that the costs for a province approximate the costs for its matched area.

In some cases, production costs are available for a country, such as Saudi Arabia, which is made up of several USGS provinces. For these cases, the costs for the country are assigned to each of the country's provinces. In other cases, several areas for which cost data are available lie in one USGS province, such as the Magallanes Basin in Southern Argentina. The future volume of that province, as given in USGS (2000), is allocated among the different areas for which

---

[9] A rate of return on invested capital is included in the *UCC* (ultimate capital cost) estimated by our cost sources. The rate of return reflects the opportunity cost of using the scarce resource capital in the production of energy rather than in another economic activity.

[10] This calculation differs from the standard method of calculating capital costs, which takes annual depreciation and divides by the volume produced in that year. For the purposes of this study, which does not require calculating capital costs in a particular year, the chosen approach is more appropriate.

[11] *URR*, a common term in the literature, is used in most of the sources from which we derive production costs. URR is used in Eq. 3 (and later in this Appendix) since this is how most sources present total volumes that will eventually be recovered. It is assumed to be a good proxy for "future volumes."

costs are available. This is done with the following equation:

$$AFV = TFV \times \frac{URR}{URRP} \quad (7)$$

where *AFV* is the allocated future volume of a given area within a USGS province in BOE, *TFV* is the total future volume (BOE) for the USGS province (e.g., the Magallanes Basin in Southern Argentina), *URR* is the ultimately recoverable resources in BOE for the given area for which cost data are available (e.g., the Tierra del Fuego area in the Magallanes Basin province), and *URRP* is the sum of the ultimately recoverable resources in BOE of all the areas for which cost data are available within the province. Repeating this procedure for every area allocates the future volume proportionally according to each area's resources.

Since costs for every USGS province are not available, not every province is listed in Tables 5 and 6. As described in section "Conventional Petroleum Resources," those provinces not individually listed are shown in the rows named "remaining provinces." There is a "remaining provinces" row for each of the eight regions. The average total production costs for a given region's "remaining provinces" is calculated using the available costs of other areas in that region. Specifically, the *simple average* of all the other areas available costs is assigned to the "remaining provinces." For example, the "remaining provinces" of the Middle East and North Africa region, for which cost data are not available, show average total production costs of US $3.92 per BOE (see row 17, column 8 of Table 5). This figure of US $3.92 per BOE is the simple average of all the other available costs for the Middle East and North Africa.

Since previous studies do not consider the unassessed provinces, their average total production costs are also unknown. These costs are estimated for each region on the basis of the known costs and the future volumes (including volumes from unassessed provinces) for the eight regions. Specifically, the average total production costs (in US$ per BOE) of the unassessed provinces within a given region (*ATCU*) are calculated using:

$$ATCU = HATC \times \left(1 + \frac{UVOL}{AVOL}\right) \quad (8)$$

where *HATC* is the highest average total production costs for an assessed USGS province within a region (US$ per BOE), *UVOL* is the estimated future volumes of unassessed provinces within the region (BOE), and *AVOL* is the estimated future volume of assessed provinces within the region (BOE).

This procedure presumes that within a region the average production costs for all the unassessed provinces are higher than those for all the assessed provinces including the assessed province with the highest costs. This is consistent with the notion that the highest-quality, least-costly resources are usually extracted first. The procedure also presumes that the difference between the production costs for the unassessed provinces and the highest cost assessed province increases as the volumes for unassessed provinces rise relative to assessed provinces.

Appendix C

Table 5 Conventional oil and NGL resources ranked by average total costs

	1	2	3	4	5	6	7	8
	Region	USGS petroleum province	Location	Future oil and[a] NGL (BBOE)	Future oil and NGL[b] plus future reserve growth (BBOE) Oil growth = 43%	Average[c] capital cost (2006 US$/BOE)	Average[c] operating cost (2006 US$/BOE)	Average total[c] production cost (2006 US$/BOE)
1	Middle East and North Africa	Mesopotamian Foredeep Basin	Iraq, Kuwait	**301.737**	**431.393**	–	–	**1.08**
2	Middle East and North Africa	Zagros Fold Belt	Iraq, Iran	**128.543**	**183.778**	–	–	**1.08**
3	Middle East and North Africa	Khleisha Uplift	Iraq	**0.715**	**1.022**	–	–	**1.08**
4	Middle East and North Africa	Widyan Basin-Interior Platform	Iraq, Saudi Arabia	**45.305**	**64.773**	–	–	**1.08**
5	Middle East and North Africa	Rutbah Uplift	Iraq	**0.155**	**0.222**	–	–	**1.08**
6	Middle East and North Africa	Anah Graden	Iraq	**0.106**	**0.152**	–	–	**1.08**
7	Middle East and North Africa	Greater Ghawar Uplift	Saudi Arabia	**118.481**	**169.392**	–	–	**2.05**
8	Middle East and North Africa	Rub Al Khali Basin	Saudi Arabia, U.A.E	**139.687**	**199.711**	–	–	**2.05**
9	Middle East and North Africa	Interior Homocline-Central Arch	Saudi Arabia	**30.665**	**43.842**	–	–	**2.05**
10	Middle East and North Africa	Qatar Arch	Saudi Arabia, Qatar	**20.912**	**29.898**	–	–	**2.05**
11	Middle East and North Africa	Fahud Salt Basin	Oman	**3.258**	**4.658**	–	–	**2.05**
12	Asia Pacific	Baram Delta/ Brunei-Sabah Basin	Malaysia Baram Delta, Offshore	**7.681**	**10.982**	1.46	1.02	**2.48**
13	Central and South America	Magallanes Basin	Argentina Faro Virgenes, On/Off Shores	**0.051**	**0.073**	0.26	2.54	**2.79**
14	Central and South America	Magallanes Basin	Argentina Tierra del Fuego, On/Off Shore	**0.293**	**0.419**	1.07	2.00	**3.07**

15	Central and South America	Magallanes Basin	Argentina Santa Cruz II	**0.194**	**0.277**	0.88	2.36	**3.24**
16	Middle East and North Africa	Sirte Basin	Libya	**30.687**	**43.873**	2.05	1.54	**3.59**
17	Middle East and North Africa	Remaining provinces		**19.019**	**27.191**	–	–	**3.92**
18	Middle East and North Africa	North Egypt Basin	Egypt West Bakr	**0.561**	**0.802**	2.91	1.16	**4.07**
19	Middle East and North Africa	Red Sea Basin	Red Sea	**12.386**	**17.708**	–	–	**4.10**
20	Middle East and North Africa	Chad	Chad Sedigui Oil Project	**0.003**	**0.004**	2.50	1.64	**4.14**
21	Sub-Saharan Africa and Antarctica	Niger Delta	Nigeria	**67.278**	**96.187**	1.19	2.97	**4.16**
22	Sub-Saharan Africa and Antarctica	West-Central Coastal	Off Shore Angola, Congo, Gabon, Eq. Guinea, Cameroon	**45.007**	**60.057**	1.19	2.97	**4.16**
23	Middle East and North Africa	Nile Delta Basin	Egypt El Mansoura	**0.646**	**0.923**	2.23	2.21	**4.43**
24	Middle East and North Africa	Abu Gharadiq Basin	Egypt Bade El Din	**1.371**	**1.960**	2.97	1.48	**4.46**
25	North America	Alberta Basin	Canada Arctic Cameron Hills	**0.101**	**0.144**	2.58	1.89	**4.47**
26	Middle East and North Africa	North Egypt Basin	Egypt Qarun	**0.684**	**0.978**	2.86	1.93	**4.79**
27	Central and South America	Magallanes Basin	Argentina Santa Cruz I	**0.211**	**0.302**	2.36	2.50	**4.86**
28	Asia Pacific	Pearl River Mouth Basin	China xijiang Area, Off Shore	**0.701**	**1.003**	3.12	2.15	**5.27**
29	Central and South America	Llanos Basin	Colombia	**8.642**	**12.355**	2.63	2.97	**5.61**
30	Central and South America	Middle Magdalena	Colombia	**1.418**	**2.027**	2.63	2.97	**5.61**
31	Central and South America	Putumayo-Oriente-Maranon Basin	Colombia, Ecuador, Peru	**6.865**	**9.815**	2.63	2.97	**5.61**

(continued)

Table 5 (continued)

	1	2	3	4	5	6	7	8
	Region	USGS petroleum province	Location	Future oil and[a] NGL (BBOE)	Future oil and NGL[b] plus future reserve growth (BBOE) Oil growth = 43%	Average[c] capital cost (2006 US$/BOE)	Average[c] operating cost (2006 US$/BOE)	Average total[c] production cost (2006 US$/BOE)
32	Central and South America	Magallanes Basin	Venezuela	**22.690**	**32.440**	2.82	2.82	**5.64**
33	Central and South America	East Venezuela Basin	Venezuela	**33.410**	**47.766**	2.82	2.82	**5.64**
34	Central and South America	Guyana-Suriname Basin	Guyana, Suriname, French Guiana	**17.623**	**25.196**	2.82	2.82	**5.64**
35	Central and South America	Campos Basin	Brazil Chinook, Offshore	**16.500**	**23.590**	2.43	3.42	**5.85**
36	Asia Pacific	Pearl River Mouth Basin	China Block 15/34 (Panya 4-2 and Panyu 5-1). Offshore	**0.603**	**0.862**	3.32	3.03	**6.35**
37	Europe	Campathian-Balkanian Basin	Romania, Bulgaria	**3.011**	**4.305**	3.18	3.28	**6.46**
38	Asia Pacific	Kutei Basin	Indonesia Makassar, Offshore	**9.181**	**13.126**	3.18	3.38	**6.57**
39	Asia Pacific	Bohaiwan Basin	China Block 11/05 (Peng Lai 19-3). Offshore	**12.653**	**18.090**	3.94	2.84	**6.78**
40	Sub-Saharan Africa and Antarctica	Remaining provinces		**1.739**	**2.486**	–	–	**6.84**
41	Central and South America	Remaining provinces		**45.684**	**65.314**	–	–	**6.94**
42	Former Soviet Union	North Caspian Basin	Kazakhstan Aktyubinsknet Contract Area	**1.647**	**2.355**	1.03	6.06	**7.10**
43	Middle East and North Africa	Trias/Ghadames Basin	Algeria EI Merk	**16.392**	**23.436**	3.00	4.13	**7.13**
44	Central and South America	Potigar Basin	Brazil Koch	**0.984**	**1.407**	–	–	**7.40**
45	Europe	Vestford-Helgeland	Offshore Norway	**15.815**	**26.900**	5.77	2.09	**7.87**

46	Europe	North Sea Graben	Northern North Sea Brent, Offshore	**41.846**	**59.827**	5.77	2.09	**7.87**
47	South Asia	Kohat-Potwar	Pakistan	**0.354**	**0.506**	–	–	**7.96**
48	Former Soviet Union	North Caspian Basin	Kazakhstan Kashagan Contract Area, Offshore	**29.532**	**42.222**	1.70	6.46	**8.16**
49	Middle East and North Africa	Chad	Chad Doba Oil Project	**0.041**	**0.059**	4.08	4.23	**8.30**
50	Central and South America	Magallances Basin	Argentina area Magallanes, Offshore	**0.143**	**0.204**	3.74	4.63	**8.37**
51	Europe	Remaining provinces		**5.868**	**8.389**	–	–	**8.48**
52	Sub-Saharan Africa and Antarctica	Senegal	Mauritania PSC B Tiof, Offshore	**0.120**	**0.172**	5.60	3.26	**8.86**
53	Asia Pacific	Remaining provinces		**26.516**	**37.910**	–	–	**9.00**
54	Asia Pacific	Bohaiwan Basin	China Caofedian Blocks 04/35 and 05/36, Offshore	**4.835**	**6.912**	3.85	5.22	**9.06**
55	Asia Pacific	Central Sumatra Basin	Indonesia Far East Offshore Shelf, Offshore	**5.862**	**8.381**	3.59	5.64	**9.23**
56	North America	Alberta Basin	Canada Arctic Norman Wells	**1.728**	**2.471**	5.18	4.07	**9.25**
57	North America	Vilahermosa Uplift	Gulf of Mexico	**28.778**	**41.144**	–	–	**9.50**
58	North America	Saline-Comalcalco Basin	Gulf of Mexico	**8.599**	**12.294**	–	–	**9.50**
59	North America	Tampico-Misantla Basin	Gulf of Mexico	**2.133**	**3.050**	–	–	**9.50**
60	Central and South America	Falklands Plateau	Islas Malvinas, Falkland Islands	**6.203**	**8.868**	–	–	**9.62**
61	North America	Remaining provinces		**15.254**	**21.809**	–	–	**9.71**

(continued)

Table 5 (continued)

	1	2	3	4	5	6	7	8
	Region	USGS petroleum province	Location	Future oil and[a] NGL (BBOE)	Future oil and NGL[b] plus future reserve growth (BBOE) Oil growth = 43%	Average[c] capital cost (2006 US$/BOE)	Average[c] operating cost (2006 US$/BOE)	Average total[c] production cost (2006 US$/BOE)
62	Central and South America	Reconcavo Basin	Brazil PetroReconcavo	**2.132**	**3.048**	–	–	**9.75**
63	Central and South America	Sergipe-Alagoas Basin	Brazil Tartaruga, On/Offshore	**2.127**	**3.041**	6.32	3.73	**10.05**
64	Central and South America	Espirito Santo Basin	Brazil Frade, Offshore	**4.610**	**6.591**	5.79	4.37	**10.16**
65	Sub-Saharan Africa and Antarctica	Senegal	Mauritania PSC B Chinguetti and Tevet, Offshore	**0.114**	**0.163**	6.46	3.72	**10.18**
66	Former Soviet Union	South Turgay Basin	Kazakhstan South Kumkol Area	**0.968**	**1.384**	0.75	9.52	**10.27**
67	Middle East and North Africa	Abu Gharadiq Basin	Egypt El Diyur	**1.090**	**1.558**	6.80	3.65	**10.45**
68	Asia Pacific	Bohaiwan Basin	China Chengdaoxi, On/Offshore	**0.559**	**0.800**	6.51	4.06	**10.57**
69	Asia Pacific	Songliao Basin	China	**4.414**	**6.311**	6.51	4.06	**10.57**
70	South Asia	Remaining provinces		**5.529**	**7.905**	–	–	**10.62**
71	Former Soviet Union	Middle Caspian Basin	Kazakhstan ADE Contract Area	**11.78**	**16.842**	1.52	9.17	**10.69**
72	Former Soviet Union	South Caspian Basin	Caspian Sea	**29.879**	**42.718**	1.52	9.17	**10.69**
73	North America		United States	**115.000**	**164.416**	–	–	**10.77**
74	Central and South America	Campos Basin	Brazil Bijupira-Salema, Offshore	**5.157**	**7.373**	7.06	3.77	**10.83**
75	Europe	Anglo-Dutch Basin	North Sea UK Sector, Offshore	**0.390**	**0.558**	5.49	6.25	**11.74**
76	Central and South America	Campos Basin	Brazil Polvo, Offshore	**2.360**	**3.374**	6.70	5.48	**12.18**

77	Former Soviet Union	Remaining provinces		**26.803**	**38.321**	–	–	**12.25**
78	Former Soviet Union	North Ustyurt Basin	Kazakhstan Karazhanbas, On/Offshore	**1.603**	**2.292**	2.64	9.81	**12.45**
79	Asia Pacific	South China continental shelf slope	China Block 16/06 (Lufeng 13-1), Offshore	**0.449**	**0.642**	7.28	5.53	**12.80**
80	Former Soviet Union	West Siberian Basin	Russia	**171.961**	**245.853**	–	–	**12.92**
81	Former Soviet Union	Volga-Ural Region	Russia	**24.786**	**35.437**	–	–	**12.92**
82	Former Soviet Union	Timan-Pechora Basin	Russia	**18.045**	**25.799**	–	–	**12.92**
83	Former Soviet Union	South Barents Basin	Russia	**3.668**	**5.244**	–	–	**12.92**
84	Former Soviet Union	North Sakhalin Basin	Russia	**7.170**	**10.251**	–	–	**12.92**
85	Former Soviet Union	Nepa-Botuoba Arch	Russia	**4.950**	**7.077**	–	–	**12.92**
86	Former Soviet Union	Azov-Kuban Basin	Russia	**−0.129**	**0.184**	–	–	**12.92**
87	Former Soviet Union	Angara-Lena Terrace	Russia	**0.805**	**1.151**	–	–	**12.92**
88	Former Soviet Union	Baykit Arch	Russia	**2.160**	**3.088**	–	–	**12.92**
89	Former Soviet Union	Cis-Patom Foredeep	Russia	**0.415**	**0.593**	–	–	**12.92**
90	South Asia	Bombay	India	**7.837**	**11.205**	8.31	4.98	**13.28**
91	Middle East and North Africa	Pelagian Basin	Tunisia Ashtart, Offshore	**3.182**	**4.549**	8.54	5.52	**14.06**
92	Former Soviet Union	North Caspian Basin	Kazakhstan Alibec South	**0.220**	**0.315**	3.03	11.11	**14.14**
93	Asia Pacific	Junggar Basin	China	**6.485**	**9.272**	–	–	**14.16**
94	Asia Pacific	Tarim Basin	China	**10.108**	**14.451**	–	–	**14.16**
95	Former Soviet Union	North Ustyurt Basin	Kazakhstan Arman, On/Offshore	**0.154**	**0.220**	5.62	9.13	**14.75**
96	North America	East Greenland Rift Basins	Offshore Greenland	**51.318**	**73.369**	–	–	**15.00**
97	Former Soviet Union	North Ustyurt Basin	Kazakhstan Buzachi North, On/Offshore	**0.651**	**0.931**	4.11	11.15	**15.27**
98								

(continued)

Table 5 (continued)

	1	2	3	4	5	6	7	8
	Region	USGS petroleum province	Location	Future oil and[a] NGL (BBOE)	Future oil and NGL[b] plus future reserve growth (BBOE) Oil growth = 43%	Average[c] capital cost (2006 US$/BOE)	Average[c] operating cost (2006 US$/BOE)	Average total[c] production cost (2006 US$/BOE)
99				**Total oil and NGL future volume (BBOE)**	**1,898**			
100								
101	*Areas not assessed in USGS (2000)*[d]:							
102								
103	Europe			**23.286**	**33.292**	–	–	**15.41**
104	Central and South America			**59.037**	**84.406**	–	–	**15.98**
105	South Asia			**35.267**	**50.421**	–	–	**17.43**
106	Middle East and North Africa			**4.569**	**6.532**	–	–	**18.45**
107	Asia Pacific			**291.571**	**416.859**	–	–	**18.58**
108	North America			**112.239**	**160.468**	–	–	**19.69**
109	Former Soviet Union			**29.984**	**42.869**	–	–	**20.04**
110	Sub-Saharan Africa and Antarctica			**37.047**	**52.967**	–	–	**26.72**
111								
112				Total	**3,561**			

[a]Future oil and NGL volumes come from USGS (2000)
[b]Future oil and NGL plus future reserve growth estimated by this study
[c]Costs estimated by this study
[d]Volumes and costs below row 101 estimated by this study

Table 6 Conventional gas resources ranked by average total costs

	1	2	3	4	5	6	7	8
	Region	USGS petroleum province	Location	Future gas[a] (BBOE)	Future gas plus[b] future reserve growth (BBOE) Gas growth = 56%	Average capital[c] cost (2006 US$/BOE)	Average operating[c] cost (2006 US$/BOE)	Average total[c] production cost (2006 US$/BOE)
1	Middle East and North Africa	Mesopotamian Foredeep Basin	Iraq, Kuwait	**61.479667**	**95.852948**	–	–	**0.72**
2	Middle East and North Africa	Zagros Fold Belt	Iraq, Iran	**100.186333**	**156.200512**	–	–	**0.72**
3	Middle East and North Africa	Khleisha Uplift	Iraq	**0.465000**	**0.724982**	–	–	**0.72**
4	Middle East and North Africa	Widyan Basin-Interior Platform	Iraq, Saudi Arabia	**17.006833**	**26.515354**	–	–	**0.72**
5	Middle East and North Africa	Rutbah Uplift	Iraq	**0.215000**	**0.335207**	–	–	**0.72**
6	Middle East and North Africa	Anah Graben	Iraq	**0.006500**	**0.010134**	–	–	**0.72**
7	Middle East and North Africa	Greater Ghawar Uplift	Saudi Arabia	**74.837667**	**116.679406**	–	–	**1.37**
8	Middle East and North Africa	Rub Al Khali Basin	Saudi Arabia, U.A.E	**99.941000**	**155.818013**	–	–	**1.37**
9	Middle East and North Africa	Interior Homocline-Central Arch	Saudi Arabia	**21.983167**	**34.273955**	–	–	**1.37**
10	Middle East and North Africa	Qatar Arch	Saudi Arabia, Qatar	**80.327833**	**125.239125**	–	–	**1.37**
11	Asia Pacific	Greater Sarawak Basin	South China Sea	**16.280000**	**25.382148**	0.97	0.68	**1.66**
12	Asia Pacific	Malay Basin	Gulf of Thailand	**12.520333**	**19.520452**	0.97	0.68	**1.66**
13	Asia Pacific	Baram Delta/Brunei-Sabah Basin	Malaysia Baram Delta, Offshore	**8.373000**	**13.054344**	0.97	0.68	**1.66**
14	Central and South America	Magallanes Basin	Argentina Faro Virgenes, On/Offshore	**0.273608**	**0.426582**	0.17	1.69	**1.86**

(continued)

Table 6 (continued)

	1	2	3	4	5	6	7	8
	Region	USGS petroleum province	Location	Future gas[a] (BBOE)	Future gas plus[b] future reserve growth (BBOE) Gas growth = 56%	Average capital[c] cost (2006 US$/ BOE)	Average operating[c] cost (2006 US$/ BOE)	Average total[c] production cost (2006 US$/BOE)
15	Central and South America	Santa Cruz-Tarija Basin	Argentina Ramos	**6.860667**	**10.696465**	0.94	1.03	**1.98**
16	Central and South America	Magallanes Basin	Argentina Tierra del Fuego, On/Offshore	**1.575451**	**2.456285**	0.71	1.33	**2.04**
17	Central and South America	Magallanes Basin	Argentina Santa Cruz II	**1.043681**	**1.627203**	0.59	1.57	**2.16**
18	Middle East and North America	Sirte Basin	Libya	**7.878833**	**12.283889**	1.37	1.02	**2.39**
19	North America	Alberta Basin	Alberta	**5.445653**	**8.490318**	–	–	**2.50**
20	Middle East and North Africa	North Egypt Basin	Egypt West Bakr	**0.212414**	**0.331174**	1.94	0.77	**2.71**
21	Middle East and North Africa	Red Sea Basin	Red Sea	**11.446667**	**17.846498**	–	–	**2.74**
22	Middle East and North Africa	Chad	Chad Sedigui Oil Project	**0.003007**	**0.004688**	1.67	1.10	**2.76**
23	Sub-Saharan Africa and Antarctica	Niger Delta	Nigeria	**36.715667**	**57.243396**	0.79	1.98	**2.77**
24	Sub-Saharan Africa and Antarctica	West-Central Coastal	Offshore Angola, Congo, Gabon, Eq. Guinea, Cameroon	**16.512167**	**25.744119**	0.79	1.98	**2.77**
25	Middle East and North Africa	Nile Delta Basin	Egypt El Mansoura	**0.684067**	**1.066528**	1.48	1.47	**2.96**
26	Middle East and North Africa	Abu Gharadiq Basin	Egypt Badr El Din	**1.387747**	**2.163636**	1.98	0.99	**2.97**
27	North America	Alberta Basin	Canada Arctic Cameron Hills	**0.680707**	**1.061290**	1.72	1.26	**2.98**
28	Middle East and North Africa	Remaining Provinces		**12.042032**	**18.774732**	–	–	**3.14**
29	Middle East and North Africa	North Egypt Basin	Egypt Qarun	**0.443000**	**0.690982**	1.91	1.29	**3.19**

30	Central and South America	Magallanes Basin	Argentina Santa Cruz I	**1.138561**	**1.775130**	1.57	1.67	**3.24**
31	Central and South America	Neuquen Basin	Argentina	**4.247167**	**6.621758**	1.57	1.67	**3.24**
32	Asia Pacific	Peral River Mouth Basin	China Xijiang Area, OFFSHORE	**0.727229**	**1.133822**	2.08	1.44	**3.51**
33	Central and South America	Llanos Basin	Colombia	**2.585333**	**4.030793**	1.76	1.98	**3.74**
34	Central and South America	Middle Magdalena	Colombia	**0.997333**	**1.554942**	1.76	1.98	**3.74**
35	Middle East and North Africa	Illizi Basin	Algeria Gassi Touil Integrated Gas Project	**12.074833**	**18.825873**	2.24	1.51	**3.75**
36	Central and South America	Maracaibo Basin	Venezuela	**7.216333**	**11.250985**	1.88	1.88	**3.76**
37	Central and South America	East Venezuela Basin	Venezuela	**33.271000**	**51.872816**	1.88	1.88	**3.76**
38	Central and South America	Tobago Trough	Caribbean Sea	**7.117167**	**11.096375**	1.88	1.88	**3.76**
39	Central and South America	Guyana-Suriname Basin	Guyana, Suriname, French Guiana	**7.009667**	**10.928771**	1.88	1.88	**3.76**
40	Central and South America	Campos Basin	Brazil Chinook, Offshore	**2.935602**	**4.576897**	1.62	2.28	**3.90**
41	Asia Pacific	Pearl River Mouth Basin	China Block 15/34 (Panyu 4-2 and Panyu 5-1), OFFSHORE	**0.260147**	**0.405595**	2.21	2.02	**4.24**
42	Former Soviet Union	Chu-Sarysu Basin	Kazakhstan Amangeldi Area	**0.708738**	**1.104993**	1.52	2.74	**4.26**
43	Europe	Northwest German Basin	Germany, Netherlands, North Sea	**12.709000**	**19.814602**	2.12	2.19	**4.30**
44	Europe	Provence Basin	Mediterranean Sea	**8.535000**	**13.306919**	2.12	2.19	**4.30**
45	Europe	Po Basin	Italy, Adriatic Sea	**4.553833**	**7.099882**	2.12	2.19	**4.30**
46	Europe	Transylvanian Basin	Romania	**1.936667**	**3.019457**	2.12	2.19	**4.30**

(continued)

Table 6 (continued)

	1	2	3	4	5	6	7	8
	Region	USGS petroleum province	Location	Future gas[a] (BBOE)	Future gas plus[b] future reserve growth (BBOE) Gas growth = 56%	Average capital[c] cost (2006 US$/ BOE)	Average operating[c] cost (2006 US$/ BOE)	Average total[c] production cost (2006 US$/BOE)
47	Asia Pacific	Kutei Basin	Indonesia Makassar, Offshore	**16.919833**	**26.379712**	2.12	2.26	**4.38**
48	Central and South America	Remaining provinces		**10.941628**	**17.059093**	–	–	**4.48**
49	Asia Pacific	Bohaiwan Basin	China Block 11/05 (Peng Lai 193-), Off shore	**2.477149**	**3.862122**	2.63	1.89	**4.52**
50	Sub-Saharan Africa and Antarctica	Remaining provinces		**3.599000**	**5.611201**	–	–	**4.56**
51	Central and South America	Santos Basin	Brazil Lagosta, OFFSHORE	**13.610000**	**21.219351**	3.29	1.33	**4.62**
52	Former Soviet Union	North Caspian Basin	Kazakhstan Aktyubinskneft Contract Area	**3.918396**	**6.109171**	0.69	4.04	**4.73**
53	Central and South America	Potigar Basin	Brazil Koch	**1.018888**	**1.588549**	–	–	**4.93**
54	North America	Alberta Basin	Alberta	**0.358267**	**0.558574**	–	–	**5.00**
55	Europe	Remaining provinces		**5.421167**	**8.452141**	–	–	**5.08**
56	Europe	North Sea Graben	Northern North Sea Brent, Offshore	**27.724833**	**43.225788**	3.85	1.40	**5.24**
57	Europe	Vestford-Helgeland	Offshore Norway	**30.143833**	**46.997251**	3.85	1.40	**5.24**
58	South Asia	Indus	Pakistan	**4.529167**	**7.061424**	–	–	**5.31**
59	Former Soviet Union	North Caspian Basin	Kazakhstan Kashagan Contract Area, Offshore	**41.260053**	**64.328549**	1.13	4.31	**5.44**
60	Asia Pacific	Remaining provinces		**8.304098**	**12.946919**	–	–	**5.52**
61	Middle East and North Africa	Chad	Chad Doba Oil Project	**0.166578**	**0.259712**	2.72	2.82	**5.54**

62	Central and South America	Magallanes Basin	Argentina Area Magallanes, Offshore	**0.767867**	**1.197181**	2.50	3.09	**5.58**
63	Middle East and North Africa	North Egypt Basin	Egypt Obaiyed	**1.068362**	**1.665682**	3.81	1.80	**5.61**
64	Sub-Saharan Africa and Antarctica	Senegal	Mauritania PSC B Tiof, OFFSHORE	**0.085516**	**0.133329**	3.73	2.17	**5.91**
65	Asia Pacific	Bohaiwan Basin	China Caofedian Blocks 04/35 and 05/36, Offshore	**0.946484**	**1.475663**	2.57	3.48	**6.04**
66	Asia Pacific	Central Sumatra Basin	Indonesia Far East Offshore Shelf, Offshore	**1.172333**	**1.827785**	2.39	3.76	**6.15**
67	Asia Pacific	North Sumatra Basin	Indonesia	**4.450833**	**6.939294**	2.39	3.76	**6.15**
68	Asia Pacific	Northwest Shelf	OFFSHORE Australia	**19.852667**	**30.952293**	2.39	3.76	**6.15**
69	Asia Pacific	Browse Basin	Offshore Australia	**6.342167**	**9.888072**	2.39	3.76	**6.15**
70	Asia Pacific	Bonaparte Gulf Basin	Northern Australia, Timor Sea	**6.132667**	**9.561441**	2.39	3.76	**6.15**
71	North America	Alberta Basin	Canada Arctic Norman Wells	**0.680707**	**1.061290**	3.45	2.72	**6.17**
72	North America	Villahermosa Uplif	Gulf of Mexico	**7.171167**	**11.180566**	–	–	**6.33**
73	North America	Remaining provinces		**21.268833**	**33.160238**	–	–	**6.41**
74	Central and South America	Reconcavo Basin	Brazil PetroReconcavo	**2.090317**	**3.2559013**	–	–	**6.50**
75	Central and South America	Sergipe-Alagoas Basin	Brazil Tartaruga, On/Offshore	**1.565500**	**2.440771**	4.21	2.49	**6.70**
76	Central and South America	Espirito Santo Basin	Brazil Frade, Offshore	**5.301901**	**8.266194**	3.86	2.91	**6.77**
77	Sub-Saharan Africa and Antarctica	Senegal	Mauritania PSCB Chinguetti and Tevet, Offshore	**0.064650**	**0.100796**	4.31	2.48	**6.79**
78	Former Soviet Union	South Turgay Basin	Kazakhstan South Kumkol Area	**1.003039**	**1.563838**	0.50	6.34	**6.85**
79	Middle East and North Africa	Abu Gharadiq Basin	Egypt El Diyur	**0.013795**	**0.021507**	4.53	2.44	**6.97**

(continued)

Table 6 (continued)

	1	2	3	4	5	6	7	8
	Region	USGS petroleum province	Location	Future gas[a] (BBOE)	Future gas plus[b] future reserve growth (BBOE) Gas growth = 56%	Average capital[c] cost (2006 US$/ BOE)	Average operating[c] cost (2006 US$/ BOE)	Average total[c] production cost (2006 US$/BOE)
80	Asia Pacific	Bohaiwan Basin	China Chengadaoxi, On/Offshore	**0.109534**	**0.170774**	4.34	2.71	**7.05**
81	Central and South America	Espirito Santo Basin	Brazil Sardinha, Offshore	**0.442599**	**0.690056**	4.12	2.96	**7.08**
82	Former Soviet Union	Middle Caspian Basin	Kazakhstan ADE Contract Area	**9.473500**	**14.770134**	1.01	6.12	**7.13**
83	Former Soviet Union	South Caspian Basin	Caspian Sea	**33.452167**	**52.155273**	1.01	6.12	**7.13**
84	North America		United States	**116.8333330**	**182.154850**	–	–	**7.18**
85	Central and South America	Campos Basin	Brazil Bijupira-Salema, OFFSHORE	**0.917026**	**1.429736**	4.70	2.52	**7.22**
86	Asia Pacific	Sichuan Basin	China Chuanzhong	**2.766833**	**4.313770**	5.77	1.52	**7.30**
87	Middle East and North Africa	Trias/Ghadames Basin	Algeria Hamra	**6.974167**	**10.873423**	5.32	2.05	**7.36**
88	Middle East and North Africa	Grand Erg/Ahnet Basin	Algeria	**19.619500**	**30.588762**	5.32	2.05	**7.36**
89	Europe	Anglo-Dutch Basin	North Sea UK Sector, Offshore	**7.642500**	**11.915422**	3.66	4.20	**7.86**
90	South Asia	Remaining Provinces		**5.007667**	**7.807453**	–	–	**7.97**
91	Former Soviet Union	Remaining Provinces		**−1.711610**	**−2.668571**	–	–	**8.04**
92	Central and South America	Campos Basin	Brazil Polvo, Offshore	**0.419372**	**0.653842**	4.47	3.65	**8.12**
93	Former Soviet Union	North Ustyurt Basin	Kazakhstan Karazhanbas, On/Offshore	**1.462552**	**2.280264**	1.76	6.54	**8.30**
94	Former Soviet Union	Amu-Darya Basin	Turkmenistan	**52.580000**	**81.977478**	1.76	6.54	**8.30**
95	Asia Pacific	South China Continental Shelf Slope	China Block 16/06 (Lufeng 13-1), Offshore	**0.483174**	**0.753316**	4.85	3.69	**8.54**
96	Asia Pacific	Yingehai Basin	Chian Yacheng 13-1, Offshore	**1.601186**	**2.496410**	5.65	2.95	**8.60**

97	Former Soviet Union	West Siberian Basin	Russia	**282.419667**	**440.320502**	–	–	**8.62**
98	Former Soviet Union	Volga-Ural Region	Russia	**14.448500**	**22.526656**	–	–	**8.62**
99	Former Soviet Union	Timan-Pechora Basin	Russia	**12.723333**	**19.836949**	–	–	**8.62**
100	Former Soviet Union	South Barents Basin	Russia	**38.479500**	**59.993388**	–	–	**8.62**
101	Former Soviet Union	North Sakhalin Basin	Russia	**13.278833**	**20.703029**	–	–	**8.62**
102	Former Soviet Union	Nepa-Botuoba Arch	Russia	**8.667667**	**13.513759**	–	–	**8.62**
103	Former Soviet Union	Azov-Kuban Basin	Russia	**3.894833**	**6.072435**	–	–	**8.62**
104	Former Soviet Union	Angara-Lena Terrace	Russia	**3.130000**	**4.879983**	–	–	**8.62**
105	Former Soviet Union	Baykit Arch	Russia	**3.672500**	**5.725795**	–	–	**8.62**
106	Former Soviet Union	Cis-Patom Foredeep	Russia	**0.910667**	**1.419820**	–	–	**8.62**
107	Former Soviet Union	Dnieper-Donets Basin	Ukraine	**7.949333**	**12.393806**	–	–	**8.62**
108	Former Soviet Union	North Barents Basin	Barents Sea	**9.970000**	**15.544227**	–	–	**862**
109	North America	Rocky Mountain Deformed Belt	Canada Arctic Kotaneelee	**0.665185**	**1.037091**	2.23	6.42	**8.65**
110	South Asia	Ganges-Brahmaputra Delta	India, Bangladesh, Bay of Bengal	**10.785167**	**16.815153**	–	–	**8.85**
111	South Asia	Bombay	India	**5.151500**	**8.031704**	–	–	**8.85**
112	South Asia	Irrawaddy	Myanmar	**4.968167**	**7.745869**	–	–	**8.85**
113	North America	Rocky Mountain Deformed Belt	Canada Arctic Nahanni South	**0.940315**	**1.466045**	2.91	6.00	**8.91**
114	Middle East and North Africa	Pelagian Basin	Tunisia Ashtart, Offshore	**3.787167**	**5.904572**	5.69	3.68	**9.37**
115	Former Soviet Union	North Caspian Basin	Kazakhstan Alibec South	**0.523550**	**0.816267**	2.02	7.40	**9.42**
116	Asia Pacific	Tarim Basin	China	**10.918500**	**17.023033**	–	–	**9.44**
117	Former Soviet Union	North Ustyurt Basin	Kazakhstan Arman, ON/ Offshore	**0.140245**	**0.218655**	3.75	6.09	**9.83**
118	North America	East Greenland Rift Basins	Offshore Greenland	**13.451500**	**20.972234**	–	–	**10.00**
119	Former Soviet Union	North Ustyurt Basin	Kazakhstan Buzachi North, On/Offshore	**0.594370**	**0.926683**	2.74	7.44	**10.18**

(continued)

Table 6 (continued)

	1	2	3	4	5	6	7	8
	Region	USGS petroleum province	Location	Future gas[a] (BBOE)	Future gas plus[b] future reserve growth (BBOE) Gas growth = 56%	Average capital[c] cost (2006 US$/BOE)	Average operating[c] cost (2006 US$/BOE)	Average total[c] production cost (2006 US$/BOE)
120			**Total gas future volume (BBOE):**	**1,665**				
121								
122								
123	*Areas not assessed in USGS(*2000*)*[d]:							
124								
125	Europe			**29.630**	**46.195634**	–	–	**10.22**
126	Central and South America			**34.038**	**53.068705**	–	–	**10.56**
127	South Asia			**9.142**	**14.252733**	–	–	**11.51**
128	Middle East and North Africa			**160.436**	**250.135434**	–	–	**12.18**
129	Asia Pacific			**36.228**	**56.482572**	–	–	**12.27**
130	North America			**50.299**	**78.421169**	–	–	**13.00**
131	Former Soviet Union			**163.048**	**254.208134**	–	–	**13.23**
132	Sub-Saharan Africa and Antarctica			**17.110**	**26.676529**	–	–	**17.65**
133								
134				Total	3,375			

[a]Future gas volumes come from USGS (2000)
[b]Future gas plus future reserve growth estimated by this study
[c]Costs estimated by this study
[d]Volumes and costs below row 123 estimated by this study

References

Adelman MA, Watkins GC (2008) Reserve prices and mineral resource theory. Energy J 29(Special Issue):1–16

Adelman MA, Houghton JC, Kaufman GM, Zimmerman MB (1983) Energy resources in an uncertain future: coal, gas, oil, and uranium supply forecasting. Ballinger Publishing Company, Cambridge

Aguilera RF (2006) Assessing the long run availability of global fossil energy resources. PhD dissertation, Colorado School of Mines, Golden, CO

Aguilera RF, Aguilera R (2008) Assessing the past, present, and near future of the global energy market. Society of Petroleum Engineers (SPE). J Petrol Technol 60(5):36–39

Aguilera RF, Balistreri EB, Cuddington JC (2007) Assessing previously unassessed petroleum provinces using the Variable Shape Distribution (VSD) model. Presented at the American Association of Petroleum Geologists (AAPG) Annual Convention and Exhibition in Denver, USA, between 7–10 June 2009

Ahlbrandt TS, McCabe PJ (2002) Global petroleum resources: a view to the future. Geotimes 47(11):14–18. http://www.agiweb.org/geotimes/nov02/feature_oil.html

AOC Petroleum Support Services (2004) Strategic significance of America's oil shale resource, vol II, Oil shale resources, technology and economics. Prepared for the Office of Strategic Petroleum Reserves, US Department of Energy

Banks FE (2005) Logic and the oil future. Nationalekonomiska Institutionen, Uppsala University, Uppsala

Barton CC (1995) A new approach to estimating hydrocarbon resources. United States Geological Survey Energy Resource Surveys Program, January. http://webharvest.gov/peth04/20041016205020; http://energy.usgs.gov/factsheets/HydroRes/estimat.html

Bentley R, Boyle G (2008) Global oil production: forecasts and methodologies. Environ Plann B Plann Des 35(4):609–626. http://www.envplan.com/abstract.cgi?id=b33063t

Bloomberg (2005) Shell, Exxon tap oil sands, gas as reserves dwindle, Feb 18. http://www.bloomberg.com/apps/news?pid=10000087&sid=a3Iz1vRFvXuI&refer=top_world_news

British Petroleum (2009) Statistical review of world energy 2009. British Petroleum, London

Campbell CJ (2000) Peak oil. Presented at the Technical University of Clausthal, Germany, December. http://www.hubbertpeak.com/de/lecture.html

Campbell CJ, Laherrere JH (1998) The end of cheap oil. Sci Am (March):78–83

Center for Global Energy Studies (2003) London, http://www.time.com/time/magazine/article/0,9171,450939,00.html

Charpentier RR (2005) Estimating undiscovered resources and reserve growth: contrasting approaches. In: Dore AG, Vining BA (eds) Petroleum geology: North-West Europe and global perspectives—proceedings of the 6th petroleum geology conference, London, pp 3–9

Chevron Texaco (2009) Heavy oil: unlocking the potential. http://www.chevron.com/documents/pdf/heavyOil.pdf

Deffeyes KS (2001) Hubbert's Peak: the impending world oil shortage. Princeton University Press, Princeton

Drew LJ (1997) Undiscovered petroleum and mineral resources, assessment and controversy. Plenum Press, New York/London

Economides MJ, Oligney RE (2000) The color of oil: the history, the money and the politics of the world's biggest business. Round Oak Publishing Company, Katy

Ecopetrol SA (2003) Annual report. http://www.ecopetrol.com.co/contenido.aspx?catID=283&conID=41869

Energy Information Administration (2006) Petroleum & Other Liquids Supply. http://www.eia.gov/pub/oil_gas/petroleum/analysis_publications/oil_market_basics/Supply_text.htm

Farrell A (2008) Energy notes: news from the University of California Energy Institute 6(2):2–3

Global Policy Forum (2002) Oil in Iraq: the heart of the Crisis, December 2002. http://www.globalpolicy.org/security/oil/2002/12heart.htm

Goldman Sachs (2005) Global energy, 100 projects to change the world. Goldman Sachs, London

Goodstein DL (2004) Out of gas: the end of the age of oil. W.W. Norton and Company Inc., New York

Hefner RA III (2002) The age of energy gases in the new millennium. Oklahoma City, OK: The GHK Company. Based on presentation at 10th repsol-harvard seminar on energy policy in Madrid, Spain, June 1999, and the Society of Petroleum Engineers (SPE) gas technology symposium in Calgary, Canada, 30 Apr 2002

Holland SP (2008) Modeling peak Oil. Energy J 29(2):61–80

Institute for the Analysis of Global Security (2003) Prospects on Russia's Stance towards OPEC. Potomac, Maryland, http://www.iags.org/n0929033.htm

Isaacs E (2005) Canadian oil sands: development and future outlook. Alberta Energy Research Institute, Calgary, Alberta. http://www.cge.uevora.pt/aspo2005/abscom/Abstract_Lisbon_Isaacs.pdf

Kaufman GM (1962) Statistical decision and related techniques in oil and gas exploration. Prentice-Hall, Englewood Cliffs

Klett TR, Gautier DL, Ahlbrandt TS (2005) An evaluation of the U.S. Geological survey world petroleum assessment 2000. AAPG Bull 89(8):1033–1042

Laherrere JH (2000) Is USGS 2000 assessment reliable? World Energy Council cyberconference, London, May. http://www.oilcrisis.com/laherrere/usgs2000/

Lynch MC (2005) Unconventional oil: filling in the gap of flooding the market? Presentation for Department of Energy National Energy Modeling System, Washington DC, April 12 http://www.energyseer.com/Unconventional%20Oil.pdf

Minerals Management Service (1996) An assessment of the undiscovered hydrocarbon potential of the nation's outer continental shelf. Report MMS 96–0034

Mommer B (2004) The value of extra heavy crude oil from the Orinoco Belt. Middle East Econ Surv 47(11):D1–D11

National Energy Board of Canada (2006) Canada's Oil Sands-Opportunities and challenges to 2015 questions and answers. Calgary, Alberta, http://www.neb.gc.ca/clf-nsi/rnrgynfmtn/nrgyrprt/lsnd/pprtntsndchllngs20152006/qapprtntsndchllngs20152006-eng.html

Odell PR (2004) Why carbon fuels will dominate the 21st century's global energy economy. Multi-Science Publishing Company, Brentwood

Oil and Natural Gas Commission of India (2004) Annual Report. http://www.ongcindia.com/financial_new.asp

Pakistan Oilfields Limited (2005) Directors report. http://www.pakoil.com.pk/investor_relations/areports.html

Porter ED (1995) Are we running out of oil? American Petroleum Institute Policy Analysis and Strategic Planning Department. Discussion Paper 81, December

Radetzki M (2002) Is resource depletion a threat to human progress? Oil and other exhaustible materials. In: 9th international energy conference & exhibition – energex 2002, Mineral and Energy Economic Research Institute of the Polish Academy of Sciences, Krakow, Poland, May 19–24

Simmons MJ (2005) Twilight in the desert: the coming Saudi oil shock and the world economy. Wiley, Hoboken

Suncor Energy Inc (2006) Oil sands tour. Fort Mc Murray, Alberta

Tertzakian P (2006) A thousand barrels a second: the coming oil break point and the challenges facing an energy dependent world. McGraw-Hill, New York

Tilton JE (2002) On borrowed time? Assessing the threat of mineral depletion. Resources for the Future, Washington, DC

Tilton JE, Skinner BJ (1987) The meaning of resources. In: McLaren DJ, Skinner BJ (eds) Resources and world development. Wiley, New York

United States Department of Energy Office of Petroleum Reserves – Strategic Unconventional Fuels (2006) Fact sheet: U.S. oil shale economics. Washington DC, http://www.fossil.energy.gov/programs/reserves/npr/Oil_Shale_Economics_Fact_Sheet1.pdf

United States Geological Survey (1995) National oil and gas assessment. CD-ROM Denver, Colorado, http://energy.cr.usgs.gov/oilgas/noga/1995.html

United States Geological Survey (2000) World petroleum assessment. CD-ROM. U.S. Geological Survey, Reston http://pubs.usgs.gov/dds/dds-060/

United States Geological Survey (2008) Circum-Arctic resource appraisal. Fact Sheet 2008–3049. Menlo Park, California, http://pubs.usgs.gov/fs/2008/3049/

Wood M (2004–2006) Upstream asset analysis reports. http://www.woodmacresearch.com/cgi-bin/wmprod/portal/energy/productMicrosite.jsp?productOID=664067

Wood JH, Long GR, Morehouse DF (2004) Long term world oil supply scenarios: the future is neither as Bleak or Rosy as some assert. Energy Information Administration, Washington, DC. http://www.eia.doe.gov/pub/oil_gas/petroleum/feature_articles/2004/worldoilsupply/oilsupply04.html

Coal: An Energy Source for Future World Needs

Thomas Thielemann

Abstract

Since 2004, international hard coal prices have been at rather expensive levels above 60 USD/t. Some argue that these higher prices might indicate the threat of a physical scarcity of fossil fuels – similar to the situation with oil and gas. This is not true. The supply situations with lignite and hard coal appear to be largely not critical. Adjusted to the rise in global coal consumption, which is expected until 2100, nature by and large can meet the world's coal demand. This is shown for lignite and for hard coal here, differentiated in space and time. The only area of potential concern is Asia (especially China). But today's and coming eager efforts in China to convert coal resources into reserves will most likely deliver the coal needed for the Chinese market. Up to the year 2100, and from a geoscientific point of view, there will be no bottleneck in coal supplies on this planet. The CO_2 emissions correlated with this coal use have to be a considerable concern if mankind wants to stabilize the world's atmospheric CO_2 concentration. A widespread deployment of CCS (carbon capture and storage) technologies is recommended to curtail the future rise in atmospheric CO_2.

Keywords

Coal • Energy supply • Reserves • Resources • Geological coal availability • Global hard coal market • CCS

T. Thielemann (✉)
RWE Power AG, Stüttgenweg 2,
Cologne D-50935, Germany
e-mail: thomas.thielemann@rwe.com

R. Sinding-Larsen and F.-W. Wellmer (eds.), *Non-Renewable Resource Issues: Geoscientific and Societal Challenges*, International Year of Planet Earth,
DOI 10.1007/978-90-481-8679-2_5, © Springer Science+Business Media B.V. 2012

Introduction

Mining raw materials has been a business for humans for more than 5,000 years. We know of the quarries that supplied the materials for Egypt's pyramids, say, or the flint pits in the Maastricht region. Depending on the raw material and the deposit, people even in those days saw deposits being depleted within a few decades or generations. As a result, they started thinking about the size and extent of a raw material deposit. Scientifically precise calculations gained weight with the emergence of mine surveying in the sixteenth century (Agricola 1556).

There are now many different analyses available on the worldwide reserves of different natural resources and for various occasions. As regards coal, for example, natural resources have been examined by the International Energy Agency (IEA 2008a, 2011), by the World Energy Council (WEC 2004), by BP (BP 2011) and by the Federal Institute for Geosciences and Natural Resources (BGR 2009, 2011). Ruppert et al. (2002) did an inventory especially for the situation in the USA.

The aim of this article is to describe the likely role of coal for world energy supply during this century. This will involve outlining the mechanisms of raw material supplies, describing the world market for coal and possible trends and, finally, describing the deployment of CCS (carbon capture and storage) technologies. In view of the complexity of the subject, though, the world's coal market and coal supplies will certainly evolve a little differently along the route we can predict today.

What Is Meant by Reserves and Resources?

Every year, the world consumes large amounts of fossil energy raw materials. If sufficient raw materials are to be available at all times, a built up of reserves for future mining is needed. For this reason, mining companies secure their annual output for some decades by exploring and developing deposits (Gerling and Wellmer 2004; Wellmer 2008). These developed quantities, i.e. economically minable using today's technology at today's prices, are called reserves.

Resources, on the other hand, include known, though at present not economically minable, quantities. Every year, coal resources move into reserves as our knowledge of coal deposits improves and new pits are developed. This conversion of resources into reserves replaces some of the reserve losses due to annual coal production. In recent decades, such conversion only accounted for 5–20% of specific world annual production (Thielemann et al. 2007). Due to big existing volumes of reserves, companies are recently in the favourable situation to only show limited exploration efforts. Measured by the pit openings and extensions planned for the next few years (Kopal 2006), annual output will not be replaced by new same-size reserves in the foreseeable future. However, if more exploration effort would be needed, all annually consumed coal reserves could be replaced by converting resources into reserves.

The sum of reserves and resources is called total resources. Adding these total resources and the coal quantities produced in recent centuries (i.e. the accumulated output), we obtain the total potential.

Approach in Calculating Reserves

Lignite and hard coal are classified into reserves and resources, using a range of cut-off parameters. Important factors include seam or coal thickness, depth, energy content, barren partings and ash content, coal rank (measured as vitrinite reflectance) and volatile matter content. Coals with a calorific value of about <6,000 kcal/kg, for example, are classified as lignites, and coals with >6,000 kcal/kg as hard coals. Depending on the national limits for the workability of the coal thickness concerned, total resources usually refer to thicknesses starting at 60 cm. Although a few deposits exist with lower thicknesses being extracted (e.g. coking coal in a Canadian seam down to 35 cm), the quantities involved are negligible. Here, coals with a maximum depth of 1,800 m are included in the total-resource calculations even if a depth of 1,500 m currently appears to be the lower limit for economic extraction. The maximum permissible barren partings content in most countries is 35% vol.

For the author, taking into account the current world market prices for coal in combination with known production costs, coal deposits developed at the moment and producing at prices below US$ 50/t count as reserves.

World Hard Coal Production

In 2008, 5.5 billion metric tons (Gt) of hard coal (=4.637 billion tce, metric tons of coal equivalent) were produced worldwide. The world's biggest producer was China, with output of 2.479 Gt or 45% of the total, followed by the USA (968 Gt, 17.6%), India (453 Mt, 8.2%), Australia (325 Mt, 5.9%) and South Africa (244 Mt, 4.4%). Fifty-seven percent of hard coal output in 2008 came from Asia, 18.7% from North America, 7.5% from the CIS, 6.5% from Oceania, 4.7% from Africa, 4% from Europe, 1.5% from Central and South America as well as negligible quantities from the Middle East.

Despite the wide regional scatter of coal deposits, consolidation trends cannot be overlooked in the companies acting on the world market. About one third of export hard coal output and just under 50% of the current expansion projects are accounted for by the Big Four. These four companies – also referred to as the RBXA group – are Rio Tinto, BHP Billiton, Xstrata/Glencore and Anglo-American. On the coking-coal market, BHP Billiton has a share of roughly 30% of the export capacities. However, rising coal prices are also encouraging small companies to boost their mining activities in this sector.

In order to serve the hard coal needs that are growing worldwide, continuous expansion of production capacities is necessary. US$ 30–60 per metric ton of annual production capacity (t p.a.) must be invested in already developed locations (brownfields) for their infrastructure. In hitherto undeveloped regions (greenfields), the required investment may reach US$ 80–120 per metric ton p.a. (IEA 2003). For the period from 2001 to 2030, the IEA (2003) calculates a lower amount of specific investment for the raw material hard coal than for oil and gas. In the case of hard coal, the costs amount to a converted US$ 3.4/tce, while they reach US$ 15.4/tce for crude oil and as much as US$ 19.6/tce for natural gas.

Once a field is developed and a mine opened, the production costs vary by coal type and by producing country. In the years 2006 and 2007, costs for producing thermal coals varied between 14 and 30 USD/t. In Australia, these costs may rise up to 42 USD/t (Ritschel and Schiffer 2007). Coking coals tend to show higher production costs worldwide, between 20 and 43 USD/t. In the Appalachian Mountains (USA), where most of the easily accessible coal seams are already mined, production of coking coals can be as expensive as up to 80 USD/t (Ritschel and Schiffer 2007). This is also reflected in the consumer prices. Coking coals are more expensive than thermal coals.

Since 1990, worldwide hard coal production capacity has expanded by 91%, from 3.5 to 6.7 Gt/a in 2010. This increase was largely due to a rise in steam-coal capacities, whereas hardly any expansion took place in coking coal (production capacities between 0.5 and 0.6 Gt/a).

World Market

There is no world market for lignite, since this fuel, in view of its low energy content, is used almost exclusively at regional level. By contrast, a functional world market for hard coal has existed since the 1920s. Maritime hard coal trade has seen an upswing since the second oil-price crisis of 1979, which directed consumers' attention to hard coal. Besides the aspect of greater security of supply, it is, above all, the price factor that speaks in favour of coal. The price for hard coal varied mainly between 62 and 78 USD/t, in the period 2004–2009, with one exceptional period. Within 2008, the coal price tripled and reached a maximum of 219 USD/t in July 2008.

In the last 10 years, the seaborne coal trade has more than doubled (Schiffer and Kopal 2005; BGR 2009). Half of this growth was accounted for by the last 5 years, during which the market expanded 5–8% p.a. In 2008, 917 million metric tons (Mt) of hard coal were traded, including 820 Mt seaborne and 97 Mt in cross border trade (IEA 2008a). In 2010, these numbers rose to 963 Mt traded hard coal, including 713 Mt thermal coal and 250 Mt coking coal. Since 1990, total world trade in thermal coal grew by about 403 Mt, from 310 to 713 Mt (VDKi 2011). Within the coking-coal market, growth was some 58 Mt in both supply and demand, from 192 to 250 Mt. In the last 15 years, supply and demand have been in long-term balance. Excess capacities such as those in the early 1990s, or the bottlenecks of the last 5 years, have been and can be offset within a few years. This proves that the world hard coal market is very competitive (Thielemann et al. 2006).

Maritime trade was divided into an Atlantic and a Pacific market. In this respect, the Atlantic market comprises hard coal exporters such as Russia, South Africa, Colombia and the USA, as well as coal consumers, mainly the EU and again the USA. The Pacific area as a regional market consists of the Pacific and Indian Oceans, with the major exporters Australia, Indonesia, China and Russia and the main importers Japan, South Korea and China. Due to especially growing Chinese imports from many sources worldwide, coal maritime trade more and more merges into one global market.

CO_2 Emissions from Coal

The consumption of 1 t of lignite leads to about 1 t of CO_2 emissions, whereas the use of 1 t of hard coal releases around 3 t of CO_2 into the atmosphere. Recalculated to energy contents, these emissions sum up to about 1,000 g CO_2/kWh$_{electric}$ for lignite and 750 g CO_2/kWh$_{electric}$ for hard coal. Recent trends in energy supply and use are unsustainable – economically, environmentally and socially. Without decisive action, energy-related CO_2 emissions will increase by 130% above 2005 levels by 2050 (IEA 2008b, 2009). We have to and can change the world's current path. But this implies a wide deployment of low-carbon energy technologies. To reach international greenhouse gas emission goals, energy efficiency, carbon capture, transport and storage (CCS), renewable energies and nuclear power would need to play a more prominent role, already near term.

IEA (2009) argues for an investment of USD 2.5–3 trillion into CCS from 2010 to 2050, which would represent about 6% of the overall investment needed to achieve 50% reduction in GHG (greenhouse gas) emissions by 2050. On the long term, the "Stern report" (prepared by N. Stern, former Chief Economist and Senior Vice-President of the World Bank from 2000 to 2003) predicted even lower electricity prices when deploying CCS widely (Stern 2006). CCS should not only be deployed to coal use. It should be more widely adopted, e.g. by gas and biomass power plants, in the oil and gas processing sector and in emission-intensive industries like iron and steel, cement, chemicals and paper. This implies also new efforts to assess the CO_2 storage capacities worldwide, especially of deep saline formations and low-costs opportunities like depleted oil and gas fields. CO_2 capture in power plants – using the different processes pre-combustion, post-combustion and oxyfuel – involves a 7–10% fall in the efficiencies of coal-fired plants (WEC 2004; BMWA 2004) so that, if ground-covering use is made of this technology, growing energy requirements will lead to even steeper increases in hard coal consumption than those described in previous scenarios.

Where Are the World's Lignite and Hard Coal Deposits?

There are coal deposits in many countries – whether states of the Organisation for Economic Co-operation and Development (OECD), economies in transition or developing countries – so that coal has an enormous geostrategic

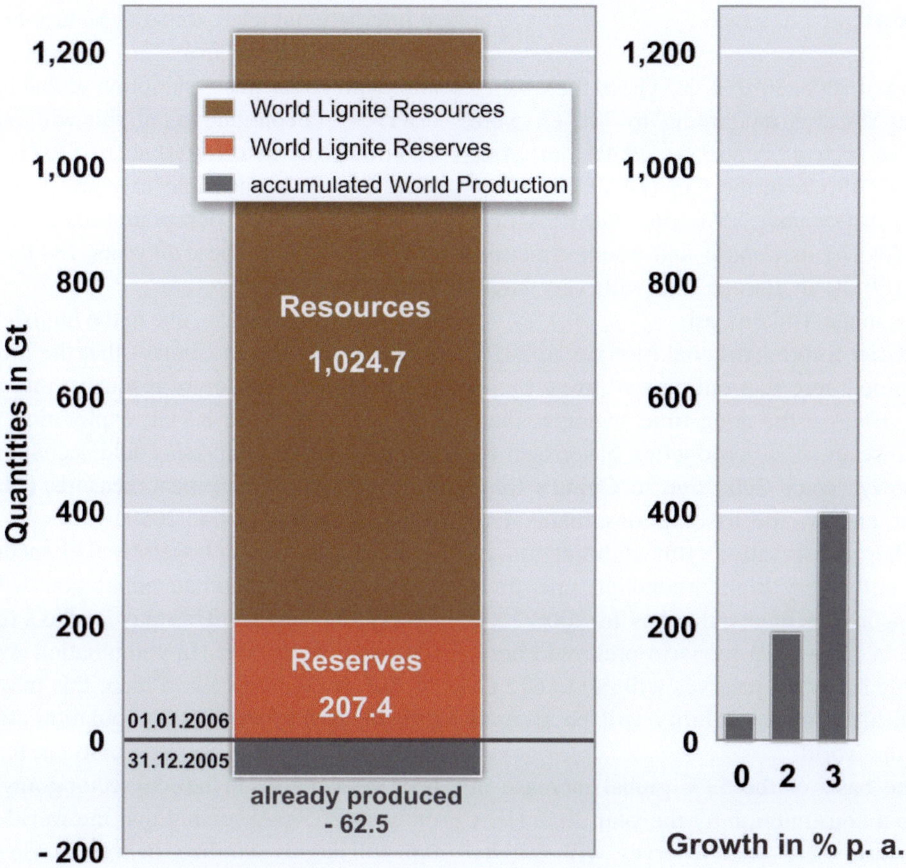

Fig. 1 Worldwide total of lignite quantities in terms of resources, reserves and accumulated production (*left*, position: 31.12.2005) and consumption of lignite reserves until 2100, assuming different scenarios (*right*, average growth in world production from 0% to 3% p. a.)

advantage compared to crude oil and natural gas, about 70% of whose reserves are concentrated on a "geostrategic ellipse" between northwest Siberia across the Caspian region to the riparian states of the Persian Gulf. Since lignite and hard coals differ both in the distribution of reserve quantities and in their so-called static lifetimes, the two raw materials will be treated separately in what follows.

Lignite

Worldwide lignite resources amount to 1,025 billion metric tons (Gt). To this, another 207 Gt of reserves can be added (Fig. 1). Since 1800, some 62.5 Gt have already been extracted and consumed. Depending on the growth scenario assumed for lignite mining (growth rates between 0% and 3%/a), different quantities of reserves will be transferred to the category "already produced" until the year 2100 (Fig. 1).

Worldwide lignite production fell from a maximum of just under 1.2 Gt in 1988 to 0.9 Gt in 1995. In the last 10 years, however, global output has been largely stable, with growth rates between 0% and 2%/a. If this trend continues, not more than already today known lignite reserves will be used up by the year 2100 (Fig. 1). This being so, lignite – viewed from a global angle – is an energy raw material with very high security of supply. No other fossil energy carrier offers lignite's availability, which is already foreseeable with such certainty today for many decades to come.

Hard Coal

At the end of 2008, 30% (223.5 Gt) of the world's hard coal reserves amounting to 746 Gt were located in North America, 29.4% in Asia (219.6 Gt), 19.1% in the CIS (142.7 Gt), 9.8% (72.9 Gt) in Oceania, 7.5% (56.2 Gt) in Africa, 2.3% (17.2 Gt) in Central and South America, 1.6% (11.9 Gt) in Europe and with very small quantities in the Middle East.

To predict future hard coal reserves in 2020, it is assumed here that output will grow by an average 3%/a – the long-time average since 1995. However, since production has grown by a good 6%/a since 2000 due to China's huge thirst for energy, the assumptions made here look rather conservative. But stabilization of this currently very high production growth is uncertain, as the financial crisis in 2008/2009 showed. In the +3%/a scenario presented here, worldwide hard coal reserves will fall to 672 Gt by 2020, although availability will be assured all over the world.

On the basis of the 53% global increase in world coal consumption by the year 2030 (IEA 2006), world hard coal reserves will fall by about 14% net to some 642 Gt. This reduction will be dampened by the conversion of resources into reserves, by an amount of approx. 69 Gt by 2030, so that, from 2009 until 2030, some 170 Gt of today's and tomorrow's reserves are likely to be consumed. North America, with 30.5% (196 Gt), will continue to have the biggest hard coal reserves. In the CIS, reserves are likely to amount to 145 Gt (22.6%) in 2030. Dramatic will be the fall in coal reserves in Asia, owing to the very high coal consumption there, to an estimated 156 Gt. Hence, these quantities will only account for 24.3% of world hard coal reserves in 2030. Then, some 70 Gt (10.9%) will be available in Oceania, 49 Gt (7.6%) in Africa, 16 Gt (2.5%) in Central and South America, 8 Gt (1.2%) in Europe and negligible amounts in the Middle East.

According to UN data, world population growth will slow down after 2030. In combination with higher efficiencies in energy generation at international level, this may reduce the rise in hard coal output from an average 3–2%/a. Although under this assumption global hard coal reserves will continue to fall, this will only be by 66 billion metric tons (Gt) net to 576 Gt in 2050, since it will be possible to convert just under 118 Gt hard coal resources into reserves between 2030 and 2050. In these 20 years, 184 Gt is likely to be consumed, therefore.

It is in this scenario, due to the high demand in Asia – especially in China – that the most comprehensive conversion of resources into reserves will be made. With 82 Gt, exploration projects and pit extensions in Asia could account for 69% of worldwide development measures (118 Gt) in the period from 2030 to 2050.

Economic wealth hopefully will encompass a broader basis in human population within the twenty-first century. This may lead to a reduction of annual birth rates. In combination with birth control in countries like China, this may reduce growth rates of the world population after 2030 (EIA 2005; IEA 2008a, 2011). So far, there have been no scenarios in the coal sector going beyond the year 2050. Assuming that the world population will largely stabilize, an average rise in world hard coal production of 1%/a is estimated for the period 2050 until 2100. On these assumptions, world hard coal reserves will fall by 255–321 Gt by 2100. Assuming that, thanks to region-specific exploration on a similar scale as in the past, resources become reserves after 2050, it is to be expected that, between 2050 and 2100, some 224 Gt of resources will be converted into reserves, including 150 Gt (67%) in Asia alone. Hard coal consumption for these 50 years could amount to some 479 Gt. For Asia, a dramatic decline in hard coal reserves must be expected since the hunger for coal will eat up much of the reserves. In North America, the hard coal needs of many new power plants (especially in the USA) will reduce reserves to a hardly lesser extent. Oceania, due to its still-strong export alignment, will have almost half of its reserves left by 2100 compared with 2050. Accordingly, the CIS (especially Russia) in the long term will have the biggest hard coal reserves.

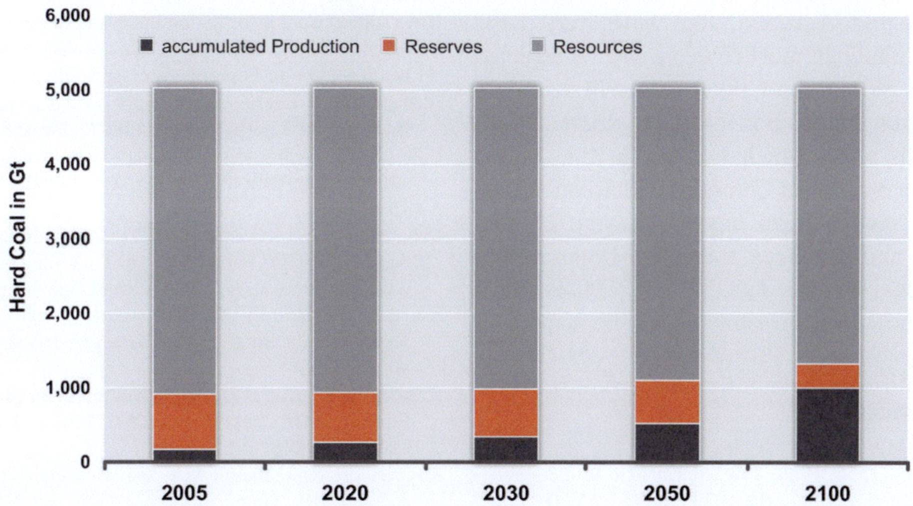

Fig. 2 Developments in resources, reserves and accumulated production of hard coal between 2005 and 2100

Should the annual rise in output be greater than 1%/a, Asia will have to convert resources into reserves on a much larger scale than presumed here. Since it will certainly be possible to cover some needs on the world market, the pressure of Asia, specifically China and India, on world coal supplies and world market prices will be much higher than today. All the same, from a raw material angle in this scenario, there will be no bottleneck in coal supplies until 2100.

The total global potential of hard coals amounts to 5,030 Gt. In the scenario presented here, these figures remain constant for the period from 2005 until 2100 (Fig. 2) because all major coal basins including their coal seams are known. However, it is assumed that resources will be converted into reserves on a continuous basis. On the assumptions made here of an initial increase in hard coal output of 3%/a, which falls to 1%/a after 2050, this transformation will amount to some 411 Gt. This is equivalent to 10% of today's existing resources. Despite these newly accessible future reserves, the size of the available reserves will fall continuously, from 730 Gt in 2009 to 321 Gt in 2100 (Fig. 2). The main reason for this is a doubling of hard coal output between 2009 and 2050. Nevertheless, the geological availability of hard coal will be restricted at no point in time until 2100.

Summary

Crude oil, uranium and natural gas reserves amount to 40–70 times the current annual consumption. For hard coal, this ratio is 150 times annual consumption, and for lignite well over 200 times.

Thanks to this favourable supply situation, coal will be able to replace some of the oil and gas in the course of this century. Today already, South Africa, for example, liquefies some 47 Mt/a of hard coal and uses it as fuel in the transport sector. A growing proliferation of the techniques for coal liquefaction and coal gasification will boost this trend. China and the USA have already launched projects for coal liquefaction (IEA 2008a, 2011). So, coal will continue to play an outstanding role in the world's energy supply.

As it was possible to demonstrate in this outlook until 2100, coal is, globally speaking, an energy raw material with very high security of supply. It is already possible to say that the availability of lignite and hard coal is assured until at least 2100. Although Asia's and, specifically, China's thirst for energy will pose huge challenges for Asia's energy supply throughout the entire century, from a deposit standpoint, it will be possible to cover the rising demand using Asian coal deposits and the world market. North

America and Australia, too, have very large reserves although most of these will be depleted by 2100. Within the CIS, Russia will acquire a much stronger position as coal exporter compared with today. Much the same is true – albeit to a lesser extent – of Colombia and South Africa. From a raw material angle, i.e. sum of coal reserves and resources, there will be no bottleneck in coal supplies until the year 2100, assumed that the necessary investments will be made.

References

Agricola G (1556, 1928, 2006 reprint) De re metallica – unchanged reprint of the 1928 edition by VDI-Verlag. Matrixverlag, Wiesbaden, 564 p

BGR (2009) Energierohstoffe 2009: Reserven, Ressourcen, Verfügbarkeit. BGR, Hannover, 286 pp. (http://www.bgr.bund.de) Accessed date 19 Jun 2010

BGR (2011) Short study "Reserven, Ressourcen und Verfügbarkeit von Energierohstoffen" 2010. BGR, Hannover, 84 pp. (http://www.bgr.bund.de) Accessed date 20 Dec 2011

BMWA (2004) Research and development concept for low emission and fossil fuelled power plants. Report of the COORETEC working group (German). BMWA report 527. Federal Ministry of Economics and Labour, Berlin, 87 p

BP (2011) BP statistical review of world energy 2006. BP plc, London, 45 pp. (www.bp.com/statisticalreview) Accessed date 2 Jan 2012

EIA (2005) International energy outlook 2005. Energy Information Administration, Washington, DC, 186 pp. (www.eia.doe.gov) Accessed date 19 Jun 2010

Gerling JP, Wellmer FW (2004) Raw material availability – with a focus on fossil energy resources. World Min Surf Undergr 56(4):254–262

IEA (2003) World energy investment outlook 2003. International Energy Agency/OECD, Paris, 516 pp

IEA (2006) World energy outlook 2006. International Energy Agency/OECD, Paris, 629 pp. (www.iea.org/books) Accessed date 2 Aug 2010

IEA (2008a) Coal information 2008. International Energy Agency/OECD, Paris, 512 pp. (www.iea.org/books, www.oecdbookshop.org) Accessed date 2 Jan 2012

IEA (2008b) Energy technology perspectives 2008. International Energy Agency/OECD, Paris, 84 p

IEA (2009) Technology roadmap, carbon capture and storage. International Energy Agency/OECD, Paris, 48 p

IEA (2011) Coal information 2011. International Energy Agency/OECD, Paris, 542 pp. (www.iea.org/books, www.oecdbookshop.org) Accessed date 2 Jan 2012

Kopal C (2006) Angebot und Nachfrage am Steinkohlenweltmarkt. Z f Energiewirtschaft (ZfE) 30(1):67–82

Ritschel W, Schiffer HW (2007) Weltmarkt für Steinkohle, Edition 2007, Report by RWE Power. Essen and Cologne, 106 p

Ruppert LF, Kirschbaum MA, Warwick PD, Flores RM, Affolter RH, Hatch JR (2002) The US Geological Survey's national coal resource assessment: the results. Int J Coal Geol 50:247–274

Schiffer HW, Kopal C (2005) Beitrag des Steinkohlen-Welthandels zur Sicherung der Energieversorgung. Z f Energiewirtschaft (ZfE) 29(3):171–180

Stern N (2006) Stern review: the economics of climate change. H.M. Treasury, London. www.hm-treasury.gov.uk/stern_review_report.htm Accessed date 2 Jan 2012

Thielemann T, Gerling JP, Rempel H, Stribrny B (2006) Entwicklung und Perspektiven des Weltsteinkohlemarktes. Energiewirtschaftliche Tagesfragen 56 (Special 1/2006), Essen:S15–S20

Thielemann T, Schmidt S, Gerling JP (2007) Lignite and hard coal: energy suppliers for world needs until the year 2100 – an outlook. Int J Coal Geol 72(1):1–14

VDKi (2011) Annual report 2011. Verein der Kohlenimporteure, Hamburg, 130 pp. (http://www.verein-kohlenimporteure.de/wDeutsch/jahresbericht/index.php?navid=19) Accessed date 20 Dec 2011

WEC (2004) Sustainable global energy development: the case of coal – a report of the World Energy Council (WEC). World Energy Council, London, 171 p. (www.worldenergy.org) Accessed date 19 Jun 2010

Wellmer FW (2008) Reserves and resources of the geosphere, terms so often misunderstood. Is the life index of reserves of natural resources a guide to the future? Z dt Ges Geowiss 159(4):575–590

Uranium and Thorium: The Extreme Diversity of the Resources of the World's Energy Minerals

Michel Cuney

Abstract

Uranium and thorium represent two elements which can be used for the production of energy. The types of uranium deposits, the mechanisms driving their genesis and their resources are relatively well known because extensive exploration programs and scientific research projects have been developed worldwide since the early 1950s. However, there are still several types of U deposits which have been underexplored such as those related to Na-metasomatism, and several others with low uranium grade but large tonnages such as phosphates, which mining has been discontinued, but which may become significant resources in the future depending of the evolution of the U prices. At the opposite, the types of Th deposits, their genesis and their resources are poorly known because of the extremely limited commercial use of this commodity. Recent evaluations of the Th resources show however that they may be as important as identified U resources, but a better evaluation of the cost of Th extraction for most of these resources is crucially needed. However, as the Th cycle does not need the costly enrichment process required for the use of uranium in most nuclear reactors and that the totality of Th can be burned in a nuclear power station, at least theoretically, the cost of thorium extraction can probably be significantly higher than that of U.

U deposits are extremely diverse and may be formed at nearly all steps of the geological cycle, whereas Th deposits dominantly result from concentration during two major types of processes: magmatic fractionation of peralkaline complexes, associated carbonatites as an incompatible element and associated fluid fractionation from these types of melt, and as heavy mineral accumulation in placer-type deposits, essentially as monazite.

Uranium is already widely used for energy production, but only a very small part of it is burned in the present nuclear reactors. A great variety of nuclear materials are produced during the U nuclear cycle: depleted U,

M. Cuney (✉)
UMR G2R 7566 CNRS – CREGU, Nancy Université,
BP 70239 – 54 506, Vandoeuvre Cedex, France
e-mail: Michel.Cuney@g2r.uhp-nancy.fr

R. Sinding-Larsen and F.-W. Wellmer (eds.), *Non-Renewable Resource Issues: Geoscientific and Societal Challenges*, International Year of Planet Earth,
DOI 10.1007/978-90-481-8679-2_6, © Springer Science+Business Media B.V. 2012

spent fuel, reprocessed U, reprocessed Pu and military highly enriched U and Pu, which can be transformed to usable fuel in the present and future generations of nuclear power stations for a sustainable use of these resources. The present review of available and potentially available nuclear fuel resources in the world shows that these resources are considerable and can be largely increased with increasing exploration, the improvement of processing of low grade – large tonnage deposits (phosphates, black shales) – of refractory ore associated with peralkaline intrusions, the spreading of recycling spent U fuel and the development of new technologies for burning 238U and 232Th in new generations of nuclear power reactors.

Keywords
Uranium • Thorium • Deposits • Resources • Sustainable use

Introduction

Uranium and thorium represent two major mineral resources for the production of energy. The use of U is already well developed whereas that of Th is still at the state of project. Moreover, until now, only a small fraction of the natural U, the isotope ^{235}U, is used in the reactors, leaving 99.3% of the natural uranium as ^{238}U unused. Also, only a small part of the nuclear material issued from nuclear fuel cycle is reprocessed to be used as a nuclear fuel. Therefore, a sustainable development of nuclear energy requires a wider development of existing technologies and the development of new ones to use more thoroughly the U and Th resources.

Uranium is enriched in the continental crust (1 ppm in average, Javoy 1999; 2.7 ppm in the upper crust) relatively to the primitive mantle (21 ppb) because of its incompatible behaviour during partial melting and fractional crystallization. Uranium deposits occur at nearly all stages of the geological cycle from surficial, sedimentary, diagenetic, metamorphic, igneous, to hydrothermal environments (Fig. 1). The about 1,200 U deposits listed in the new UDEPO database (OECD/NEA & IAEA 2009) have been grouped into 15 major types mainly based on their geological setting according to the classification developed by the IAEA–NEA in 1988–1989 and Dahlkamp (1993). However, in this chapter, we use a classification based on the geological, physical and chemical conditions controlling their formation as proposed by Cuney (2009, 2010) (Table 1). The same type of classification has been developed here to present the Th deposits, but the variety of processes at their origin is much more limited.

The main aim of this contribution is to present a synthetic review of the world U and Th resources in the two first parts of the chapter. A third part is devoted to a summary of present and future programs devoted to the sustainable use of the U resources.

Uranium Resources

Three types of deposits: Olympic Dam, diagenetic-hydrothermal deposits and those related to meteoric water infiltration represent over 75% of identified world U resources. In the 2007 edition of the Red Book, published each 2 years by the IAEA, identified resources represent the sum of the reasonably assured resources for which grades and tonnages are estimated with the highest degree of confidence and of the inferred resources for which further exploration work is needed to confirm their estimation. In the 2007 edition, world identified resources recoverable at less than \$130US/kg U correspond to 5.469 Mt U (Table 2).

Very large low-grade resources (several millions t U) are also associated with phosphates (Morocco has about 70% of the resources of this type), with grades of 50–350 ppm U, and with

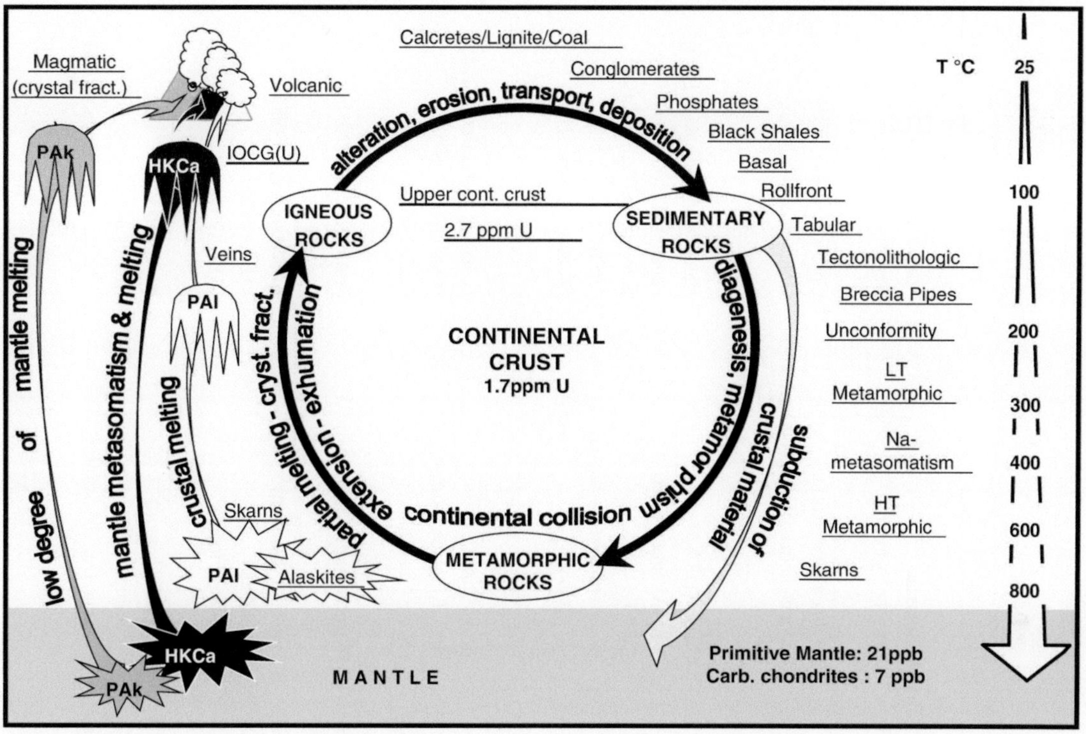

Fig. 1 Position of the major uranium deposit types along the geological cycle, the major uranium reservoirs of the Earth and the major uranium-rich magmatic associations: *PAk* peralkaline, *HKCa* high-K calc-alkaline, *PAl* peraluminous and alaskites (From Cuney 2009, modified)

black shales, the best example being the Cambrian alum shales of southern Sweden. Including speculative (a resource that is expected to occur in geological provinces that may host U deposits according to the IAEA classification) and higher-cost resources, such as phosphates and black shales, the 2007 IAEA Red Book assesses the world global U resource at 11 Mt.

The different types of U deposits will be presented as far as possible from the highest to the lowest temperature fractionation processes, that is from magmatic processes to hydrothermal processes to sedimentary and surficial processes.

Magmatic Deposits

Related to Crystal Fractionation

Crystal fractionation is one of the major primary processes leading to U enrichment but is the most efficient in peralkaline magmatic systems. As a result of their peralkalinity and high temperature, peralkaline melts have a very low degree of polymerization which permits high solubilities of the large, highly charged elements like U, Th, Zr, REE and Nb and thus causes their simultaneous and continuous enrichment until the most extreme stages of magmatic fractionation. Hence, deposits associated with peralkaline magmas are always hosted by the last and most fractionated magma batch at the roof or at the margin of the peralkaline plutonic complexes. When these melts crystallize, abundant and complex Th–Zr–REE–Nb silicates, phosphates and/or oxides are formed, and U is distributed among these mineral phases. These minerals are too refractory to allow for economic recovery of U. Only covaluation of the associated metals may allow future economic mining of such deposits.

The best example of this type of mineralization is the Kvanefjeld deposit (83,320 t U at 260 ppm U) in the Ilimaussaq plutonic complex in southern Greenland which is hosted by layered

Table 1 Genetic classification of the uranium deposits (From Cuney 2010)

	Deposit type	Size	Grade%	Type example	Reference
M	Magmatic				
MCF	Crystal fractionation	Small to medium	0.04	Ilimaussaq	Sørensen et al. 1974)
MPM	Partial melting	Large	0.01–0.05	Rössing (Namibia)	Berning et al. (1976)
H	Hydrothermal				
HV	Hydrothermal-volcanic	Small to very large	0.05–0.2	Streltsovkoye (Russia)	Chabiron et al. (2003)
HG	Hydrothermal-granitic	Small to large	0.1–0.6	Niederschema-Alberoda (Germany)	Golubev et al. (2000)
HD	Hydrothermal-diagenetic				
HDIa	*Intraformational redox*				
HDIaTb	*Tabular*	Small to large	0.1–0.2	Grants (USA)	Hansley and Spirakis (1992)
HDIaTl	*Tectonolithologic*	Small to large	0.1–0.4	Arlit–Akouta (Niger)	Pagel et al. 2005)
HDIaCb	*Collapse breccia*	Small	0.1–0.9	(USA)	Wenrich and Sutphin 1989
HDBb	*Basin/basement redox*	Small to very large	0.2–20%	McArthur River (Canada)	McGill et al. (1993)
HDIf	*Interformational redox*	Small to medium	0.1–0.4	Oklo (Gabon)	Gauthier-Lafaye (1986)
HMp	Metamorphic-hydrothermal				
HMt	Metasomatic-hydrothermal				
HMtAm	*Alkali-metasomatism*	Small to large	0.1–0.25	Michurinka (Ukraine)	Scherbak and Bobrov (2005)
HMtSk	*Skarns*	Small	0.1	Mary Kathleen (Australia)	Maas et al. (1987)
E	Evapotranspiration	Small to large	0.014–0.1	Yeliree (Australia)	Carlisle et al. (1978)
S	Syn-sedimentary				
SMs	*Mechanical sorting*	Large	0.02–0.15	Witwatersrand (South Africa)	Frimmel et al. (2005)
SRt	*Redox trapping*				
SRtm	*Marine (black shales)*	Very large	0.005–0.03	Randstatt (Sweden)	Andersson et al. (1985)
SRtc	*Continental (coal, lignite…)*	Small to medium	0.001–0.05	Kazakhstan	IAEA (2005)
SCcr	*Crystal–chemical/redox (phosphates)*	Very large	0.005–0.03	Morocco, Florida	IAEA (2001)
M	Meteoric water infiltration				
MB	*Basal-type*	Small to large	0.01–0.1	Blizzard (Canada)	Boyle (1982)
MRf	*Roll-front*	Small to very large	0.01–0.2	Wyoming	Adams and Cramer (1985)
O	Others	Very large	0.01–0.04	Olympic Dam (Australia)	Hitzman and Valenta (2005)

Table 2 World uranium identified resources (reasonably assured + inferred) at less than 130 US$/kg (IAEA 2007) and presumed additional potential

Country	Resource	Potential
Australia	1,243,000	High
Kazakhstan	817,300	High
Russian Federation	545,600	High
South Africa	435,100	Low
Canada	423,200	High
United States	339,000	Moderate
Brazil	278,400	*Very high*
Namibia	275,000	High
Niger	274,000	High
Ukraine	199,500	Moderate
Jordan	111,800	Moderate
Uzbekistan	111,000	Moderate
India	72,900	Moderate
China	67,900	Moderate
Mongolia	62,000	High
Denmark	32,300	Moderate
Algeria	19,500	High
Argentina	12,000	High
Central African Republic	12,000	Moderate
France	11,700	High
Malawi	11,600	
Spain	11,300	High
Sweden	10,000	Moderate
Somalia	7,600	
Turkey	7,300	
Portugal	7,200	Moderate
Germany	7,000	Moderate
Greece	7,000	
Romania	6,700	
Japan	6,600	Low
Vietnam	6,400	
Italy	6,100	Moderate
Gabon	5,800	High
Indonesia	5,800	
Slovenia	5,500	
Peru	2,900	Moderate
Congo Democratic Republic	2,700	*Very high*
Mexico	1,800	Moderate
Iran	1,600	
Chile	1,500	
Zimbabwe	1,400	
Finland	1,100	Moderate
Czech	700	Moderate
Total	5,468,800	

lujavrites, a highly fractionated quartz undersaturated peralkaline syenite. At Kvanefjeld, U is hosted mainly by steenstrupine, a complex U–Th–REE silicophosphate, disseminated in the lujavrites (Sørensen et al. 1974).

Magmatic fluids rich in fluorine are always involved in formation of U deposits related to extreme fractionation of peralkaline magmas because F and U are simultaneously enriched in highly fractionated melts (Linnen and Cuney 2005). Hence, some U and Th are fractionated from the silicate melt into F-rich magmatic fluids and transported over a few 100 m as at Bokan Mountain in Alaska, USA (Staatz 1978), or several kilometres from the intrusion in the Th veins of the Front Range, USA (Staatz et al. 1979). Also, some minor mineralization occurring in volcanic–plutonic environment can be related to such fluid/melt partitioning. Possible examples include the early Th–U–Zr–REE mineralization with Th-bearing uraninite in the Morro do Ferro deposit hosted by the Poços de Caldas peralkaline caldera in Brazil, the early U–Th–Mo mineralization in the Zoujiashan deposit hosted by rhyolitic lava and rhyodacites, in the western part of the Xiangshan caldera, China (Jiang et al. 2005), and the Rexpar U–Th–REE deposit in British Columbia, Canada (Morton et al. 1978). However, as the fluid/peralkaline melt partition coefficients of U are very low (Peiffert et al. 1996), the amount of U able to fractionate outside the intrusion is low, and related U deposits of this type are generally non-economic.

Extreme degree of crystal fractionation of highly potassic magmas may also lead to the formation of slightly peraluminous granites (Grosse 2007) or pegmatites (Lentz 1996) with a few to some hundreds ppm U, but typically, these bodies are too small to represent economic mineralization.

Related to Partial Melting

Partial melting in continental crust may also lead to the formation of U deposits. This type of deposit comprises low-grade disseminated uraninite concentrated in granitic pegmatoids occurring in migmatitic domains and represents the other major type of deposit dominantly related to magmatic processes. The type example is the Rössing deposit in Namibia whose mineralization is hosted by granitic pegmatite sheets and small plutonic bodies, called alaskites or pegmatoids (Berning et al. 1976; Cuney 1980, 1982). They typically intrude into epicontinental sediments (arkoses, quartzites, black shales, marls, limestones), possibly associated with acidic volcanics metamorphosed to a high grade with evidence of partial melting. This type of environment is generally enriched in U especially in arkoses and/or acidic volcanics and in organic matter- or phosphate-bearing lithologies. The Rössing U deposit in Namibia (246,500 t) is one of the lowest-grade (300 ppm) U deposits ever mined. Similar U mineralization is known in the vicinity of Rössing at Goanikontes, Husab, Valencia and Rössing South.

The large accumulation of alaskite dykes with relatively high U grade at Rössing results from the combination of the following parameters: (1) a U-rich source represented by intracontinental platform sediments and probably by the existence of synchronous felsic volcanism developed in rift structures; (2) a low degree of partial melting of the quartz-feldspar-rich part of the volcanosedimentary sequence that explains their weakly peraluminous character and their U enrichment; (3) structural control of alaskite emplacement linked to the late kinematic evolution of the Rössing Dome (Basson and Greenway 2004); (4) a chemical barrier represented by the Rössing formation marbles, stopping the rise of the alaskitic melts as the result of skarn-formation reactions with production of carbon dioxide shifting the solidus of the melts to a higher temperature and finally leading to their accumulation in the vicinity of the marble layers; (5) a reducing barrier represented by the graphite and sulphides from the Rössing formation, which has prevented the fractionation of U in magmatic fluids at this level, and the trapping of the U from magmatic fluids deriving from alaskites crystallized at deeper structural levels under oxidized conditions and (6) favourable climatic conditions allowed the oxidation of uraninite in the weathering zone and precipitation of uranophane in fractures enriching the upper part of the deposit.

Two hypothesis have been proposed for the genesis of the primary U mineralization in the pegmatoids: (1) extreme fractionation of a deeper seated granite body, as proposed for the Grenville pegmatites in Ontario (Lentz 1996) and for Rössing by Bowden et al. (1995), or (2) partial melting of U-rich meta-sediments or meta-volcanics, as proposed for Rössing and the Mont Laurier pegmatoids, Québec (Cuney 1982). The partial melting model for the origin of the mineralized leucocratic granite dykes of the Rössing type is supported by a series of features summarized by Cuney (2010).

Hydrothermal Deposits

Volcanic-Hydrothermal Systems

Uranium deposits associated to volcanic and volcanic–plutonic complexes correspond most typically to U veins and disseminations formed in geothermal systems developed shortly after the eruption of felsic volcanic rocks. The main U districts are hosted by caldera complexes, filled with alternating mafic and felsic volcanic rocks, and subordinate amounts of sedimentary layers. Highly fractionated felsic volcanic units generally represent the largest proportion of the volcanic succession. As the pyroclastic material can be transported in the atmosphere over several tens to hundreds of kilometres, this type of volcanism may provide U enrichement within the sedimentary basins located far away from these volcanoes. The pyroclastic rocks represent particularly favourable U source rocks because they are relatively permeable, and a high proportion of the U is hosted by the glassy matrix from which it can be easily mobilized during devitrification.

The genesis of large deposits requires also the existence of a relatively shallow magmatic chamber (a few kilometres to about 5 km) that is active over several million years to provide the heat flux and the fracturing necessary to promote focused and protracted convective fluid circulations. As a consequence, the rocks within the caldera and underlying basement are deeply altered, and U is intensively mobilized from these rocks (Chabiron et al. 2003). The U ore bodies are mainly trapped along basement structures reactivated after the caldera filling and may extend from the volcanic pile several hundreds of metres down into the basement, as at the Antei deposit in the Streltsovkoye caldera in Russia (Chabiron et al. 2003). They may also occur along the ring structures of the caldera as at McDermitt, USA (Castor and Henry 2000), at the margin of resurgent domes as at Xiangshan, China (Hu et al. 2009) or along sedimentary layers within the caldera as at the Dornot deposit in Mongolia (Petrov et al. 2003).

Peralkaline volcanic rocks represent by far the best source for U deposit related to volcanism. The Streltsovkoye caldera, (Transbaikalia, Russia), which hosts the largest U ore field of this type in the world (18 deposits, 250,000 t U at 0.20%), shares all the preceding favourable characteristics (Ishukova et al. 1991). The caldera is very large (140 km^2), filled with a up to 1-km-thick pile of alkali basalts, trachydacites and liparitic tuffs with interlayered sedimentary horizons, and the magmatic activity lasted from 170 Ma to 142 ± 7 Ma. The early phengite–illite alteration stage (133 ± 5 Ma) and pitchblende deposition (133 ± 4 Ma) (Chernyshev and Golubev 1996) immediately follow the last magmatic event.

The exceptional size of the resources at Streltsovkoye also results from the juxtaposition of four main U sources: (1) liparitic tuffs which represent 30–35 vol.% of the volcanic pile, (2) Variscan high-K calc-alkaline granitoids in the basement (Chabiron et al. 2003), (3) Ordovician U mineralization in the basement (Chernyshev and Golubev 1996) and (4) fluids expelled from the volcanic melts or from underlying magma chamber. This last parameter is however minor because the U fluid/peralkaline melt partition coefficient is strongly in favour of the melt (K_{DU} fluid/melt = $3 \cdot 10^{-2}$ to $4 \cdot 10^{-2}$, Peiffert et al. 1996).

A series of similar deposits are known in Asia: in the Dornot district in Mongolia, with similar ages to Streltsovka; the Permian–Jurassic Chatkalo-Kumarinski district and the Silurian-Devonian Karamazar district in Uzbekistan; the Betpakdalinskii district in the Balkhash Lake area and the Kokchetov area in Kazakhstan; the Tchouckotka region in eastern Siberia and the Xiangshan Cretaceous caldera in China (Jiang et al. 2005). Similar but smaller U deposits, are also known in other parts of the world such as the ones hosted by the Tertiary McDermitt caldera in Nevada (Castor and Henry 2000).

High-K calc-alkaline meta-luminous volcanic rocks are a less favourable U source because a significant but variable portion of the U in these rocks tends to be trapped in accessory minerals (Leroy and George-Aniel 1992). Most deposits related to this type of volcanism have a relatively small size.

Highly peraluminous acidic volcanics, mineralized in U, are essentially known in the Macusani district, Peru. Pitchblende and autunite occur in sub-vertical to sub-horizontal fractures in the top tens of metres of Pliocene crystal-rich flows/tuffs. A resource of 10,000 t U has been estimated for the Chapi site and 30,000 t U to the whole Macusani district at an average grade of 0.1% U (IAEA 2009).

Granite-Hydrothermal Systems

Uranium deposits related to post-magmatic high-level hydrothermal systems associated with granites are dominantly associated with two mica peraluminous leucogranites. The largest U district of this type, from which about 300,000 t U have been extracted, is the mid-European Variscan U province, which extends for more than 2,000 km from Spain–Portugal to Bohemia, through Cornwall, Brittany, French Massif Central, Vosges and the Black Forest. The mineralization may occur within the granites (the so-called intragranitic deposits) as is the case for most deposits from the French Massif Central or within enclosing metamorphic rocks (the so-called peribatholitic deposits) as the deposits from the Erzgebirge. The mineralized veins extend over a length of a few metres to several hundreds of metres and occasionally up to 1 km. The thickness of the ore bodies is commonly 1–2 m but ranges from a few centimetres to 15 m, and their vertical extension varies between tens of metres to hundreds of metres. The grade ranges from a few hundreds ppm to several% U and averages 0.1–0.3% U. Local U enrichment occurs at the intersection with lamprophyre dykes.

Other districts with U mineralization related to peraluminous leucogranites include the Yenshanian belt (upper Jurassic to Cretaceous) in South–Eastern China and the Achala batholith in Central Argentina.

In the Variscan belt, primary U deposition occurred at the Stephanian–Permian transition (270 ± 15 Ma), 30–50 Ma after the emplacement of the granites (Holliger and Cathelineau 1986; Křibek et al. 2009), during a rifting event. This event results from the late Variscan collapse of the orogenic edifice, with the formation of numerous intracontinental Permian basins. Minor U occurrences have given slightly older ages at 425 and 330–340 Ma in Vendée (Cathelineau et al. 1990). In the Saint Sylvestre granite, which hosts the major uranium resources of the French Massif Central, the richest hydrothermal mineralization occurs where brittle fracturing was superimposed to late magmatic shear zones controlling the emplacement of the late generation of U-rich fine-grained leucogranites (Cuney et al. 1990). Highly fractionated melts and fluids, enriched in U, expelled from these granite injections, were emplaced in the late magmatic shear zones. This phenomenon has enhanced the enrichment in magmatic-hydrothermal uraninite in the coarse-grained granites in the vicinity of the late fine-grained leucogranites (Cuney et al. 1990).

Most U deposits have vein-type morphology, but de-quartzified granites (episyenites) also represent a trap for many deposits in the French Massif Central (Leroy 1984; Cathelineau 1986). Episyenite formation is associated with meteoric fluid infiltration (Turpin et al. 1990) induced by the decompression resulting from a sudden increase in the rate of basement exhumation during late orogenic uplift at about 305 Ma (Scaillet et al. 1996). During their infiltration, the meteoric fluids (0.4 wt% NaCl) are heated up to 350–400°C. When the fluids moved back towards the surface, the decompression led to their boiling and then to their condensation at about 370°C ± 25°C and 200 ± 50 bars which led to the dissolution of quartz (El Jarray et al. 1994). The vertical extension of episyenite bodies may reach at least 1 km as observed at the Bernardan deposit (6,610 t U, 0.56%) in Western Marche.

Then, the hydrothermal fluids responsible for the leaching of uraninite are low-salinity and low-temperature (160–180°C) aqueous solutions (Dubessy et al. 1987). Uranium deposition resulted from the mixing of an oxidized meteoric fluid able to leach uraninite from the enclosing leucogranites and of a connate "heavy" fluids with $\delta^{18}O = +8‰$ to $+15‰$ and $\delta D = -45‰$ to $-30‰$. The connate fluid that probably derives from an overlying Permian basin (Turpin et al. 1990; Křibek et al. 2009) may have provided the reductants (hydrocarbons and/or H_2S) for U deposition. The $\delta^{13}C$ values of the carbonates of $-17.6‰$ indicate an organic origin for the carbon (Turpin et al. 1990).

Hydrothermal U deposits may also present different types of relations with highly fractionated high-K calc-alkaline granites, but they are less common and generally of smaller size. At Hotagen in Sweden, the deposits occur within a Palaeoproterozoic high-K calc-alkaline granite, but they were formed during a Caledonian tectonic-hydrothermal event. The Bois Noirs–Limouzat deposit in France (Cuney 1978), with 6,920 t U at grades of 0.27% U, is hosted within a high-K calc-alkaline granite, but drilling at depth has intersected uraninite-rich peraluminous leucogranites which probably represent the major U source for the system (Poty et al. 1986). The Pribram deposit in the Czech Republic, with 42,000 t U (Křibek et al. 1999), is hosted by Late Proterozoic and Cambrian weakly metamorphosed sediments, at the south–western margin of the high-K meta-luminous to slightly peraluminous calc-alkaline granites, granodiorites and monzonites of the Blatna suite which belongs to the Carboniferous Central Bohemian Pluton. A last but the major example of relation between highly potassic calc-alkaline magmatism and uranium deposit is the Olympic Dam deposit in Australia which is hosted by the highly fractionated Roxby Down granite. This deposit will be discussed in the last section dealing with other types of uranium deposits.

Some other U deposits related to hydrothermal systems located at high structural level do not appear to be associated to U-rich granites.

For example, for the Upper Cretaceous Schwartzwalder deposit, Colorado, USA (Wallace and Whelan 1986), (3,600 t U, 0.37%), the U source remains unknown.

Diagenetic-Hydrothermal Systems

Diagenetic-hydrothermal systems correspond to fluid circulations through porous fluvial, lacustrine, deltaic to near shore siliciclastic sediments, more rarely limestones during or after their burial. This leads to the formation of a large variety of epigenetic U deposit types. The diagenetic fluids are brines which may mix with infiltrated meteoric fluids. According to the location of the redox control within the basin and their depositional or structural environments, these deposits are subdivided into five subtypes.

Tabular Intraformational Redox Control Deposits

They are transitional with the syn-sedimentary deposits, because U may begin to precipitate shortly after sedimentation and burial, but is dominantly concentrated during diagenesis. Ore bodies are conformable with the stratigraphy and form within coarse sandy layers intercalated between less permeable clay-rich horizons, generally at the margins of palaeochannels. Volcanic ashes within the sandstones commonly represent a major source of U and other elements (Mo, Zr). Uranium is generally leached under oxidizing and slightly acidic conditions, but interaction with volcanic ashes can make the solutions more alkaline. The fluids transporting U are generally low-temperature basinal brines. In the Grants region, USA, Hansley and Spirakis (1992) proposed that U was transported by brines expelled from evaporites.

Reduction of U in sandstone is controlled by the organic matter present as debris of fossil plants or layers of authigenic, or migrated organic material (humates), either with bacteria as a catalyst or through the production of biogenic hydrogen sulphide (Spirakis 1996). In sandstones, relatively poor in organic material, the reduction of U may be caused either by H_2S (biogenic and/or non-biogenic) produced from the interaction of oxidized groundwaters with pyrite or by H_2S and/or hydrocarbons injected along structures and deriving from deep hydrocarbon reservoirs (Spirakis 1996). Major tabular deposits are those from the Grants region, Colorado, where over 240,000 t U at 0.09–0.21% have been mined. A few are known in Australia (Beverley, Honeymoon, Four Miles…), but many more have to be expected in this country with further exploration.

Tectonolithologic Intraformational Redox Control Deposits

The ore bodies are controlled both by the host lithology, permeable palaeochannels rich in reducing organic compounds as in the preceding deposit types, but together with a strong tectonic control. Typical examples are the deposits from the Arlit–Akouta area in Niger (Pagel et al. 2005). Two types of U sources can be considered: (1) the erosion of the U-rich volcanic and granitic complexes of the Aïr massif and (2) the acidic peralkaline volcanic tuffs dispersed in the different sedimentary formations from Devonian to Jurassic (Forbes et al. 1984).

In tectonolithologic deposits, mineralizing fluids are also diagenetic brines. For the Akouta deposit in Niger, temperatures of 85–175°C and salinities of 15–22 wt% NaCl have been estimated (Forbes 1989). The fluids were alkaline, as attested by the abundance of carbonate and analcime in the sediments, and oxidizing. The processes controlling ore deposition are similar to those described for tabular deposits and for some of the roll-front deposits. Uranium mineralization occurs in coarse-grained immature and permeable sandstones deposited in a fluvio-deltaic environment of the Guezouman and Tarat formations. At a more detailed scale, the mineralization is preferentially located in bottom-set layers and along basal fore-set layers of palaeochannels where the organic matter is preferentially accumulated and replaced by pyrite providing abundant reductants (Sanguinetti et al. 1982).

The Arlit–Akouta uranium district in Niger is one of the largest one with tectonolithologic deposits with about 115,000 t U already mined and more than 70,000 t U as identified resources at 0.2–0.4% U.

Diagenetic-Hydrothermal Karsts

They are generally called solution collapse breccia pipes. They are mainly known in the Grand Canyon region in Arizona, USA, but similar occurrences have been reported in China. In the Grand Canyon region, they form near-vertical, 30–175 m large, cylindrical pipes, located in flat-lying upper Palaeozoic to Triassic marine platform sediments (Wenrich and Titley 2008). Solution (karstification) of the Mississippian Redwall Limestone started mainly along and at intersections of fracture sets. Overlying strata have collapsed up to a vertical distance of 1,000 m. Thousands of pipes are known but only about 100 of them are variably mineralized. Average ore grades vary from 0.4% to 1% U, and resources from some hundreds to 1,500 t U. Pitchblende associated with Cu-sulphide, arsenide and arsenic–sulphosalt minerals and locally bitumen occurs as disseminated replacements and minor fracture fillings.

The fluids in ore-related sphalerite, dolomite and calcite with homogenization temperatures of 80–173°C and salinities of 4–17 wt% equiv. NaCl were of probable diagenetic origin, derived from deeper part of the basin (Wenrich and Titley 2008). Temperatures, salinities, the Cu > Zn > Pb (± Ag, Co, Mo, Ni, V) sulphide paragenesis associated with the U ores, the ages of deposit formation and the presence of sulphur-rich bitumen are similar to those of Mississippi Valley–type (MVT) deposits. Two mineralization events have been dated at 260 and 200 Ma (Ludwig and Simmons 1992), a timing corresponding to a period of stability in the Pangea, that allowed many MVT deposits to form throughout the world.

In South China, Devonian–Carboniferous unmetamorphosed carbonate rocks host several deposits (Sanqilinyi, Sanbaqi and Saqisan), sharing many of the characteristics of the Arizona breccia pipes (Min et al. 2002). They occur in solution collapse breccias and solute on-fault breccias mineralized with pitchblende, coffinite and Fe, Cu, Pb, Zn and Ni sulphides. The matrix of the breccia is composed of organic matter, clays and limestone fragments. The deposits seem to result from the superimposition of an early syn-palaeokarstic U mineralization associated with organic matter and a saline basinal brine percolation event at 110–251°C driven by a late Yanshanian tectonic event, enriching the U ore.

Basin/Basement Redox-Controlled Deposits

They are usually termed unconformity-related deposits because they are generally located in the vicinity of an unconformity between a metamorphic basement and an undeformed clastic sedimentary cover. The basement usually comprises Archaean granitic domes rimmed by U-rich and graphite-rich metamorphosed Palaeoproterozoic epicontinental sediments intruded by granitic bodies. The base of the basins consists of oxidized continental fluviatile, aeolian to marginal marine, highly mature, siliclastic sediments. The two main provinces hosting this type of deposit are the Athabasca Uranium Province, Saskatchewan, Canada and the East Alligator River Uranium Province, Northern Territory, Australia. Other U deposits which share some of the characteristics basin/basement redox-controlled deposits are those related to the Proterozoic basins of Thelon in Nunavut (Fuchs and Hilger 1989); Otish in Northern Québec (Gatzweiler 1987); Pasha Ladoga in Karelia, with the Karku deposit (Velichkin et al. 2005); and the metamorphosed Coolbro Sandstone in Australia, with the Kintyre deposit (Hanly 2005).

Possible U sources of unconformity-related deposits were the basement rocks and the sandstone cover. Archaean rocks underlying the Athabasca Basin are mainly composed of U-poor tonalites and cannot have represented a U source, but in the Northern Territory, high-K–Th–U Archaean granites do exist (Nanambu Complex). Palaeoproterozoic sediments which rim the Archaean domes in both provinces consist of epicontinental clastic to chemical deposits. Calc-silicates and meta-arkoses, as well as leucogranites and pegmatoids derived by partial melting of the Palaeoproterozoic meta-sediments, are enriched in uraninite and represent particularly easily leachable U sources (Sibbald et al. 1976; Annesley et al. 2003). Additional potential U sources are the Palaeoproterozoic U–Th-rich high-K calc-alkaline granitoids. Also, late Hudsonian vein-type U deposits (e.g. Beaverlodge, Gunnar) demonstrate the ability of the Palaeoproterozoic basement to represent a favourable U source to form these deposits, but they may also themselves have contributed as U sources to the unconformity-related deposits.

Another alternative proposed by several authors is that the sandstone cover represents the main U source. However, the average U content of the regional sandstones is below 1 ppm, and 50–80% of this U is presently bound to zircon. Only the monazite destruction may have represented a significant U source, as in the Franceville basin, Gabon, with the extraction of about 75% of its U (Mathieu et al. 2001; Hecht and Cuney 2000). The average Th content of the sandstones being generally below 10 ppm, the amount of U deriving from monazite alteration, with a Th/U ratio of about 20, was limited. Moreover, a significant part of the liberated U was trapped in altered zircons (U increases from a few hundred to several thousand ppm), but not leached out as proposed by Fayek and Kyser (1997), and in Fe–Ti oxides deriving from detrital Fe–Ti oxide alteration. Therefore, the strong U enrichment of the Palaeoproterozoic basement is preferred as the major source of U for the genesis of such large high-grade deposits.

Uranium solubility was favoured by the high fO_2, acidity, and chlorinity (up to 6 molal) of basinal diagenetic brines. High fO_2 resulted from the lack of organic matter when the sandstones were deposited in continental environment during the lower Palaeoproterozoic. The pH, controlled by the kaolinite–illite–quartz paragenesis, was acidic. Detrital feldspar was lacking or has been altered during diagenesis. Early diagenetic brines in detrital quartz overgrowths are Na rich and derive from evaporitic layers which should have been present in the upper parts of the basins, whereas the brines trapped later in pervasively silicified zones and drusy quartz, close to the mineralized zones, were enriched in Ca and Mg through their interaction with Ca-Mg rich basement lithologies (Derome et al. 2003, 2005). Laser ablation ICP-MS analyses (Richard et al. 2010) reveal that the Ca-rich brines are the richest in U with contents reaching several hundreds of ppm. Temperature and pressure close to the unconformity decrease from the early diagenetic (160–220°C, 1–1.25 kbar) to the ore stage (140–160°C, 0.6 kbar), (Derome et al. 2003, 2005).

The oldest U mineralization age obtained so far is ca 1,590 Ma, from LA-ICP-MS U–Pb dating of uraninite and ^{40}Ar–^{39}Ar dating of syn-ore illite, for both basement- and sandstone-hosted deposits (Alexandre et al 2009). This event and younger U–Pb ages also obtained on uraninite, interpreted as remobilization events, are attributed to several episodes of hydrothermal circulation resulting from far-field, continent-wide tectonic events. However, it is still uncertain if all the U has been deposited during a single event or if accretion of multiple stages of U deposition have occurred.

Most unconformity-type deposits are related to tectonic contraction, leading to reverse faulting along Hudsonian graphite–sulphide-bearing basement faults. If faulting is important to rejuvenate the permeability of the structures, the magnitude of the displacements is not always sufficient to create significant openings. Quartz dissolution represents another major way of creating a huge space permitting the deposition of massive U mineralization (Lorilleux et al. 2002).

To explain U reduction, as well as the source of Mg and B in the alteration envelops and of the metals (Ni, Co, Cu, Zn, Au) in the polymetallic deposits, a basement-derived reduced fluid has been proposed by various authors. However, direct evidences for such a reduced fluid are still very weak.

In conclusion, if the degree of U and graphite enrichment of the basement and the intensity of quartz dissolution were some of the key factors explaining the genesis of the high grade – large-size unconformity-related – deposits, the extreme reactivity of the oxidized, acidic, highly saline, Ca–Na brines generated at the base of thick Proterozoic sandstone basins (free of organic matter and K-feldspar) able to scavenge U even from very refractory minerals seems to be the most important key factor.

Interformational Redox-Controlled Uranium Deposits

They are characterized by the location of the redox barrier controlling U deposition to the upper part of an oxidized sandstone formation below an oil generating shale formation as exemplified by the deposits (Oklo, Mounana, Okelonbondo, …) of the Early Proterozoic Franceville Basin in Gabon (Gauthier-Lafaye 1986). The Francevillian series correspond to a 4–5-km-thick, unmetamorphosed, sedimentary sequence deposited at about 2.1 Ga. At the base of the basin, the FA formation (500–1,000 m thick) is mainly composed of conglomerates and fine- to coarse-grained, fluvial to deltaic sandstones (Weber 1968). The overlying FB formation consists mainly of organic matter–rich marine shales. All the U deposits of the Franceville basin are located in the upper part of the FA formation, immediately below the redox boundary represented by the FB black shales (Gauthier-Lafaye 1986). Uranium deposition has been dated by a U–Pb discordia at 2.05 ± 0.03 Ga (Gancarz 1978). The ore forming fluids are similar to those of unconformity-related deposits with 28.7 wt% NaCl to 30 wt% $CaCl_2$ eq. and have been trapped at 135–155°C and 1 kbar (Mathieu et al. 2000). Detrital monazites are also widely altered in the FA formation. The amount of U liberated during monazite alteration is here largely sufficient to account for the U in all the U deposits (27,635 t U at 0.38%) of the Franceville basin (Cuney and Mathieu 2000). Then, U transported by the diagenetic brines has precipitated at the redox interface between the FA and FB formations.

Metamorphic Systems

After diagenesis, with increasing temperature and pressure, U deposits can also be formed during the circulation of metamorphic fluids in association with dehydration, folding, faulting and/or thrusting of the rocks. The lowest grade of metamorphism associated with the most important release of fluids from the sediments is the most favourable for the formation of U deposits related to such processes. More specifically, during the metamorphism of epicontinental platform sediments, oxidized brines expelled from evaporitic layers, extremely efficient for transporting U, may mix with hydrocarbons produced by black shales, for the precipitation of U. Uranium deposited by metamorphic fluids generally occur in veins. Typical examples are the Mistamisk U veins from Labrador, Canada (Kish and Cuney 1981), and some, if not all, U deposits of the Copper Belt, as exemplified by the Shinkolobwe deposit (25,500 t U, 0.40%) (Ngongo-Kashisha 1975; Audeoud 1982), in the Democratic Republic of Congo and deposits from Zambia (Meneghel 1979) such as the Kansanshi deposit (Kříbek et al. 2005). They have all been formed during very-low-grade metamorphism at 350°C ± 50°C of sediments deposited in epicontinental setting.

At higher temperature, the metamorphic origin of the deposits is rarely well constrained. Many of them, if not all, are in fact metamorphosed deposits. For example, the Tiraun deposit in Switzerland considered as syn-metamorphic by von Pechmann and Bianconi (1982) corresponds in fact to a metamorphosed U mineralization. The Mary Kathleen skarn-hosted U deposit (10,176 t U at 0.11%), Australia, initially attributed to contact metamorphism (Cruikshank et al. 1980), is in fact temporally disconnected from the granite intrusions (1,740–1,740 Ma) and has been related to a later metamorphic event (at 1,550 ± 15 Ma). Conversely, the Lagoa Real U district (100,000 t U at 0.12%), Brazil, which genesis was related to the Brazilian metamorphic event (at about 550 Ma) by Lobato et al. (1983), is in fact much older, 1,450 Ma according to Turpin et al. (1988), or even older, at 1,870 Ma according to Chavès et al. (2007), and in fact represents a U deposit metamorphosed by the Brazilian event.

Metasomatic Systems

Na-Metasomatism-Related Deposits

Na-metasomatism is a widespread geological process that occurs in a large variety of conditions, but only some specific occurrences are associated with U ore genesis. Na-metasomatism generally affects large volume of rocks, within which the mineralization occurs in a core of the alteration zones showing the most advanced stage of alteration. Na-metasomatism is generally developed in felsic rocks to form albitites, but it may occur within meta-volcanic or meta-sedimentary rocks. In felsic rocks, K-feldspar and plagioclase are replaced by albite, quartz is dissolved and newly formed albite and other Na and/or Ca minerals are deposited in the vugs left by quartz together with the deposition of U. In the largest mineralized districts of this type, the albitites form discontinuous zones extending over several tens of kilometres. Individual Na-metasomatic zones are normally several metres wide and several hundreds to several thousands of metres long. This alteration is associated with deep regional structures as best exemplified in the U district of Central Ukraine (Belevtsev and Koval 1968). Fluids associated with these processes derived from either or both a basinal brine or a magma as proposed for the Valhalla U deposit in Queensland, Australia, (Polito et al. 2007), or from the surface, as in the central Ukrainian U deposits (Cinelu 2008). Some authors consider that these deposits are related to metamorphic processes (Kushev 1972; Lobato et al 1983), others to post-granite emplacement hydrothermal circulation (Kazansky and Laverov 1978; Turpin et al. 1988). The geodynamic conditions and drivers responsible for the development of the largest U ore fields associated with Na-metasomatism are still poorly recognized.

The Michelin deposit (36,800 t U at 840 ppm) from Central Labrador, Canada, is hosted by deformed and recrystallized ash flow tuffs and porphyritic rhyolites with interbedded clastic sedimentary rocks and basalts belonging to the Palaeoproterozoic Aillik Group, but the major U concentration step is associated with typical Na- and Ca-metasomatism (Gandhi 1978).

Worldwide, it is remarkable that U mineralization associated with Na-metasomatism seems to occur mainly during one major period of Earth history, between 2.0 and 1.5 Ga with the central Ukraine district (180,000 t U) (Scherbak and Bobrov 2005), Kurupung (2,750 t U) in Guyana (Cinelu and Cuney 2006), Skuppesavon (690 t U at 700 ppm) (Smellie and Laurikko 1984) and Pleutajok (4,000 t U at 0.10%) (Adamek and Wilson 1977) in Sweden, Morocco (Idir and Renard 2002), Liangshanguan (3,000 t U) in China (Jiarong and Zhutian 1988) and Lagoa Real (100,000 t U at 0.12%) in Brazil (Chavès et al. 2007). A far less important Na–U event is associated with the Pan-African–Brazilian orogenesis (500 ± 50 Ma) with deposits of: Espinharas (8,000 t U) (Ballhorn et al. 1981), Itataia, for the early hydrothermal phase affecting granite sheets (67,700 t U at 0.08%) (Netto 1983), both to the north-east of Brazil, and Kitongo (11,000 t U at 0.09%) in Cameroon (Vels and Fritsche 1988). The Coles Hill deposit in Virginia (45,000 t at 0.05%) in the USA is probably of Mesozoic age (Jerden and Sinha 1999).

Uranium deposits associated with Na-metasomatism represent a largely underexplored model. Vast domains of Palaeoproterozoic crust have not been evaluated for this type of deposit. Moreover, resource estimations from some districts like Lagoa Real in Brazil and Michelin in Canada remain underestimated because they have been evaluated by relatively shallow drilling (some hundreds of metres), whereas the Ukrainian district shows that such type of mineralization may extend down to 2 km or more.

Skarn-Related Deposits

They are associated to metasomatic processes generally occurring during contact or regional metamorphic events. Therefore, they can be also classified as metamorphic U deposits. The Mary Kathleen U–REE skarns in Australia are hosted in Palaeoproterozoic amphibolite-grade metamorphosed calc-silicate, mafic to intermediate igneous and sedimentary rocks. The sediments correspond to impure carbonates, sandstones and evaporates deposited in a shallow water shelf setting. A first phase of metamorphism and deformation (1,750–1,730 Ma) was accompanied by emplacement of granitic and mafic bodies and rhyolitic dykes. Skarns result from an extensive hydrothermal system by interaction between the volatile- and U–Th-rich high-K calc-alkaline Burstall Granite emplaced at 1,737 ± 15 Ma and the enclosing calc-silicates, and highly saline fluids derived from evaporates (Oliver et al. 1999).

A first phase of U and REE enrichment is presumed to have occurred in these skarns, at or near the present orebody (Maas et al. 1987). During the second phase (1,550–1,500 Ma), upper amphibolite facies conditions were reached (600–650°C, 3.5–4 kb) with a new phase of highly saline hydrothermal activity producing intense scapolitization in the sediments and the main phase of U–REE mineralization dated at 1,550 ± 15 Ma by Page (1983). Disseminated minute grains of uraninite enclosed in allanite or along veins are exclusively associated with the zones of retrogressed massive garnet ± diopside skarns (McKay and Miezitis 2001). Fluid inclusions in apatite recorded temperatures up to 500°C with 30–50% dissolved solids. The U–REE mineralization (8,550 t U at 0.11%) at Mary Kathleen may correspond to a metamorphosed deposit with recrystallization of an older, granite-related U–REE mineralization, upgraded during a later metamorphic hydrothermal event.

The uranothorianite-bearing pyroxenites of Tranomaro, Madagascar, represent an extreme type of U deposit, hosted by skarns (Moine et al. 1985), but in fact, they correspond more to a Th deposit because the Th–U ratios higher than 1 and thus will be described in the Th resources section.

Syn-sedimentary Deposits

Syn-sedimentary U deposits in marine environments are formed during sedimentation in shelf environments and essentially comprise U-rich black shales and phosphorites. These syngenetic U enrichments do not always reach a grade sufficient to make economically exploitable deposits at the present time, but they constitute potential resources for the future. They also represent an extremely favourable environment for the subsequent development of other types of U mineralization through additional concentration steps, for example partial melting or hydrothermal remobilization in a large variety of T-P-X conditions. Three major types of syn-sedimentary U deposits can be distinguished according to the main mechanism controlling U deposition during sedimentation: (1) mechanical sorting, (2) redox trapping and (3) crystal-chemical and redox trapping.

Mechanical Sorting

This is the main process controlling the genesis of the quartz-pebble conglomerates (QPC) deposits. They represent the first U deposits formed on the Earth. The earliest ones occur in South Africa and are hosted by the Dominion Group which has been deposited between 3,086 ± 3 Ma and 3,076 ± 6 Ma (Frimmel and Minter 2002); the latest ones are hosted by the Elliott Lake Group in Eastern Canada and were deposited between 2,496 and 2,215 Ma (Corfu and Andrews 1986). The two endmember models proposed for their genesis vary from a purely syngenetic origin, the placer model, with a detrital deposition of uraninite in fluvial to fluvio-deltaic environments (Frimmel et al. 2005), to an epigenetic origin with U precipitating from hydrothermal fluids (Barnicoat et al. 1997) or a combination of both in the modified placer theory (Pretorius 1961). The controversy is of crucial importance, the placer model being one of the major arguments for the existence of very low oxygen levels in the atmosphere prior to about 2.3 Ga. A series of features such as the high Th contents of the uraninites and their magmatic REE patterns strongly support a detrital origin of the primary uranium concentration (Cuney 2010). Later remobilization of the primary uraninite has occurred, but is not significant for the ultimate origin of the metal accumulation. Historic production from this type of deposit is 246,000 t U, and identified resources reach 547,900 t U.

Redox Trapping

In shallow water marine environments, large amounts of U may accumulate adsorbed on organic materials and clay minerals in organic- and pyrite-rich black shales, also named oil shales. The largest deposits are the Cambro–Ordovician alum shales of the Ranstad area in Sweden (Andersson et al. 1985) and the Chattanooga and the New Albany shales in the USA (Nash et al. 1981), but the Silurian graptolitic black shales of the Ronneburg-Gera deposits in Germany are the only deposits of this type which have been mined to any large extent because of their higher grades (169,230 t U at 850–1,700 ppm U) which result from a combination of syngentic, hydrothermal and supergene enrichments (Lange and Freyhoff 1991). The Cambrian–Ordovician alum shales in southern Sweden are devoid of secondary enrichments. They represent U resources over 1 Mt U, and the Ranstad deposit alone contains ~254,000 t U at 170–250 ppm U. The alum shale has been deposited in shallow marine waters of the Iapetus Ocean. It contains up to 25 wt% organic carbon, and U and associated trace elements (Mo, V, Ni, Zn, Cu) are clearly syngenetic. The preferential U enrichment in the vicinity of the palaeoshores is interpreted to reflect a more vigorous bottom water circulation that promoted higher rates of mass transfer across the sediment/water interface relatively to the sediments deposited farther away. In the most enriched Upper Cambrian biozone (Peltura scarabaeoides Zone), the U enrichment is inversely correlated to the thickness of the zone, reflecting the low rate of detrital particle supply. The Kolm beds, formed by resuspension of the sediments in an anoxic water column that favoured U adsorption during diffusive exchanges between suspended particles and seawater, may reach 1,000–8,000 ppm U (Schovsbo 2002).

In continental environments, coal, lignites, organic-rich bog peats, swamps, closed anoxic lakes, and also karst caverns represent other environments prone to concentrating surficial U accumulations generally in the order of a few tens of ppm (Morales et al. 1985) which may still be forming today. In such organic-rich environments, U can be deposited by adsorption on the organic material and/or by reduction of UO_2^{2+} by the anaerobic bacterial activity. Uraniferous lignite deposits in Kazakhstan and China contain from 20,000 to 50,000 t of U with U contents between 500 and 2,000 ppm. Several of these deposits have been mined. China is presently testing U extraction from coal or lignite ash.

Crystal-Chemical and Redox Trapping

These are the main mechanisms of U fixation in the sedimentary phosphates. The sedimentary phosphates probably represent the largest (possibly up to 15–22 Mt U) low-grade U repository which may become a resource in a near future. It would be more sustainable to extract U during phosphoric acid fabrication rather than to spread fertilizers with 10–200 ppm U in the agricultural fields (Yamazaki and Geraldo 2003). Phosphorite deposits formed along shallow continental shelf from upwelling of nutrient-rich marine waters promoting intense development of living organisms onto a shallow continental shelf with restricted circulation. The organic matter and biologic activity has an important role in accumulation of U by oceanic phosphorites for its reduction (Baturin and Kochenov 2001). U^{4+} is a proxy for Ca^{2+} in the apatite structure, but a fraction of the U may also precipitate outside of the apatite structure together with other metals. The largest episode of phosphate deposition occurred during the late Cretaceous to Eocene (90–45 Ma) period, under the same palaeolatitude (8–15°N), along the southern margin of the Tethys Ocean. The phosphorite belt extends from Turkey to Morocco and beyond across the Atlantic to Colombia and Venezuela. The global Late Cretaceous sea-level rise, the circum-equatorial Tethyan oceanic circulation, associated with a northward Eckman offshore transport of surface waters arising from dominant easterly winds, have resulted in upwelling of nutrient-rich waters along the southern Tethys shelves and deposition of an organic and phosphate-rich sequence (Soudry et al. 2004). Morocco hosts a large part of the world resources of this type (6.9 Mt U, with average U contents of 50–150 ppm, IAEA 2001). However, most of the world production from phosphorites was extracted from Miocene–Pliocene phosphorites of Florida (17,150 t U) (Cathcart 1978) and in Kazakstan (40,000 t U). Small amounts of U have also been extracted from imported phosphates in Belgium (686 t U).

Meteoric Water Infiltration

Basal-Type or Sealed Palaeovalleys Deposits

They are generally located in colluvium between a U-rich granitic basement and a basalt cover and are named palaeovalley or infiltration-type in Russian literature (e.g. Vitim deposits, Transbaïkalia, Russia, Kondrat'eva et al. 2004). The basal-type U deposits are probably the equivalent of these deposits in the western countries. The colluvium is composed of poorly consolidated, highly permeable, fluvial to lacustrine, organic matter–bearing gravels and sands deposited in palaeovalleys incised in igneous and metamorphic rocks. It is capped by a low permeable cover consisting of plateau basalts or sediments. Uranium is leached from the granitic basement and precipitates by reaction with the organic matter during groundwater percolation in permeable sediments, between two aquitards: the basement and the basalt cover. The Blizzard deposit in Canada, with resources of 4,000 t U at 0.21% (Boyle 1982), the Tono deposits in Japan (5,000 t U at 500 ppm) and the Saint Pierre du Cantal deposit in the French Massif Central, which has produced 1,304 t U (Carré 1979), are typical examples of this type.

Roll-front deposits represent the best example of epigenetic U deposition at a redox interface (front). Host rocks are younger than Ordovician and were deposited in fluvial to lacustrine environments in intermountain basins, as in the Tertiary Powder River Basin in Wyoming (Adams and Cramer 1985), or in marine marginal plains, in channel, lagoonal and beach-bar settings on the marine marginal plains (Finch and Davis 1985). The source of U may be U-rich granites or volcanics external to the sediments and/or syn-sedimentary U deposited in reducing conditions or intercalated U-rich volcanic ashes. The reduced sandstone is buff, grey, grey-green or black coloured due to disseminated organic matter and/or pyrite and occurs down the hydrologic gradient relative to the oxidized sandstone. The reduction may be either of syn-sedimentary origin, resulting from the maturation of detrital plant debris during diagenesis, or may be epigenetic, induced by the infiltration of hydrocarbons or H_2S migrated from deep oil reservoirs along faults (South Texas, Goldhaber et al. 1978; Kazakhstan, Aubakirov 1998; Ordos basin, China, Cai et al. 2007). The migration of hydrocarbons in the Ordos basin, China, is evidenced by the presence of oil/gas inclusions in the calcite cement and healed fractures in quartz grains. Bacterial sulphate reduction may also be involved in some roll-front deposits (Dongsheng district, north of the Ordos basin, China, Cai et al. 2007) as pyrite aggregates intimately intergrown with coffinite have large variations in $\delta^{34}S$ values ($-34‰ < \delta^{34}S < +18‰$) typical of bacterial reduction. In most proposed models, U is transported by oxidized low-temperature meteoric waters infiltrated through permeable sandstones confined between two aquitards corresponding to mudstone. The reduced sandstone is oxidized and takes a pink-red colour due to hematite, or a yellow-orange colour due to limonite, or even may become white grey when iron has been totally leached out. Uranium deposition is controlled by the interplay of the permeability of the sediments, the distribution of reducing components and for some of them by bacterial activity. In some other occurrences, interbedded mafic volcanics may also play a role in the reduction processes. The resulting orebodies exhibit C-shapes in cross section and are sinuous along the roll-front interface. In the Inkai deposit, Kazakhstan, they may extend laterally up to more than 100 km. Uranium ore minerals are typically pitchblende or/and coffinite. Resources associated with this type of deposit are very important and reach over 1 Mt U for Kazakhstan.

Evapotranspiration

Calcretes

These deposits are known from Tertiary to present time, in surficial continental environments. Most commonly, they are hosted by surficial sediments deposited in fluviatile to lacustrine/playa systems, which have not been subjected to significant burial and may be cemented by a variety of minerals: calcite, gypsum, dolomite, ferric oxide, strontianite or halite and uranyl minerals. The cement results of the fluctuation of the groundwater level and of evapotranspiration in arid to semi-arid climatic conditions. The economically important calcrete are referred to as valley calcretes by Carlisle et al. (1978), but the ore bodies are hosted in fact by highly immature, porous, fluviatile valley-fill sediments.

The U source is most commonly U-rich granites or acidic volcanics which are weathered in large catchment areas, and the associated V is generally derived from mafic rocks or shales. Uranium is leached and transported by slightly alkaline, oxidizing surface waters or shallow groundwaters and carried as uranyl carbonate complexes. During downflow migration, water alkalinity and salinity increase due to progressive evaporation. In valley calcrete deposits, basement highs along the valley represent barriers to the groundwater flow, directing it towards the surface, where evaporation sharply increases, inducing CO_2 degassing, and the breakdown of the uranyl carbonates. The predominance of vanadates (carnotite and tyuyamunite) in these deposits is due to their solubility, two orders of magnitude lower than that of other uranyl minerals, in the pH range 5–8.5 (Langmuir 1978). Mann and Deutscher (1978) propose that V transported as V^{4+} complexes in slightly reduced groundwater is oxidized to V^{5+} by mixing with highly oxidized groundwater transporting U in uranyle carbonates, thus resulting in carnotite precipitation. Increasing K^{+}, V^{5+} and U^{6+} activities in the solution during evaporation of upwelling groundwaters also helps for carnotite deposition.

Calcrete deposits are known in Australia, Namibia, South Africa, Mauritania, Somalia, Jordania, USA (called caliches), Argentina and China. The largest ones are the Langer Heinrich (63,520 t U at 510 ppm) and Trekkopje (42,360 t U at 100 ppm) deposits in Namibia and those of Yeelirrie (44,520 t U at 1,270 ppm), Lake Maitland (9,190 t U at 260 ppm) and Lake Way (3,900 t U at 460 ppm) in Australia.

Other Types

Uranium mineralization associated with IOCG deposits, mainly represented by the Olympic Dam iron oxide Cu–Au (U–Ag–REE) deposit (IOCG deposit) in South Australia, are classified in this category because too little information is available about the U mineralization process and the mechanism of U deposition within the deposit, from studies of the author, does not seem to result from a single process. Mineralization occurs in a zoned hematite–breccia complex hosted by the Roxby Downs Granite, one of the most fractionated high-K calc-alkaline granite of the Hiltaba suite, and dated at 1,588 ± 4 Ma (Creaser 1996). The breccia complex extends approximately over 5–7 km along a NW–SE direction and over more than 1 km at depth. Granite emplacement is contemporaneous with the extrusion of the Gawler Range Volcanics (Reynolds 2000). Dating indicates that magmatic activity, brecciation and Cu–Au mineralization are synchronous (Johnson and Cross 1995). The Olympic Dam ore genesis is related to the unmixing of a hot, high-saline fluid from a granitic magma which then mixes with oxidized meteoritic fluids (Hitzman et al. 1992). The high Cl–S ratios of the fluids compared to other hydrothermal systems are the reason for very low enrichments in Pb and Zn, elements weakly soluble in S-poor fluids.

Compared to other IOCG deposits such as those of the Conclurry district in Queensland, Olympic Dam is entirely hosted within U–Th-rich high-K calc-alkaline granites and volcanics. In the mineralized breccias, U minerals are represented both by finely dispersed euhedral uraninite crystals, typical of a high-temperature mineralization, and by much more abundant botryoidal pitchblende, coffinite and U–Ti oxides in veins corresponding to lower temperature deposition. Hitzman and Valenta (2005) proposed that leaching of U from the wall rocks by the hydrothermal fluids produced the enrichment in the IOCG deposits. Uranium enrichment in the ore is 10–40 times larger than in unaltered host rocks. The only available age determined on pitchblende at 1,400 Ma, is from an unpublished report of Trueman et al. (1988). This means either that pitchblende have been re-equilibrated during a later hydrothermal event with radiogenic Pb loss, or has been deposited 190 Ma after the formation of the diatreme. The last interpretation would mean that the main stage of U deposition may be much younger than the IOCG system. Therefore, a more detailed study of the U mineralization process is clearly needed for IOCG deposits and especially in the Olympic Dam deposit, by far the largest U deposit of the world, with resources of 1.9 Mt U at 340 ppm, but U is produced as a by-product of Cu, Au and Ag ores.

A series of other deposits, such as those located in limestone, marbles or related to K-metasomatism need further research to be classified according to the present genetic scheme. The Tumallapalle deposit hosted by metamorphosed stromatolitic dolomitic limestone is being mined in India in the Proterozoic Cuddapah basin in Andhra Pradesh state; mineralization has been dated by Pb–Pb isotopes at 1,756 ± 29 Ma (Zachariah et al. 1999). The non-metamorphosed equivalent of this type of mineralization is possibly represented by the Mesozoic Todlito lacustrine limestone, New Mexico, USA (Gabelman and Boyer 1988). The U mineralization seems to be epigenetic and is controlled primarily by permeable zones developed during folding and secondarily by joints.

The Itataia deposit hosted in graphitic marbles and alaskitic dykes also remains quite unique (Netto 1983). The first ore stage, occurring in the alaskites, shares the same characteristics as Na-metasomatism-related deposits, but the vugs left by quartz dissolution are filled with xenomorphic zircon aggregates, coffinite, sulphides, organic matter and idiomorphic apatite. The main ore stage corresponds to a hydrothermal karst developed in the marbles and filled with massive pink U-bearing collophane.

The Elkon deposit, located in the Sakha Republic of eastern Asian Russia, is also an unusual type of U deposit with vein stockworks associated with K-metasomatism and gold mineralization (Kazanskii 2004 and references therein). Total resources of the Elkon district are estimated in excess of 324,000 t U, at an average grade of 0.1–0.15% U, and with gold values ranging from less than 1 ppm to several ppm (wise-uranium.org/). Host rocks are Archaean to Palaeoproterozoic high-grade metamorphic rocks. The mineralization is controlled by NW–SE striking and steeply SW dipping faults reactivated during the Mesozoic and by pyrite–carbonate–potassium feldspar altered rocks. Three main stages of alteration separated by brecciation episodes have been recognized: (1) pyrite (Au, Ag)–ankerite–orthoclase, (2) pyrite(Au, Ag)–dolomite–orthoclase and (3) calcite–adularia, all associated with dequartzification and Na loss. The only primary U mineral is brannerite; Boitsov and Pilipenko (1998) distinguish three U ore varieties: gold–brannerite, gold–uraninite and brannerite–silver–gold mineralization. Temperatures of the U mineralization were 160°C and 250°C according to fluid inclusion temperatures. U–Pb isotopic dating gives ages of 135–130 Ma for the primary brannerite which correlates with the emplacement of Early Cretaceous monzonite and nepheline syenite intrusions (Boitsov and Pilipenko 1998).

Unconventional Resources

According to the IAEA, unconventional resources represent resources from which uranium can only be recovered as a minor by-product, such as U associated with phosphorites, non-ferrous ores, carbonatites, black shales, lignite and seawater. However, this definition may evolve with the uranium prices and technological improvements, and some of these resources, like uranium from black shales or from phosphorites, may become a major resource in the future.

Other significant non-conventional resources which have to be considered are the following:

- Several projects are being developed in the world (many in South Africa and also in the Czech Republic, Kyrgyzstan and Tajikistan) for reprocessing the tailings produced during previous uranium or other metal extraction. For example, the company Rand Uranium, created by Harmony Gold and USA-backed Pamodzi Resource Fund, is currently estimating the feasibility of reprocessing tailings for Au and U extraction in the Randfontein/Westonaria region, Witwatersrand, South Africa.
- About 1,100 t U have been already recovered from lignite ash from 1964 to 1967 produced in North Dakota, USA. Recently, an agreement has been signed in 2007 between Sparton and the Xiaolongtang Guodian Power Company of Yunnan, China, to test the extraction of U from coal ash produced by the burning of lignite coal. This coal has a high ash content (20–30%) and an average U content of 65 ppm (20–315 ppm). With an average U content of 125 ppm, annual coal ash produced from three power stations

contains about 150 t U. Assuming a U recovery rate of 70%, 105 t U/year could be produced.

- U may also be extracted from monazite recovered from sand placers if REE and Th production from this resource restart in the future. Monazite from sand placers typically contains several thousands ppm U.
- U has been recovered from porphyry copper operations from ores from the USA and Chile with very-low-grade U contents (tens of ppm). It is probable that several presently mined ore deposits contain significant levels of U which may have not been checked to avoid additional regulations specific for radioactive materials. Recently, the Talvivaara Ni–Zn mine, with 15–20 ppm in the ore, announced (www.talvivaara.com) a production of about 350 t U/year from the leach solution with 25 mg/l U.
- Some tens of tons of U are produced each year from water treatment processes associated with the management of former U mines and tailings. For example, about 38 t of U has been produced in 2007at the Königstein mine in Germany from cleaning of mine water (www.world-nuclear.org).

Thorium Resources

Introduction

The average abundance of Th in the Earth's crust is about 5.6 ppm, but Th is enriched in the upper crust with 10.5 ppm. At high temperatures, Th commonly has the same behaviour as U and is deposited together with rare earth elements (REE) in a great variety of minerals. The most important Th minerals are monazite ([(Ce, La, Nd, Th)PO_4] with a few hundreds of ppm up to 26.4% Th), uranothorite ([(Th,U)(SiO_4)] with up to 38% UO_2) and thorianite [ThO_2] which forms a continuous solid solution with uraninite [UO_2]. Other minerals that contain smaller amounts of Th are allanite ([(Ce, Ca, Y, Th,U)$_2$(Al, Fe, Mg)$_3$(SiO_4)$_3$(OH)] with up to 4.35% Th), bastnäsite ([(Ce, Th, La, Y, Ca)(CO_3)F] with up to 41% Th), betafite–euxenite–pyrochlore [(Y,Ca,Ce,U,Th)(Nb,Ta,Ti)$_2$$O_6$], brockite [(Ca, Th, Ce)PO_4,H_2O], xenotime ([(Y,U,Th)PO_4] with up to 2.2% Th), fluorapatite [$Ca_5(PO_4)_3F$] and zircon [$ZrSiO_4$]. Thorogummite is a product of the alteration of thorite caused by hydration induced by its metamictization. Many of the Th-bearing minerals are remarkably resistant to weathering because ThO_2 has a very low solubility at low temperature; therefore, Th tends to become enriched in the oxidized zones of most Th deposits. Its solubility strongly increases with decreasing pH. At a pH of 3, Th solubility reaches 100 ppm as a sulphate complex. For example, Th is deposited in acid hot spring concretions in Th-rich quaternary alkali basalts in Italy. At low temperature, Th is strongly adsorbed on limonite and clay minerals. At high temperature, Th solubility is enhanced primarily as an F-complex, but the Cl, PO_4 and CO_2 anions may also markedly increase its solubility. In addition, there are a number of hydroxide, silicate and phosphate complexes (Langmuir and Herman 1980). Thorium occurs in a large variety of deposit types:

1. Placer deposits which corresponds to sands enriched in heavy minerals
2. Carbonatites (e.g. Fen in southern Norway)
3. Alkaline and peralkaline igneous complexes (e.g. Ilimaussaq in Greenland)
4. Pegmatites (e.g. pegmatoids from anatectic domains as in the Mont Laurier district, Canada)
5. Thorium-bearing veins and lodes (e.g. the Lehmi Pass veins in the USA)
6. Quartz-pebble conglomerates (e.g. Witwatersrand and Blind River-Elliot Lake deposits)
7. Skarns and hornfels deposits (e.g. the Mary Kathleen deposit, Tranomaro district, Madagascar)

The main source of the world's Th until recently was the monazite. Monazite is recovered as a by-product of processing heavy-mineral sands mainly for the extraction of ilmenite, rutile, leucoxene and zircon. Thorium itself is a by-product of refining monazite for its REE content. However, future production will probably come from bastnaesite in the carbonatites and from thorite in the Th veins.

Early use of Th mainly resulted from the very high melting point of ThO_2 (3,300°C). In non-energy applications, Th was used as chemical catalyst and in the light bulb elements, lantern mantles, arc-light lamps, welding electrodes and heat-resistant ceramics. The production reached about 1,000 t in the 1970s and has decreased thereafter. A continuous oversupply of Th on the market was caused by the production of REE from monazite. Increased costs related to increasing regulatory constraints have progressively limited monazite commercial value, and non-radioactive substitutes have been developed (Hedrick 2008). For example, between 1980 and 1995, 165,000 t of monazite were mined in Western Australia, and most of the production was exported to France for extraction of REEs, but the monazite plant in France has closed because its operators were unable to obtain a permit for a disposal site for Th and its decay products. The total amount of Th which has been produced worldwide until 1987 is estimated at 39,000 t (IAEA 1987): 17,000 t in the USA, 8,000 t in Australia, 6,000 t in China, 4,000 t in India, 2,000 t in Germany and 2,000 t in France as purified Th nitrate. Some quantities have been also produced in Canada, Malaysia, Brazil, Turkey and the former USSR. The production of Th is presently of some hundred tonnes per year (IAEA 2002).

World Resources

It is generally believed that Th resources should be three to four times larger than the U resources because the Clarke concentration of Th is three to four times larger than that of U. However, such a statement does not take into account the differences of chemical properties between U and Th. Uranium has two valence states, U^{4+} and U^{6+}, and is very soluble under the oxidized state, as UO_2^{2+}. Thorium has only one valence state, Th^{4+}, and has a low solubility at low to intermediate temperature. This property has several major consequences: (1) Th minerals can be concentrated physically in placer-type deposits as resistates, (2) most other Th deposit types are formed in magmatic and high-temperature hydrothermal environments, and (3) metallurgical extraction of Th from ores requires more costly processes than for U.

Most Th resource estimations correspond to potential resources because specific exploration and comprehensive resource estimation of Th deposits have never been really undertaken. For example, in China, a substantial amount of Th is produced annually as a by-product of the huge REE production at Bayan Obo, in Inner Mongolia, but the only available resource estimation dates back to 1984. The estimates of the Th resources mainly date from exploration made in the 1950s–1980s for U and thus are uncertain. Hence, detailed investigations of the resources as well as ore processing tests should be performed before it can be decided whether the Th resources can be defined as an economical asset. Thorium resources are found in a variety of countries and are compiled in Table 3.

Thorium resources are dominantly associated with peralkaline magmatic complexes: first in carbonatites, then in vein-type occurrences generally deriving from magmatic fluids expelled from carbonatites and in peralkaline rock complexes (Table 4, Fig. 2). Placer-type deposits with monazite represents one quarter of the world resources.

Thorium as a Nuclear Fuel

More than 99.99% of natural Th is ^{232}Th, the rest are ^{230}Th and ^{228}Th. Although not fissile itself, ^{232}Th is fertile and absorbs slow neutrons to produce ^{233}U, which is fissile. The main factors generating interest in Th as a potential fuel in nuclear reactors with respect to U result from the fact that (1) the Th fuel cycle produces much less Pu and other transuranic elements and thus may present less problems in waste disposal; (2) a large part of the highly radioactive actinides are "burnt" because the Th fuel may remain in reactor for up to 9 years and thus will decay to background levels in about 100 years; (3) it would be more proliferation resistant; (4) Th fuel can be used in current light water U reactors; (5) virtually all of the Th is potentially useable in a reactor, compared with the 0.72% ^{235}U of natural U and (vi) ^{233}U is a better fuel than ^{235}U and ^{239}Pu because of

Table 3 World thorium resources by country

Country	Global resource	Identified (10^3 t Th)	Prognosticated (10^3 t Th)
Australia[a]	489	13	476
USA[b]	432	440	274
China[c]	–	–	380
Turkey	880	344	400–500
India	319	319	–
Brazil	1,306	221	329–700
Venezuela	–	300	–
Norway[d]	180	132	132
Egypt	295	100	280
Russia		75	–
Greenland		54	32
Canada	173	44	128
South Africa	115	18	130
Malaysia[b]		–	4.5
CIS[e]	1,650	–	–
Others	388	33	81
Total	*6,227*	*2,053*	*2,655–3,126*

Sources: IAEA (2006)
Identified Resources (*RAR* reasonably assured resources + *EAR-1* estimated additional resources) are estimated at a cost of USD 80/kg
[a]http://www.australianminesatlas.gov.au/aimr/commodity/thorium_09.jsp, taking into account an arbitrary 10% loss during the mining and processing of Th from an initial estimate of 543,700 t Th
[b]USGS commodities evaluation 2010
[c]World Mining. Annual review. 1984
[d]USGS has deleted in 2010 the resources for Norway in its report because of the poor evaluation of the occurrences
[e]CIS = Commonwealth of Independent States. Among the other countries having Th resources mainly as placer deposits are Madagascar, Liberia, Burma, Thailand, Indonesia, Sri Lanka and Bangladesh. Data from many countries of the world are lacking, and many of the figures are based on assumptions. USGS data are reported in ThO_2 whereas all others are given in Th with the same numbers!

Table 4 Thorium resources (in 1,000 t Th) classified according to major deposit types (IAEA-NEA 2007)

Deposit	Resource	%.
Carbonatite	1,900	31.26
Placer	1,500	24.68
Vein-type	1,300	21.39
Alkaline rocks	1,120	18.43
Other	258	4.24
Total	*6,078*	*100.00*

Including prognosticated resources, the total reaches over 6 Mt Th. The differences between with the estimations of Table 3 result from the different costs and degrees of geological assurance used

its higher neutron yield per neutron absorbed. So, theoretically, one unit mass of natural Th could produce about 40 times more energy than the same amount of natural U (i.e. $U^{238}+U^{235}$) (UIC Nuclear Issues Briefing Paper # 67, 2004). 1970s data indicated that a 1,000 MWe Th fuel reactor requires an initial loading of ~40 t Th and ~10 t of highly enriched U (~90% U^{235}) and annual reloading of ~10 t Th. Compared to the present needs of 67,000 t of natural U for the 435 U-fuelled reactors, 500 Th-fuelled 1,000-MWe reactors would have an annual demand of about 5,000 t of Th only. Hence, 3-Mt of Th would be enough to last 500 Th-fuelled reactors 600 years.

Only a few commercial-scale Th-fuelled nuclear reactors have been developed in Germany and the USA in the 1970s–1980s but shut down in late 1980s. Presently, the Moscow's Kurchatov Institute is developing a Th–Pu fuel for feeding existing Russian Vodo-Vodyanoi Energetichesky (VVER)-1,000 reactors, partly to burn up stocks of Pu, in a joint program with the US company

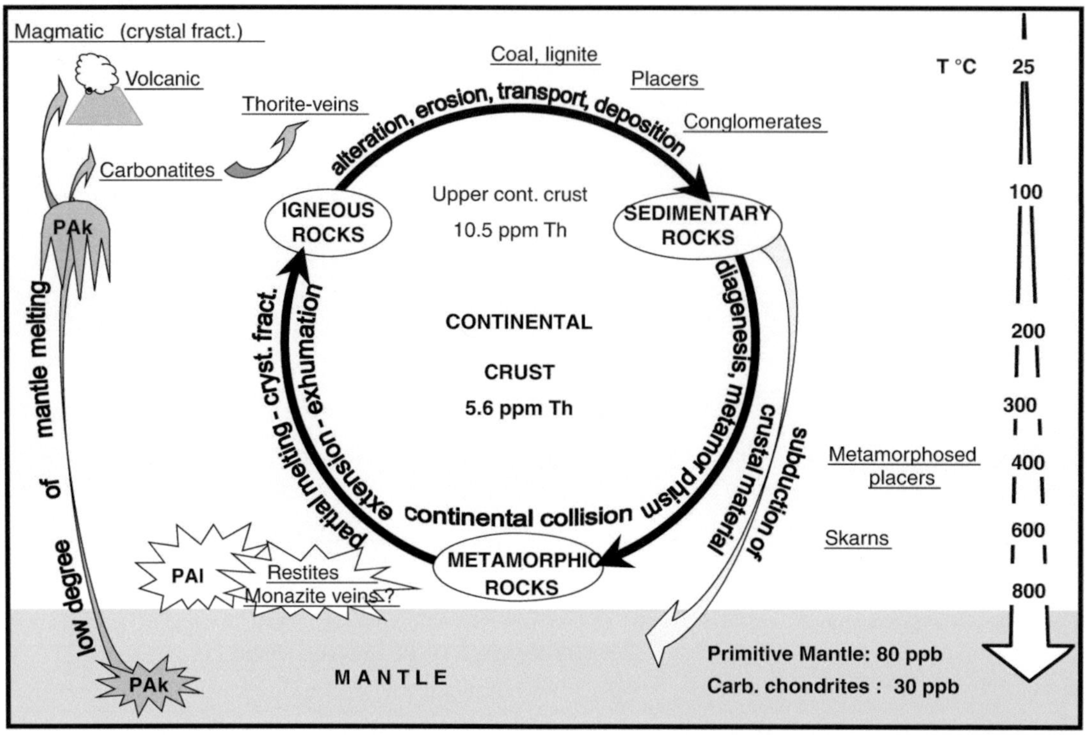

Fig. 2 Position of the major thorium deposit types along the geological cycle, the major thorium reservoirs of the Earth and the major magmatic associations associated with Th deposits: *PAk* peralkaline, *HKCa* high-K calc-alkaline, *PAl* peraluminous (Cuney)

Thorium Power and US government funding. Fissile ^{239}Pu fuel is placed in the centre of the reactor and the Th around it (http://www.world-nuclear.org/info/info.html, May 2007). India is also building a 470-MWe fast breeder reactor that will use a Th-fuelled blanket with anticipated completion in 2010. India is also currently testing components for a 300-MWe technology demonstrator Th-fuelled reactor for completion towards 2020. The development of a full scale Th-fuelled commercial Th reactor in India is scheduled for 2030.

Thorium Deposits

During partial melting and fractional crystallization in the mantle, Th has a strongly incompatible behaviour similar to that of U and other highly charged and/or large ionic radius elements (U, REE, Zr, Nb, …). This behaviour is particularly enhanced in peralkaline magmas because of their high degree of depolymerization resulting from the high temperature of these melts and their excess in alkalis with respect to the quantity of alumina necessary to make the feldspars (Peiffert et al. 1996; Cuney 2009). The excess of alkalis appears as feldspathoids, sodic-pyroxenes and amphiboles and other alkali-rich phases. Hence, Th, U, REE, Zr and Nb are enriched together in peralkaline silicate melts, and the initial Th–U chondritic ratio of 3–4 tends to be preserved (Cuney and Kyser 2008). Thus, in ore deposits resulting from the fractional crystallization of peralkaline magmas, Th will tend to be three to four times more abundant than U. Like for U resources in this type of magmatism, Th is bound to refractory minerals together with the REE, Zr and/or Nb. As a consequence of these properties, a large part of Th deposits are directly or indirectly related to peralkaline magmatism: (1) carbonatites which generally represents the last phase of peralkaline complexes, (2) magmatic fluids mainly deriving from carbonatites leading to Th deposition in metasomatized rocks (fenites) or in veins associated or not with U, REE, Zr and Nb and (3) quartz saturated or undersaturated peralkaline acidic plutonic, pegmatitic or volcanic rocks (Fig. 2).

Deposits Related to Crystal Fractionation

Peralkaline igneous rocks generally originate from low degree of partial melting of the mantle followed by more or less extensive fractional crystallization. The U-, Th-, REE-, Zr- and Nb-mineralized bodies represent always the most fractionated parts of plutonic complexes and are located at their margin or their apical part (Cuney and Kyser 2008). Peralkaline complexes are known all over the world (Fitton and Upton 1987; Woolley 1987, 2001; Kogarko et al. 1995). They are typically anorogenic, zoned or layered, enriched in Na or K. The mineralized intrusions contain an extremely large variety of Th–U-bearing minerals.

Acidic Plutonic, Pegmatitic or Volcanic Rocks

Some peralkaline granites, pegmatites and syenites may be strongly enriched in Th and associated elements. They represent a large reserve of Th and U for the future. The Bokan Mountain peralkaline granite in the USA and the silica undersaturated Ilimausacq peralkaline complex in Greenland, already described as a U resources in the preceding section, represent also two to three times larger Th resources than U resources, the average chondritic ratio being essentially preserved during fractionation. Hence, the U resources of Ilimaussaq being estimated at 85,000 t U and using an average Th–U ratio of 3 for the deposit, a resource of 255,000 t Th is obtained.

The Lovozero massif in Russia's Kola Peninsula, emplaced during Upper Devonian at the same time as the Khibiny massif, is one of the world's largest peralkaline complexes, with an area of 650 km^2. It is unique regarding the abundance and diversity of Th mineralization (Ermolaeva et al. 2007). In the peralkaline pegmatites and veins of Lovozero, Th has segregated into complex but specific Th minerals such as umbozerite $Na_3Sr_4Th(Mn,Zn,Fe,Mg)(Si_8O_{24})(OH)$, Ti–Th silicates $Na_{0-7}Sr_{0-1}ThTi_{1-2}Si_8O_{22-23}(OH),nH_2O$ and Na–Th silicates $(Na,K)_4Th_3[Si_8(O,OH)_{24}],nH_2O$, thorite, steenstrupine–thorosteenstrupine series members and Th phosphate $(Th,Na,K,Ca,Mg,U,Sr,Ba)[(P,Si,AlO_4)],nH_2O$.

The Thor Lake Be, Y, REE, Nb, Ta, Zr and Ga deposits in the Great Slave Lake district, Canada, are related to the nearly circular (16 by 33 km) Blatchford Lake peralkaline complex (Davidson 1982). The Blatchford Lake complex consists of a suite of gabbro, quartz syenite and granite, intruded by the hypersolvus reibeckite Grace Lake Granite and then by the hypersolvus reibeckite Thor Lake Syenite. The Thor Lake Syenite has been dated by U–Pb in zircon at 2,094 ± 10 Ma (Bowring et al. 1984). Principal radioactive minerals are zircon and bastnaesite with minor REE-rich apatite and monazite. Felsic dykes within the Grace Lake Granite and Thor Lake Syenite are one order of magnitude richer in U and Th than the host rocks. Most mineralized zones occur within the western part of the Thor Lake Syenite. The deposits are related to albitite or greisens. Accessory minerals include zircon, uranothorite, allanite, magnetite, hematite, columbite, pyrochlore, carbonates, bastnaesite and lanthanite. U occurs in an east–west trending pegmatite dyke complex that can be traced over a strike length of 2.5 km and widths of up to 400 m. An indicated resource of 4.06 Mt at 183 ppm Th (743 t Th) and an inferred resource of 122.3 Mt ore at 173 ppm Th (21,158 t Th) have been estimated from the data published by www.avalonraremetals.com (March 2009) taking the lowest cut off at 1.6% REEO.

The Sæteråsen trachyte (Permian Oslo graben, Norway) has high concentrations of Zr, Nb, Y, REE and Th (490 ppm) in euxenite, pyroklore, chevkinite, fergusonite and apatite (Ihlen 1983). Mineral separation performed in 1980 showed, however, that the trachyte was difficult to exploit with respect to these metals as it is very fine-grained. A preliminary estimate gives 8 Mt of ore at 490 ppm Th.

In the Høgtuva Palaeoproterozoic granitic orthogneiss occurring in basement windows of the Caledonides in Norway, another potential resource for Be, Zr, REE, U and Th has been identified. The resource has been drilled for Be and contains 0.5 Mt of rocks with 500 ppm Th (Lindahl 2007).

The Brockman Zr, Hf, Nb, Ta, Y, REE, Ga and Be deposit in Western Australia is hosted in the "Niobium Tuff," a fluorite-bearing rhyolitic volcaniclastic unit of the Halls Creek Group (Taylor et al. 1995). Up to 371 ppm Th has been reported by Ranstead (1994), but no resource evaluation is available.

Carbonatites

With 1.9 Mt Th, more than 31% of the world Th resources are concentrated in carbonatites. Carbonatite occurrences have been reviewed in North and South America by Woolley (1987), in the former USSR by Kogarko et al. (1995) and in Africa by Woolley (2001). Carbonatites generally contain some tens of ppm Th; however, some may reach several hundreds of ppm, such as the carbonatites from the Gardar Province in Greenland contain with up to 843 ppm Th (Upton and Emeleus 1987). An extremely large variety of Th minerals can be present: pyrochlore, monazite, apatite, perovskite, titanite, zircon and baddeleyite and also less frequently thorianite, thorite, betafite, cerite, bastnäsite, xenotime, allanite and aeschynite. However, in the present market conditions, carbonatites do not represent a commercial source of Th. Carbonatite-related deposits can be subdivided into deposits generated by magmatic and/or magmatic-hydrothermal processes and those enriched by supergene processes.

Magmatic carbonatite-related deposit is best represented by the *Araxá carbonatite*, in Brazil, a Cretaceous ring complex of approximately 4.5 km in diameter, which not only represent the largest Nb deposit of the world (495,000 t at 2,5% Nb_2O_5), but probably also the largest Th deposit. Thorium resources have been estimated to more than 1,200,000 t ThO_2 mainly hosted by pyrochlore in the Barreiro deposit with additional resource in the Área Zero deposit (30,000 t Th, IAEA 2002). The magmatic incompatible element enrichment has been strongly enhanced by carbonate dissolution during lateritic weathering.

The Mountain Pass deposit in California/USA is associated with a peralkaline intrusive complex composed of eight 100/2,000-m-long plugs of alkaline intrusions (shonkinites, syenites and carbonatites) and about 200 dykes of carbonatite, emplaced at 1.4 Ga. The Sulphide Queen carbonatite, which represents the main intrusion of this type in the complex, is tabular. The reserves were estimated at 29 Mt at 8.9% REO, mostly hosted by bastnaesite, the second in importance after Bayan Obo. The Th resource of the Mountain Pass deposit is not known, the bastnasite has relatively low Th (about 100 ppm), but thorite has been also described. The carbonatite is likely derived from an enriched mantle source (Castor 2008).

The Late Cretaceous Poços de Caldas intrusion in Brazil is one of the largest peralkaline complexes of the world, about 28 km in diameter, emplaced between 78 and 76 Ma (Schorscher and Shea 1992). Similar to Ilimaussaq, it is composed of a silica-undersaturated highly potassic igneous suite. The igneous complex forms a nearly perfect circular ring structure of caldera type. A Th–REE mineralization exists within the caldera at Morro do Ferro, with about 30,000 t of Th and 100 t of U, as uranothorite and fluorcarbonates. The deposit results from the weathering in lateritic conditions of a carbonatite body (Waber 1992).

The Fen Complex in Telemark (600 Ma), Norway, mainly consists of søvite (calcite carbonatite), rauhaugite (dolomite carbonatite), rødberg (hematite–calcite–carbonatite) and fenite (alkali-metasomatized granitic gneiss). The highest concentrations of Th are found in the iron-rich rocks rauhaugite and rødberg, with concentrations ranging from 0.4 to 0.13 wt% Th. The US Geological Survey has estimated reserves of about 170,000 t Th and resources of 150,000 t Th. However, the scientific basis for these old estimates is somewhat unclear, and they have not been confirmed by the Norwegian Geological Survey. Moreover, separation of very fine-grained ore minerals is a great challenge for processing the ore. Thus, the Fen carbonatite rather represents a chemical reserve.

The peralkaline Palabora intrusive complex was emplaced at about 2030 ± 18 Ma in north–eastern Transvaal, South Africa (Eriksson 1984).

In the central part of the complex, phoscorite and carbonatite form the Loolekop pipe (1.4 by 0.8 km). This pipe has been intruded by a dyke-like banded carbonatite. The main U- and Th-bearing mineral is uranothorianite, $(U, Th)O_2$, which is recovered as a heavy mineral together with baddeleyite and processed for U production (Verwoerd 1986). With an average grade of 0.01% Th, the carbonatite should contain about 10,500 t Th for every 100 m of depth (Verwoerd 1986).

The Bayan Obo deposit in Inner Mongolia, China, is a giant polymetallic REE–Fe–Nb ore deposit of hydrothermal origin which hosts 70% of the world's REE resources (Yuan et al. 1992; Chao et al. 1992). It is located on the northern edge of the North China craton, near the suture of the Caledonian subduction of the Mongolian plate beneath the craton. It is hosted in marble and quartzite that were metamorphosed prior to mineralization. Replacement textures provide evidence for the epigenetic, hydrothermal and metasomatic origin of the ore deposit. Hydrothermal activity evidences range from 1.26 Ga to beyond 343 Ma. Monazite and bastnaesite U–Pb ages indicate that the episodes of REE mineralization lasted about 150 Ma (555–395 Ma). The most abundant REE minerals at Bayan Obo are monazite, bastnäsite and huanghoite $[Ba(Ce,La,Nd)(CO_3)_2F]$. The main Th minerals include thorianite and thorite. Thorium also occurs as isomorphic substitution in the REE–Nb minerals, REE–fluorocarbonate minerals and aeschynite-type minerals (Luo et al. 2010). However, the Th content and a recent evaluation of the resources of this deposit are lacking. The monazites and bastnaesites of Bayan Obo ores are typically very low in U (<5 ppm) but have up to 7,000 ppm Th. Pb and Nd isotopes from Bayan Obo ore and gangue minerals indicate a crustal origin for the REEs (Wang et al. 1994). Chao et al. (1995) propose that the REEs and associated elements were scavenged from the underlying crust by acidic hydrothermal solutions activated by Caledonian subduction.

Several carbonatite occurrences in Australia, such as the Mount Weld, Cummins Range and Yangibana carbonatite complexes, are also considered as potential Th resources. The mineralization results from primary magmatic enrichment of the incompatible elements further upgraded by supergene alteration in the case of the Mount Weld and Cummins Range occurrences. The 2.0 Ga Mount Weld carbonatite complex is a large, high-grade Y, Nb, Ta, P, Zr, Ti and REE resource (Duncan and Willett 1990; Lottermoser 1990) where the REE are contained in secondary monazite relatively poor in Th. The typical Th content of the deposit is about 600 ppm (Mernagh and Miezitis (2008). In the Cummins Range carbonatite complex, the main Th minerals are monazite, pyrochlore, bastnäsite, aeschynite, baddeleyite, pandaite and thorianite (Andrew 1990; Richards 1985). About 500 ppm Th has been measured in the top 48 m of weathered zone in one drill hole (Richards 1985). The Yangibana ferrocarbonatite–magnetite–REE-bearing dykes in the Gascoyne Province of Western Australia occur as lenses and pods from 2 to 25 m in width and traceable for up to 25 km (Pearson et al. 1996). They are typically the last stage of a peralkaline complex fractionation enriched in REEs, fluorite and U–Th. Grab sample values vary from 1,062 to 5,230 ppm Th. Th is mainly hosted by monazite and bastnäesite.

Deposits Related to Partial Melting

During partial melting of the continental crust of peraluminous material, at low temperature, the produced magmas are strongly polymerized, and this severely limit the dissolution of large and highly charged elements (U, Th, LREE, Zr) in the silicate melts (Peiffert et al. 1996; Montel 1993). Therefore, the most abundant large and highly charged elements (Th, LREE, Zr) will be already saturated during partial melting and enter into the structure of accessory minerals (monazite, zircon, uranothorite). These accessory minerals will concentrate into the restitic material. Generally, these accumulations are of limited size and essentially correspond to centimetres to metres long biotitic selvages within migmatites in high-grade metamorphic domains. Thorium concentrations may reach several tens to hundreds of ppm.

During the extraction of a silicate melt, the accessory minerals may be transported as a suspension through channelways essentially concordant with the local metamorphic foliation on variable distances. Then, they may locally form larger accumulations in veins, concordant or discordant with the local metamorphic foliation when the most mobile fraction of the silicate melts depleted in these elements escapes towards higher structural levels along tectonic structures.

Such occurrences are known in many high-grade metamorphic domains such as the Mont Laurier pegmatoidic veins of the Grenville belt in Canada (Cuney 1982), the Th- and U-bearing leucocratic granitic rock and pegmatite–aplite dykes from the Bancroft area of Ontario, in the Namib desert of Namibia, in the Charlebois Lake area of northern Saskatchewan and on the north shore of the St. Lawrence River in Québec (Cuney and Kyser 2008). Similar occurrences occur at Crockers Well in South Australia where thorianbrannerite occurs in sodic granitic and felsic gneisses (Ashley 1984).

Deposits Related to Metasomatic Systems

Metasomatic Th deposits occur in marbles in high-grade metamorphic domains generally at the contact with granitic intrusions but sometime with no visible intrusion at least at the level of observation and may result from magmatic or metamorphic fluid infiltration.

The uranothorianite-bearing pyroxenites of Tranomaro, South–East Madagascar, represent an example of a Th and U deposit hosted by skarns (Moine et al. 1985). A metasomatic zonation may be developed between granite injections and marbles, dominantly controlled by a bimetasomatic exchange of Ca and Si between the two lithologies, but the granites are not always outcropping. The main mineralization is associated with skarns composed of non- or poorly mineralized meionite-bearing endoskarns and highly mineralized Al-diopside–spinel–corundum exoskarns. Uranothorianite is the main ore mineral. A later episode of mineralization, of minor importance, corresponds to fluorphlogopite – fluorpargasite – uranothorianite – REE-rich hibonite veins. Fluid circulation responsible for metasomatism and U–Th mineralization is synchronous with granulitic metamorphism from 850°C to 800°C and 5–3 kbar. Although CO_2 was a major fluid in the metasomatic reactions, fluorine has played a major role in the complexing and transport of Th, U, REE and Zr in the skarns (Moine et al. 1998). Most of the Th–U ratios of the mineralized rocks are between 2 and 3, close to the average crustal ratio and thus indicate a simultaneous enrichment in U and Th. It is proposed that the fluids involved in the metasomatic reactions and Th–U–REE–Zr mineralization are derived from the oversaturation of the granitic melts injected in the Tranomaro metamorphic series and from the devolatization reactions occurring at depth during granulitic metamorphism (Boulvais et al. 1998). From 1955 to 1968, the French Atomic Energy Commission (CEA) and private companies have produced 1,030 t of U and 3,218 t of Th, thus corresponding to an ore with an average Th–U ratio of about 3.

The Mary Kathleen deposit is hosted in the Corolla Palaeoproterozoic formation close to the contact with the foliated, coarse-grained Th–U-rich Burstall Granite. The Corolla formation (1,780–1,760 Ma) is composed of metamorphosed evaporates, calc-silicate rocks, marble, meta-pelites, meta-psammites and minor meta-volcanics. To its western margin, the Burstall Granite is cut by a network of microgranite and rhyolite. Two main phases of hydrothermal activity and associated metasomatism were recognized. Phase 1 (1,750 and 1,730 Ma) corresponds to the emplacement of granitic and mafic bodies and the formation of large garnet/pyroxene skarns with minor U and REE (±Th) mineralization (Holcombe et al. 1992). During phase 2 (1,550–1,500 Ma), upper amphibolite facies was reached with intense scapolitization of the Corolla formation (skarns 2) together with the main U, REE and Th mineralization (Oliver et al 1999). The hydrothermal fluids involved in the two phases were highly saline and are assumed to be derived from the evaporitic Corolla formation sediments (Oliver et al. 1999). The mineralization which consists essentially of allanite and uraninite is disseminated throughout the allanite zones. Mary Kathleen is considered to be a skarn-hosted metamorphic–metasomatic-hydrothermal deposit. The global resource within the orebody has been estimated at 10,200 t U at 0.11% U, with 3.0% REOs and 220 ppm Th. Hence, the resource of the Mary Kathleen deposit can be estimated at about 2,040 t Th.

Deposits Related to High-Temperature Hydrothermalism

Uranothorite Veins

These deposits are widely distributed throughout the world but are relatively rare. They are localized at shallow crustal levels, in shear zones, faults, breccia zones and diatremes in meta-sedimentary and meta-volcanic rocks, and generally associated with carbonatites, but such rocks have not always been identified in the vicinity of the veins. Narrow fenitized (alkali-metasomatized) wall-rock selveges may occur adjacent to veins and dykes. The main Th-bearing minerals in the deposits are thorite, thorogummite, monazite and brockite. They are frequently associated with various REE minerals (allanite, bastnäsite, xenotime, cenosite, florencite and synchysite). Quartz, hematite and/or magnetite, feldspar, micas, carbonates, barite, apatite, fluorite and rutile are the common gangue minerals. Some deposits show transitional features between the primary magmatic concentrations disseminated within the most fractionated part of peralkaline magmas or carbonatite and vein systems such as at Eskisehir in Turkey, at Bayan Obo in China and at Mountain Pass in California, USA.

The Nolans Bore deposit in the Northern Territory, Australia, has world-class characteristics, in terms of size and grade of REE, P, U and Th. It occurs as veins and lodes, localized in shear zones, fractures, stockworks, breccia zones and in two zones over a strike length of about 2 km. Four styles of REE mineralization have been recognized (http://www.arafuraresources.com.au/nol_geo.html): (1) massive fluorapatite dykes with 4–6% REE, (2) high-grade REE ore with 10–20% REE found in cheralitic apatite-poor rocks, (3) calc-silicate-hosted apatite–allanite–epidote REE mineralization and (4) zones of low-grade REE mineralization in gneisses (Hussey 2003). The average Th content of Nolans Bore fluorapatite is 0.23% Th (0.07–0.59%) (Hussey 2003). According to Mernagh and Miezitis (2008), this high Th content suggests that the Nolans Bore deposit is not related to a carbonatite, or to fluids derived from them, but is more likely to hydrothermal fluids associated with an NYF-type pegmatite.

The Lemhi Pass District, Idaho and Montana, is characterized by quartz–thorite–hematite veins, shears and brecciated zones and monazite–thorite–apatite-bearing shears and replacements with biotite and alkali feldspar. Allanite and monazite are locally abundant, and brockite and xenotime may also occur. They are typically fine-grained and coloured in pink to brown as a result of the abundant hematite and thorite. Within the Lemhi Pass district, Staatz (1979) mapped 219 Th–REE veins. The veins range from 1 m to over 1,325 m in length and from a few centimetres to 12 m in width. The Last Chance vein (1,325 m long and 3–8 m wide) represents the largest individual Th resource in the district. The ten largest veins, with an average grade of 0.43% ThO_2, represent 95% of the district's identified Th resources.

Mineralization has not initially been linked to any specific intrusives in the district. However, the possible presence of an underlying buried peralkaline complex possibly associated with carbonatites is indicated by the occurrence of: (1) carbonate minerals in many of the veins, (2) one breccia pipe with a calcite matrix adjacent to a mafic dyke, (3) a few carbonate-bearing mafic dykes and (4) a syenite and an ultramafic sill of possible carbonatitic affiliation (Gillerman 2008). A direct relation of Th deposits, characterized by the iron oxides, apatite, REE and F association, with alkaline intrusive or carbonatite complexes is observed in the Powderhorn district in Colorado (Van Gosen and Lowers 2007). Numerous other Th vein occurrences exist in the USA: Wet Mountains district in Colorado, Capitan Mountain and Laughlin Peak in New Mexico and Bokan Mountain district in Alaska.

In the Eskisehir-Sivrihisar district, in Turkey, the Kizilcaören F–Ba–Th–REE deposit is mainly hosted by anchimetamorphic sediments and rarely trachytic tuffs and breccia pipes adjacent to peralkaline porphyritic trachyte and phonolite. Five breccia pipes occur in the area. Two different phase of carbonatite injection during late Oligocene have been distinguished: a first one occurring as dykes up to 1.5 m in width and a second one occurring as dykes ranging in size 1–10 cm cutting the breccia pipes and veins. Fracturing, brecciation and mineralization are associated with the carbonatite intrusions. The mineralization infills radial circular and funnel-shaped fractures and breccia pipes. During the first stage, predating the brecciation event, fluorite and barite are mainly deposited. The second stage corresponds to Th-rich bastnaesite with minor fluocerite, brockite, florencite and monazite deposition after the brecciation. The deposit has been interpreted as formed by the upwelling of carbonatitic magmatic fluids guided by circular, radial and funnel-shaped fractures. Temperatures of 550–300°C and 300–190°C have been obtained for the first and second stage, from fluid inclusion studies. Reserves are of 380,000 t of ore at 0.2% ThO_2 and 3% REE (Stumpfl and Kirikoglu 1985). Additional Th reserves also occur at Kuluncak, Hekimhan and Malatya (Gültekin et al. 2003), and total resources may reach 880,000 t Th.

Monazite Veins

The Mesoproterozoic Steenkampskraal monazite vein in South Africa is the largest of a number of vein-type monazite–apatite–chalcopyrite–magnetite deposits occurring in the Namaqualand high-grade metamorphic complex in South Africa (Andreoli et al. 1994), a totally different setting than the thorite veins always emplaced at shallow crustal levels. The vein intrudes the Roodewal Suite composed of charnockites, tonalites and enderbites. The deposit is tabular in shape with a thickness varying from 30 to 90 cm and has been traced down dip at least to 450 m. The monazite lode itself is dark grey to brown, is fine-grained and massive and has sharp contact with the wall rock. The ore body contains monazite, quartz, apatite and magnetite with small amounts of zircon, pyrite, chalcopyrite, galena and ilmenite. The monazite display high ThO_2 contents (8–8.8 wt%) and gives a U–Pb age of 1,180 ± 40 Ma. At an average grade of 17% REEOs and 2.5% ThO_2, Steenkampskraal is the highest grade REE deposit in the world. From the age relationships and the existence of orthopyroxene, Andreoli et al. (1994) propose that the monazite veins of the Steenkampskraal district formed in granulite facies conditions (800–860°C, 5–6 kbar). An origin by protracted fractionation of a rare metal–enriched magma, yielding anorthositic cumulates and P-rich immiscible liquids, is proposed for the deposit by Andreoli et al. (1994).

Syn-sedimentary Deposits

Because of its resistance to weathering and its high specific weight, monazite tends to accumulate as heavy mineral layers with other high-density minerals such as ilmenite, rutile, xenotime, zircon, etc. Water currents or waves may concentrate heavy mineral grains in a river bar (alluvial placers) or on a beach (coastal placers), respectively. Most of the heavy mineral sand deposits in the world are concentrated by the action of the waves, but they may also result from sorting in river valleys or deltas (alluvial or stream placers), or from aeolian sorting (aeolian placers) as extensive dune systems, as at the North Stradbroke Island in Queensland, Australia. Most heavy mineral sand concentrations occur onshore, but offshore deposits are also known. Most of the deposits are recent, but fossil more or less consolidated or even metamorphosed placers do also exist.

Necessary conditions for an efficient concentration of heavy minerals include: (1) low rates of clastic sediment supply and long periods of weathering and abrasion of the source area to create a mature heavy mineral suite, (2) an energetic swell wave climate driving large sand fluxes onshore and along shore and (3) changing sea levels, especially marine transgressions that have the effect of moving heavy minerals from the shelf onto the present coast. Additional factors which may also play a role in the formation of economic heavy mineral deposits include: (1) episodic fluctuations in storminess, (2) alternation of steep headlands and relatively flat embayments in coast and shelf morphologies, (3) diverse depositional environments with contrasting sediment budgets (e.g. sand shelf bodies and barriers) and (4) subtle tectonic effects that promote reworking of the sands (Roy 1999). For getting placers with high Th resources, the degree of enrichment of the source area in monazite is certainly a necessary additional condition, as observed especially in India (Kerala) and in Sri Lanka.

Monazite placers are reported all over the world: Australia, India, Bangladesh, Sri Lanka, Burma, Malaysia, Thailand, Indonesia, China, Brazil, Madagascar, Egypt, Liberia and the USA (Jayaram 1987).

Coastal Placers

In India, coastal placers are best developed along the beaches of the south–west coast of India, in Kerala and south–western Tamil Nadu (Jayaram 1987). They are essentially beach or barrier deposits with development of dunes produced by aeolian action which is prominent during the dry months. The beach sands of the Chavara bar (Kollam, Kerala) contain 0.5–1% monazite at 9.0% ThO_2. In Manavalakurchi, south–western Tamil Nadu, the monazite content may reach 3–4% at 9–10% ThO_2. On the east coast of India, monazite concentrations consist of extensive dunes fringing the coast in the Andhra Pradesh state and the eastern part of the Tamil Nadu and the Orissa state. At Cuttak and Ganjam in the Orissa state, the thickness of the placer is about 30 cm with a monazite content of 2.5%. The higher percentage of monazite in this area is attributed to the high density of intrusion of the pegmatites and leptynites in the hinterland. Elsewhere in India, monazite sand deposits occur in other parts of Andhra Pradesh. Monazite content of placers rarely exceeds 3%, and the maximum Th concentration is about 3,000 ppm in the sediment, with an average concentration of 60 ppm.

In Sri Lanka, relatively rich deposits occur along the N and N–W coast. Alluvial deposits also occur in the lower valley of Kaleganga in the Ratnapura district. The largest placer deposit (12,000 t monazite) near Pulmoddai is 3-km long and 50-m wide and contains 3 Mt of sand at 0.4% monazite.

In Southern China, monazite placers occur between Chianhua Hsien in south–western Hunan and Kung Chen Hsien in north–eastern Kwangsi and extensive deposits in the Kwangtung province.

In Australia, shoreline deposits are located on the east coast, northern NSW and southern Queensland and south–west WA (Mernagh and Miezitis 2008). About 165,000 t of monazite was produced from placers and exported from 1980 to 1995. About 159,500 t was produced from deposits in Western Australia and the remaining 5,500 t from New South Wales and Queensland. Most available resources in shoreline deposits in Australia are mined out; remaining resources are in national parks and built-up areas. Presently, monazite which is no more extracted is dispersed back through the original host sand to avoid the concentration of radioactivity. Hence, it becomes no more economic to recover the dispersed monazite for its REE and Th content from these sands. In order to estimate Australia's Th resources in the heavy mineral sand deposits, Geosciences Australia has used a value of 7% Th from limited historic data on the Th content in monazite. Sircombe (1997) gave a median value of 6.7% Th and a mean of 7.2% Th from the analysis of 250 monazite grains. The maximum Th value in this suite of monazites was 27%, but 95% of the grains had less than 14.4%.

In Brazil, monazite occurs associated with ilmenite and zircon in coastal placers along the eastern and south–eastern Atlantic coast. *In the USA*, the Florida beach placers are estimated to contain 0.3–1% monazite.

Dunes

Monazite placer deposits associated with the formation of dunes have been reported in India and in Australia. Near Manavalakurchi, in India, at the mouth of the Vailiyar river, monazite placers are formed by aeolian action on the beaches, especially during the dry season. They contain

3.3% monazite with 9–10% ThO_2. In Australia, dune placers occur in Queensland with resources of about 25,000 t Th. Most of them are located in national parks or conservation areas, except the North Stradbroke mine, and their monazite content is generally low, only 0.2–0.3% of the heavy mineral fraction.

Offshore Placers

Offshore placers are known in India and Australia but should exist in other countries. In India, they occur 600 m off from the Neendakarai–Kayankulam bar over a length of 22 km with 0.05% monazite at 9.5% ThO_2. In Australia, they occur off the southern Queensland and northern New South Wale coast.

Stream Placers

In stream placers, heavy minerals are concentrated in sand and gravel by the action of rivers. They occur along well-defined channels or benches. In the Cooljarloo region, Western Australia high-grade sections 40 m thick with 4.4% heavy mineral concentrations exist. In the USA, they are known from southern Virginia to central North Carolina. These deposits were formed at the intersection of shorelines and major palaeorivers, during multiple transgressive/regressive events from Cretaceous to Tertiary time.

Fossil Placers

They are known in sandstones, conglomerates and arkoses from the Proterozoic to the late Tertiary. Some of these placers may contain up to several thousands ppm Th, but most have less than 100 ppm. In Australia, the Murray Basin is a major fossil monazite placer in New South Wales (with a resource of about 27,000 t Th), Victoria and South Australia with variable monazite content (Mernagh and Miezitis 2008). In Western Victoria, the placers were formed in an offshore environment with monazite contents of 1–1.5% and are relatively large (232,000 t Th). Other fossil shoreline deposits are in the Eneabba and Cooljarloo areas north of Perth, in the Yoganup area south of Perth, in south–west Western Australia and in the Eucla Basin in South Australia and Western Australia.

Metamorphosed Placers

Metamorphosed placers are known from low-grade to high-grade domains. Examples are the low-grade metamorphosed extensive Ordovician–Silurian quartzites of the Despenaperros zone in Spain (Alia 1956) or the high-grade Palaeozoic monazite, sillimanite, staurolite and kyanite schists, gneisses and migmatites in the Piedmont of the eastern USA (Twenhofel and Buck 1956).

Coal, Lignite

Th adsorption on organic matter, clay minerals and Fe-oxides increases with increasing pH (Syed 1999). Th^{4+} is strongly absorbed on carbonaceous materials (coal, lignite, petroleum, bitumen and thucolite). However, most peat bogs have low Th contents (~10 ppm), but Th content may increase two- to tenfold in the vicinity of Th-rich rocks (Boyle 1982). Some US coal may contain up to 100 ppm Th (Facer 1979). Hence, Th extraction from ash from coal and lignite power plants has been considered. For example, 13 t of Th per year are produced in a typical 1,000 MWe coal power plant in the form of ash. One pound of Th contains the energy equivalent of 5,000 t of coal (Ragheb and Tsoukalas 2010). Hence, extraction of Th from coal would yield 31 times the energy equivalent of the coal. It is interesting to note that in the USA alone, the ashes accumulated only for the year 1981 from 154 coal-fired power plants of 1,000 MW contain 801 t of U and 1,971 t of Th, assuming U and Th concentrations of 1.3 and 3.2 ppm, respectively, in the coal (Gabbard 1993).

Towards a Sustainable Use

In the present generation of nuclear reactors (second- or third generation), only ^{235}U is burned, which represents 0.72% of the natural U (1/140). In fact, even less than that is really used because part of ^{235}U remains in depleted U after the enrichment process (0.25–0.30% ^{235}U), and a significant proportion of ^{235}U remain in the spent nuclear fuel (0.8–1% ^{235}U). The fraction of ^{235}U which is burned (3–4%) in the nuclear reactors produces a

Table 5 Inventory of recyclable materials (www.world-nuclear.org)

	Quantity (t)	Natural U equivalent (t)
Plutonium from reprocessed fuel	320	60,000
Uranium from reprocessed fuel	45,000	50,000
Military plutonium	70	15,000
Military highly enriched uranium	230	70,000
Depleted uranium tails	$1{\cdot}6{\cdot}10^6$	–

series of highly radioactive nuclides which represent the high-level wastes. Besides, about 1% of ^{238}U is transformed to ^{239}Pu (50%) and ^{241}Pu (15%) by neutron capture during the nuclear reactions. Only ^{239}Pu is fissile. About 70 t Pu contained in used fuel is removed each year in the world when refuelling reactors. A large part of the information summarized here comes from the website of the World Nuclear Organization (www.world-nuclear.org) and from OECD/NEA (2007).

During the enrichment process, every tonne of natural U gives about 130 kg of enriched fuel with 3.5% ^{235}U. This means that the production of 1 t of enriched U requires about 7.7 t natural U, and thus about 6.7 t of depleted U is produced. Therefore, huge amounts and variable types of nuclear materials are produced along the U nuclear fuel cycle (Table 5). Many processes have been developed for a sustainable use of these different types of nuclear materials: depleted U, spent fuel, Pu and even high-level nuclear wastes. Since the signature of the non-proliferation treaty between Russia and the USA (known as the Megatons to Megawatts program), a program has been set up to convert highly enriched U (HEU) from nuclear weapons into fuel for electricity production.

Depleted Uranium

Uranium enrichment leaves depleted U tails with 0.2–0.3% ^{235}U. Depleted U tails are either stored as UF_6 or de-converted back to U oxide, especially in France. Every year, over 50,000 t of depleted U is produced in the USA, Europe and Russia. The world stock in 2009 is about 1.6 Mt. With the decrease of the activities for military enrichment, a lot of enrichment capacity remains unused, particularly in Russia. Consequently, an agreement has been signed in 1995 (which should cease in 2010) between Tenex (Russia) and Areva and Tenex and Urenco for re-enrichment of part of the depleted U tails. The tails resulting from that secondary re-enrichment, with only about 0.10% ^{235}U, becomes the property of Tenex.

During de-conversion of depleted U, UF_6 is reacted with hydrogen at 700°C to yield HF which represents a by-product for sale and U-oxide powder which is packed for storage. About 300,000 t of depleted uranium was de-converted by the beginning of 2007. The main de-conversion plant is run by Areva NC at Tricastin, France. Another plant operating at Zelenogorsk in Russia has been developed based on the French technology. The construction of another plant is planned in the USA (New Mexico). The resulting depleted U oxide is either considered as a waste or as a long-term strategic resource which can be used in the fast neutron reactors of the fourth generation.

Spent Fuel

In some countries (e.g. France, UK, Japan), spent fuel is reprocessed to separate highly radioactive nuclides to recover its U and Pu, which represent about 95% of the volume of the spent fuel, and thus reduce by about 20 times the final volume of high-level wastes.

Pu is recycled by mixing it with depleted U to produce a mixed-oxide fuel (MOX) which can be used in conventional nuclear power stations. MOX fuel with of about 7% Pu is equivalent to U-oxide fuel enriched to about 4.5% ^{235}U. If weapons-grade Pu is used (over 93% ^{239}Pu), only about 5% Pu is needed; 260 t weapons-grade Pu exists in the world. If all this Pu was used in MOX fuel, it would be equivalent to a little over one year's world U production. A single recycle of Pu in the MOX fuel increases the energy derived from natural U by some 12%, and if the

U is also recycled, this reaches about 22% (based on light water reactor fuel with burn-up of 45 GWd/tU). In 2006, about 180 t of MOX fuel was loaded into over 30 reactors (mostly PWR) in Europe, providing about 2% of the nuclear fuel. So far, about 2,000 t of MOX fuel has been loaded into power reactors.

The U recovered from reprocessing used nuclear fuel is called reprocessed uranium (RepU). The re-use of RepU is not straightforward, essentially because two new U isotopes ^{232}U and ^{236}U are formed by or following neutron captures. ^{232}U is essentially a decay product of ^{236}Pu, and its abundance increases with storage time in spent fuel, peaking at about 10 years. As both isotopes decay much more rapidly than ^{235}U and ^{238}U, they are much more radioactive, and especially ^{232}U. Strong shielding is required to handle such material. Also, ^{236}U is a strong neutron absorber which slows down the chain reaction that entails the use of higher enrichment levels of ^{235}U of the new fuel to compensate for stronger neutron adsorption. Moreover, during the re-enrichment process for production of the new fuel, the lighter isotopes ^{232}U and ^{236}U tend to be further enriched. This problem limits the used of RepU. This process has been mainly used in UK, where 16,000 t of RepU from Magnox reactors has been converted to about 1,650 t of enriched fuel, but this recycling was stopped in 1996 due to economic factors. Much smaller quantities have been converted, in France and Japan. Some RepU from Swiss, Germany and Russia is presently re-enriched by blending with military highly enriched U.

Military Uranium

Since 1987, the USA and countries of the former USSR have signed a series of disarmament treaties to reduce the nuclear arsenals by about 80%. The 2,000-t HEU existing in US and Russian weapons and other military stockpiles are equivalent to about 12 years of world mine production. The contract signed between the US Enrichment Corporation (now USEC Inc) and Russia's Technabexport (Tenex) in 1994 permits USEC to purchase a minimum of 500 t of weapons-grade HEU (equivalent to around 20,000 bombs) over 20 years to 2013, at a rate of up to 30 t/year from 1999. HEU which contains at least 20% ^{235}U and usually about 90% can be diluted with depleted U, natural U or partially enriched U. This weapons-grade material is diluted by a factor of about 25:1 with depleted U, or 29:1 with depleted U that has been enriched slightly (to 1.5% ^{235}U) to minimize levels of radiogenic ^{234}U deriving from the natural decay of ^{238}U. In Russia, HEU is diluted to 15,259 t U enriched at 4.4% ^{235}U, using depleted U tails re-enriched at 1.5% ^{235}U. This quantity of diluted HEU is equivalent to about 137,000 t of natural U. Presently, dilution of HEU stockpiles is equivalent to 9,000 t U production from mines each year, and meets about 13% of annual world reactor requirements.

Conclusions

Uranium deposits are known in nearly all part of the geological cycle from high-temperature, magmatic, hydrothermal or metasomatic processes to low-temperature hydrothermal or sedimentary processes (Fig. 1). Identified U resources can satisfy U needs at the present rate of consumption for about 80 years. According to the number of new reactors which will be connected to the grid during the next years, this estimate may decrease. However, increased exploration efforts particularly in strongly underexplored countries as in Africa and South America (Table 2) may counterbalance this evolution. The main problem for satisfying the U needs will more probably result from the difficulty of opening new uranium mines due to technical, regulatory, financial, environmental and political problems. Besides, increasing burn-up and thermal efficiency in existing and new reactors, reducing the amount of ^{235}U in the depleted U, may reduce the demand and recycling of nuclear material resulting from the industrial and military U cycle (depleted U, Pu, spent fuel) and will continue to represent an additional significant source of fuel (MOX, RepU) for the nuclear power stations. In a relatively near future (2030?), reactors of the fourth generation should be able to burn most of the ^{238}U, which may increase the duration of the U resources by a factor of at least 60, i.e. about 5,000 years at the present rate of consumption.

Thorium represents an additional radioactive fuel source which should be used in conventional or specific reactors in a near future. Thorium resources occur in a large variety of geologic environments, but are essentially associated either with highly fractionated peralkaline magmatism (especially with carbonatites) and associated magmatic fluids, or placer-type concentrations. Thorium contained in the ash derived from the burning of coal may represent an additional significant resource in the future if efficient extraction processes are developed. However, the present knowledge of the Th resources in the world is relatively poor because of inadequate exploration efforts arising out of insignificant demand. In a near future, with the possible development of Th-fuelled reactors and the urgent needs of renewed REE resources, in response to the recent policy of China concerning restrictions on REE exports, Th could be first obtained as a by-product of the REE–Th-bearing minerals, the REE covering most of the extraction costs. Then, with further development of Th reactors in the future, other ore minerals with higher Th contents, such as thorite, would be more likely sources.

Uranium and thorium cannot be considered as renewable resources, because at each nuclear cycle, part of these elements are burnt out to produce energy, but as additional resources are created during each fuel cycle, U and Th can be considered as highly sustainable resources.

References

Adamek PM, Wilson MR (1977) Recognition of a new uranium province from the Precambrian of Sweden. In: Proceedings of a technical commity. IAEA, Vienna, pp 199–215

Adams SS, Cramer RT (1985) Data-process-criteria model for roll-type uranium deposits. In: Geological environments of sandstone-type uranium deposits, IAEA-TECDOC-328. IAEA, Vienna, pp 383–400

Alexandre P, Kyser K, Thomas D, Polito P, Marlat J (2009) Geochronology of unconformity-related uranium deposits in the Athabasca Basin, Saskatchewan, Canada and their integration in the evolution of the basin. Miner Deposita 44:41–59

Alia M (1956) Radioactive deposits and possibilities in Spain. In: Proceedings of the international conference on the peaceful uses of atomic energy, New York, United Nations. Geol Uranium Thorium 6:196–197

Andersson A, Dahlman B, Gee DG, Snäll S (1985) The Scandinavian alum shales. Sveriges Geologiska Undersoekning, Serie Ca: Avhandlingar och Uppsatser I A4, NR 56, 50

Andreoli MAG, Smith CB, Watkeys M, Moore JM, Ashwal LD, Hart RJ (1994) The geology of the Steenkampskraal monazite deposit, South Africa; implications for REE-Th-Cu mineralization in charnockite-granulite terranes. Econ Geol 89:994–1016

Andrew RL (1990) Cummins range carbonatite. In: Hughes FE (ed) Geology of the mineral deposits of Australia and Papua New Guinea. Austral Inst Mining Metallurgy, Melbourne, pp 711–713

Annesley IR, Madore C, Hajnal Z (2003) Wollaston-Mudjatik transition zone: its characteristics and influence on the genesis of unconformity-type uranium deposits. In: Cuney M (ed) Uranium geochemistry 2003, conference proceedings, Nancy, pp 55–58

Ashley PM (1984) Sodic granitoids and felsic gneisses associated with uranium-thorium mineralisation, Crockers Well, South Australia. Miner Deposita 19:7–18

Aubakirov KhB (1998) On the deep origin of ore-forming solutions in the uranium deposits in platform sequence of depressions (with Chu-Sarysu Province as an example). Geol Kazakhstan 354:40–47

Audeoud D (1982) Les minéralisations uranifères et leur environnement à Kamoto, Kambove et Shinkolobwe (Shaba, Zaïre). Pétrographie, géochimie et inclusions fluides. Unpublished PhD thesis, Lyon University (France)

Ballhorn RK, Thakur VK, da Fonte JEC, Suckau V (1981) Geology of the uranium deposit. Brazil. In: Uranium deposits in Latin America – geology and exploration. IAEA Proceedings, Vienna

Barnicoat AC, Henderson IHC, Knipe RJ, Yardley BWD, Napier RW, Fox NPC, Kenyon AK, Muntingh DJ, Strydom D, Winkler KS, Lawrence SR, Cornford C (1997) Hydrothermal gold mineralization in the Witwatersrand basin. Nature 386:820–824

Basson IJ, Greenway G (2004) The rössing uranium deposit: a product of late tectonic localization of uraniferous granites in the central zone of the Damara Orogen, Namibia. J Afr Earth Sci 38:413–435

Baturin GN, Kochenov AV (2001) Uranium in phosphorites. Lithol Miner Res 36:303–321

Belevtsev YN, Koval VB (1968) Genesis of uranium deposits associated with sodium metasomatism in crystalline rocks of shields. Geologicheskiy Zhurnal (Kiev) 28:3–17

Berning J, Cooke R, Hiemstra SA, Hoffman U (1976) The Rössing uranium deposit, South West Africa. Econ Geol 71:351–368

Boitsov VE, Pilipenko GN (1998) Gold and uranium in mesozoic hydrothermal deposits of Central Aldan, Russia. Geol Rudnykh Mestorozhdenii 40:354–369

Boulvais P, Fourcade S, Gruau G, Moine B, Cuney M (1998) Persistence of pre-metamorphic C and O isotopic signatures in marbles subject to Pan-African granulite-facies metamorphism and U–Th mineralization. Tranomaro. Southeast Madagascar. Chem Geol 150:247–262

Bowden P, Herd D, Kinnaird JA (1995) The significance of uranium and thorium concentrations in pegmatitic leucogranites (alaskites), Rössing mine, Swakopmund, Namibia. Commun Geol Surv Namibia 10:43–49

Bowring SA, Van Schmus WR, Hoffman PF (1984) U-Pb zircon ages from Athapuscow aulacogene, East arm of Great Slave Lake, N.W-T., Canada. Can J Earth Sci 2:1315–1324

Boyle DR (1982) The formation of basal-type uranium deposits in South Central British Columbia. Econ Geol 77:1176–1209

Cai C, Li H, Qin M, Luo X, Wang F, Ou G (2007) Biogenic and petroleum-related ore-forming processes in Dongsheng uranium deposit, NW China. Ore Geol Rev 32:262–274

Carlisle D, Merifield PM, Orme AR, Kohl MS, Kolker O (1978) The distribution of calcretes and gypcretes in southwestern United States and their uranium favorability based on a study of deposits in Western Australia and South West Africa (Namibia), US Department of Energy. Open File Report GJBX-29

Carré JL (1979) Les minéralisations uraniferes des dépôts oligocènes de Saint Pierre du Cantal dans leur cadre géologique régional et local. Unpublished PhD thesis, INPL, Nancy, 121 p

Castor SB (2008) The mountain pass rare-earth carbonatite and associated ultrapotassic rocks, California. Can Mineral 46:779–806

Castor SB, Henry CD (2000) Geology, geochemistry, and origin of volcanic rock-hosted uranium deposits in north-western Nevada and south-eastern Oregon, USA. Ore Geol Rev 16:1–40

Cathcart JB (1978) Uranium in phosphate rock. U.S. Geological Survey Professional Paper 988A

Cathelineau M (1986) The hydrothermal alkali metasomatism effects on granitic rocks: quartz dissolution and related subsolidus changes. J Petrol 27:945–965

Cathelineau M, Boiron MC, Holliger P, Poty B (1990) Metallogenesis of the French part of the Variscan orogen. Part II: Time-space relationships between U, Au and Sn-W ore deposition and geodynamic events – mineralogical and U-Pb data. Tectonophysics 177:59–79

Chabiron A, Cuney M, Poty B (2003) Possible uranium sources for the largest uranium district associated with volcanism: the Steltsovka caldera (Transbaïkalia, Russia). Miner Deposita 38:127–140

Chao ECT, Back JM, Minkin JA, Ren Y (1992) Host-rock controlled epigenetic, hydrothermal metasomatic origin of the Bayan Obo REE-Fe-Nb ore deposit, Inner Mongolia, P.R.C. Appl Geoch 7:443–458

Chao ECT, Back JM, Minkin JA, Tatsumoto M, Junwen W, Conrad JE, McKee EH, Zonglin H, Qingrun M (1995) The sedimentary carbonate-hosted giant Bayan Obo REE-Fe-Nb ore deposit of inner Mongolia, China: a cornerstone example for giant polymetallic ore deposits of hydrothermal origin. US Geol Surv Bull 2143, 65 p http://pubs.usgs.gov/bul/b2143/

Chavès AO, Tubrett M, Rios FJ et al (2007) U-Pb ages related to uranium mineralization of Lagoa Real, Bahia – Brazil: tectonic implications. Rev Geol 20:141–156

Chernyshev IV, Golubev VN (1996) The Strel'tsovskoe deposit, Eastern Transbaikalia: isotope dating of mineralisation in Russia's largest uranium deposit. Geokhimia 10:924–937 (in Russian)

Cinelu S (2008) Métasomatose sodique et minéralisations uranifères associées: Exemples du batholite du Kurupung (Guyana), du district de Kirovograd – Novoukrainsk (Ukraine) et du gisement d'Espinharas (Brésil). Unpublished PhD thesis, Henri Poincaré – Nancy I Univ

Cinelu S, Cuney M (2006) Na-metasomatism and U–Zr mineralization: a model based on the Kurupung batholith (Guyana). Goldschmidt Conference, Melbourne, Australia. Geochim Cosmochim Acta 70:A103

Corfu F, Andrews AJ (1986) A U–Pb age for mineralized Nipissing diabase, Gowganda, Ontario. Can J Earth Sci 23:107–109

Creaser RA (1996) Petrogenesis of a Mesoproterozoic quartz latite-granitoid suite from the Roxby Downs area, South Australia. Precamb Res 79:371–394

Cruikshank BI, Ferguson J, Derrick GM (1980) The association of uranium and skarn development in the Mary Kathleen area, Queensland. In: Ferguson J, Goleby AB (eds) Uranium in the Pine Creek geosynclines. IAEA, Vienna, pp 698–706

Cuney M (1978) Geologic environment, mineralogy, and fluid inclusions of the Bois Noirs-Limouzat uranium vein, Forez, France. Econ Geol 73:1567–1610

Cuney M (1980) Preliminary results on the petrology and fluid inclusions of the Rössing uraniferous alaskites. Trans Geol Soc S Afr 83:39–45

Cuney M (1982) Processus de concentration de l'uranium et du thorium au cours de la fusion partielle et de la cristallisation des magmas granitiques. In: Les méthodes de prospection de l'uranium. OCDE, Paris, pp 277–292

Cuney M (1990) Contrôles magmatiques et structuraux de la métallogenèse uranifère tardi-hercynienne; exemple du district de la Crouzille (Haute-Vienne). Chron Rech Min 499:9–17

Cuney M (2009) The extreme diversity of uranium deposits. Miner Deposita 44:3–9

Cuney M (2012) Key genetic processes that form uranium deposits throughout the geological cycle: working towards a genetic classification. Econ Geol (in revision)

Cuney M, Kyser K (2008) Recent and not-so-recent developments in uranium deposits and implications for exploration. Mineralogical Association of Canada, short course series 39: Mineralogical Association of Canada, Quebec, 257 p

Cuney M, Mathieu R (2000) Extreme light rare earth element mobilization by diagenetic fluids in the geological environment of Oklo natural reactor zones, Franceville basin, Gabon. Geology 28:743–746

Cuney M, Friedrich M, Blumenfeld P, Bourguignon A, Boiron M-C, Vigneresse JL, Poty B (1990) Metallogenesis in the French part of the Variscan orogen. Part I. U-preconcentrations in the pre-Variscan and Variscan formations – a comparison with Sn, W and Au. Tectonophysics 177:39–57

Dahlkamp FJ (1993) Uranium ore deposits. Springer, Berlin, 460 p

Davidson A (1982) Petrochemistry of the Blachford Lake complex near Yellowknife, Northwest Territories. In: Maurice YT (ed) Uranium in granites. Geological Survey Canada Paper 81:71–80

Derome D, Cuney M, Cathelineau M, Dubessy J, Bruneton P, Hubert A (2003) A detailed fluid inclusion study in silicified breccias from the kombolgie sandstones (northern territory, Australia): application to the genesis of middle-proterozoic unconformity-type uranium deposits. J Geoch Explor 78–79:525–530

Derome D, Cathelineau M, Cuney M, Fabre C, Lhomme T, Banks DA (2005) Mixing of sodic and calcic brines and uranium deposition at McArthur river, Saskatchewan, Canada: a Raman and laser-induced breakdown spectroscopic study of fluid inclusions. Econ Geol 100:1529–1545

Dubessy J, Ramboz C, Nguyen Trung C, Cathelineau M, Charoy B, Cuney M, Leroy J, Poty B, Weisbrod A (1987) Physical and chemical control (fO2, T, pH) of the opposite behaviour of U and Sn-W as exemplified by hydrothermal deposits in France and Great Britain, and solubility data. Bull Minér 110:262–281

Duncan RK, Willett GC (1990) Mount weld carbonatite. In: Hughes FE (ed) Geology of the mineral deposits of Australia and Papua New Guinea. Austral Inst Mining Metal, Melbourne, pp 711–713

El Jarray A, Boiron MC, Cathelineau M (1994) Percolation microfissurale de vapeurs aqueuses dans le granite de Pény (Massif de Saint Sylvestre, Massif Central): relation avec la dissolution du quartz. C R Acad Sci Paris 318:1095–1102

Eriksson SC (1984) Age of carbonatite and phoscorite magmatism of the Phalaborwa Complex (South Africa). Chem Geol 46:291–299

Ermolaeva VN, Pekov IV, Chukanov NV, Zadov AE (2007) Thorium mineralization in the hyperalkaline pegmatites and hydrothermalites of Lovozero pluton, Kola Peninsula. Geol Ore Deposits 49: 758–775

Facer JF (1979) Uranium in coal, Rep. GJBX-56(79). USDOE, Grand Junction Office, Colorado

Fayek M, Kyser K (1997) Characterization of multiple fluid events and rare-earth-element mobility associated with formation of unconformity- type uranium deposits in the Athabasca Basin, Saskatchewan. Can Miner 35:627–658

Finch WI, Davis JF (1985) Sandstone-type uranium deposits. An introduction. In: Geological environments of sandstone-type uranium deposits. IAEA-TECDOC-38. IAEA, Vienna pp 11–20

Fitton JG, Upton BGJ (1987) Alkaline igneous rocks. Geol Soc Lon Spec Publ 30:544

Forbes P (1989) Rôle des structures sédimentaires et tectoniques, du volcanisme alcalin régional et des fluides diagénétiques-hydrothermaux pour le formation des minéralisations à U-Zr-Zn-V-Mo d'Akouta, Niger. Géol GéochimUranium Mém, Nancy 7, 376 p

Forbes P, Pacquet A, Chantret F, Oumarou J, Pagel M (1984) Marqueurs du volcanisme dans le gisement d'uranium d'Akouta (République du Niger). C R Acad Sci Paris 298:647–650

Frimmel HE, Minter WEL (2002) Recent developments concerning the geological history and genesis of the Witwatersrand gold deposits, South Africa. Soc Econ Geol Spec Publ 9:17–45

Frimmel HE, Groves DI, Kirk J, Ruiz J, Chesley J, Minter WEL (2005) The formation and preservation of the Witwatersrand goldfields, the largest gold province in the world. In: Hedenquist JW, Thompson JFH, Goldfarb RJ, Richards JP (eds) 100th Anniversary volume, Society of economic geologists 9. Society of Economic Geologists, Littleton, pp 769–797

Fuchs H, Hilger W (1989) Kiggavik (Lone Gull): an unconformity related uranium deposit in the Thelon Basin, Northwest Territories, Canada. Uranium resources and geology of North America. In: Proceedings of a technical committee meeting, Saskatoon, Saskatchewan. IAEA-TECDOC 500. International Atomic Energy Agency, Vienna

Gabbard A (1993) Coal combustion: nuclear resource or danger. Oak Ridge Natl Lab Rev 26:25–33

Gabelman JW, Boyer WH (1988) Uranium deposits in Todilto limestone, New Mexico: the Barbara "J" No. 1 mine. Ore Geol Rev 3:241–276

Gancarz AJ (1978) U-Pb age (2.05 × 109 years) of the Oklo natural fission reactors 4851 uranium deposit. In: Proceedings of the natural fission reactors. IAEA, Vienna, pp 513–520

Gandhi SS (1978) Geological setting and genetic aspect of the uranium occurrences in the Kaipokok Bay – Big River area. Econ Geol 73:1492–1522

Gatzweiler R (1987) Uranium mineralization in the Proterozoic Otish Basin, Central Québec, Canada. Gebruder Borntraeger, Berlin/Stuttgart. Monogr Ser Miner Deposits 27:27–48

Gauthier-Lafaye F (1986) Les gisements d'uranium du Gabon et les réacteurs d'Oklo. Modèle métallogénique de gites à fortes teneurs du Protérozoique inférieur. Mém Sci Géol, Strasbourg Univ 78, 206 p

Gillerman VS (2008) Newly discovered intrusives at the Lemhi Pass thorium – REE iron oxide district, Idaho: Cambrian syenite and mystery ultramafics – signatures of a buried alkaline complex or two systems ? 104th Cordilleran and 60th Rocky Mountain sections joint meeting, paper no. 9–4

Goldhaber MB, Reynolds RL, Rye RO (1978) Origin of a South Texas roll-type uranium deposit: II. Sulfide petrology and sulfur isotope studies. Econ Geol 73:1690–1705

Golubev VN, Cuney M, Poty B (2000) Phase composition and U-Pb isotopic systems of pitchblende of quartz-calcite-pitchblende veins at Schlema-Alberoda deposit (Erzgebirge). Geol Ore Deposits 42:761–773

Grosse P (2007) Los granitos porfíricos y orbiculares del sector Centro-oriental de la Sierra de Velasco: génesis y significación regional. Unpublished Ph.D. thesis, Salta University, Argentina 285

Gültekin AH, Örgün Y, Suner F (2003) Geology, mineralogy and fluid inclusion data of the Kizilcaoren fluorite-barite-REE deposit, Eskisehir, Turkey. J Asian Earth Sci 21:365–376

Hanly A (2005) Evolution of Mesoproterozoic basins and their economic potential. PhD thesis, Queen's University, Canada, 276 p

Hansley PL, Spirakis CS (1992) Genesis of tabular uranium-vanadium deposits in the Morrison formation, Colorado Plateau. Econ Geol 87:352–365

Hecht L, Cuney M (2000) Hydrothermal alteration of monazite in the Precambrian basement of the Athabasca: implications for the genesis of unconformity related deposits. Miner Deposita 35:791–795

Hedrick JB (2008) "Thorium" in 2007 minerals yearbook – Thorium (advance release), U.S. Geological Survey, U.S. Department of the Interior 76.1–76–4

Hitzman MW, Valenta RK (2005) Uranium in iron oxide-copper-gold (IOCG) systems. Econ Geol 100:1657–1661

Hitzman MW, Oreskes N, Einaudi MT (1992) Geological characteristics and tectonic setting of Proterozoic iron oxide (Cu-U-Au-REE) deposits. Precamb Res 58:241–287

Holcombe RJ, Pearson PJ, Oliver NHS (1992) Structure of the Mary Kathleen fold belt. In: Stewart AJ, Blake DH (eds) Detailed studies of the Mount Isa Inlier. Aust Geol Surv Organ Bull 243:257–287

Holliger P, Cathelineau M (1986) Le chronomètre U-Pb en milieu uranifère: application aux gisements hydrothermaux d'U associés au batholite de Mortagne (Vendée, France). Chron Rech Min 485:33–43

Hu RZ, Burnard PG, Bi XW, Zhou MF, Peng JT, Su WC, Zhao JH (2009) Mantle-derived gaseous components in ore-forming fluids of the Xiangshan uranium deposit, Jiangxi province, China: evidence from He, Ar and C isotopes. Chem Geol 266:86–95

Hussey KJ (2003) Rare earth element mineralisation in the eastern Arunta Region. Australia, Northern Territory Geological Survey Report 2003–004, 20 p

IAEA (1987) Thorium-based nuclear fuel: current status and perspectives, technical report IAEA-TECDOC-412. International Atomic Energy Agency (IAEA), Vienna

IAEA (2001) Analysis of uranium supply to 2050, STI-PUB-1104. IAEA, Vienna

IAEA (2002) Thorium fuel utilization: options and trends. IAEA-TECDOC-1319. IAEA, Vienna, 376 p

IAEA (2006) Global and identified resources. Nuclear development forty years of uranium resources production and demand in perspective, "The red book retrospective". NEA/OCDE/IAEA, Paris, 279 p

IAEA (2005) Uranium 2005: resources, production and demand. OECD, Paris, 403 p

IAEA (2009) Uranium 2009: resources, production and demand. NEA/OCDE/IAEA, Paris, 454 p

IAEA-NEA (2007) Uranium 2007: resources, production and demand 2008. OECD, Paris

Idir EH, Renard JP (2002) Synthèse des résultats de l'exploration de l'uranium dans l'Anti-Atlas (Maroc). Mines Géologie et Energie Rabat 58:69–94

Ihlen PM (1983) Geologiske og petrokjemiske resultater fra diamantboring på Sæteråsen niob-forekomst. Norg Geol Unders, Fagrapport, 1800/76B, 39 p

Ishukova LP, Modnikov IS, Citchev IV (1991) Uranium ore-forming systems of continental volcanism regions. Geol Rudn Mestorozhdenye 3:16–25 (in Russian)

Javoy M (1999) Chemical earth models. C R Acad Sci Paris 329:537–555

Jayaram KMV (1987) An overview of world thorium resources, incentives for further exploration and forecast for thorium requirements in the near future. In: IAEA-TECDOC—412, Thorium-based nuclear fuel: current status and perspectives. International Atomic Energy Agency, Vienna, 158 p

Jerden JL Jr, Sinha AK (1999) Geology, mineralogy and geochemistry of the Coles Hill uranium deposit, Virginia: an example of a structurally controlled, hydrothermal apatite-coffinite-uraninite orebody. Geol Soc America, Abstracts with Programs 31:A-69

Jiang Y, Ling H, Jiang S (2005) Mantle derived fluid and uranium mineralization: evidence from the world-class Xiangshan uranium deposit, SE China. In: Proceedings of 8th SGA meeting, Beijing, China. Mao J, Bierlein FP (eds) Mineral deposit research: meeting the global challenge. Springer, Berlin/Heidelberg/New York, 269–271

Jiarong Z, Zhutian G (1988) The geological characteristics and metallogenetic control factors of the Lianshanguan uranium deposit, Northeast China. Precamb Res 39:51–64

Johnson JP, Cross KC (1995) U-Pb geochronological constraints on genesis of the Olympic Dam Cu-U-Au-Ag deposit, South Australia. Econ Geol 90:1046–1063

Kazanskii VI (2004) The unique Central Aldan gold-uranium ore district, Russia. Geol Rud Mestorozhdenii 46:195–211

Kazansky VI, Laverov NP (1978) Uranium. In: Smirnov VI (ed) Ore deposits of the USSR. Nedra, Moscow, pp 327–333 (in Russian)

Kish L, Cuney M (1981) Uraninite – albite veins from the Mistamisk Valley of the Labrador Trough, Québec. Miner Mag 44:471–483

Kogarko LN, Kononova VA, Orlova MP, Woolley AR (1995) The alkaline rocks and carbonatites of the world. Part II: Former USSR. Chapman & Hall, London

Kondrat'eva IA, Maksimova IG, Nadyarnykh GI (2004) Uranium distribution in ore-bearing rocks of the Malinov deposit: evidence from fission radiography. Lithol Mineral Res 39:333–344

Kříbek B, Žák K, Spangenberg JE, Jehlicka J, Prokes S, Kominek J (1999) Bitumens in the late Variscan hydrothermal vein-type uranium deposit of Pribram, Czech Republic; sources, radiation-induced alteration, and relation to mineralization. Econ Geol 94:1093–1114

Kříbek B, Knésl I, Pasava J, Caruthers H, Sykorová I, Jehlicka J (2005) Hydrothermal alteration of the graphitized organic matter at the Kansanshi Cu (Au-, U-) deposit, Zambia. In: Proceedings of 8th SGA meeting Beijing, China. Mao J, Bierlein FP (eds) Mineral deposit research: meeting the global challenge, Springer, Berlin/Heidelberg/New York, pp 277–280

Kříbek B, Žák K, Dobeš P, Leichmann J, Pudilová M, René M, Scharm B, Scharmová M, Hájek A, Holeczy D, Hein UF, Lehmann B (2009) The Rožná uranium deposit (Bohemian Massif, Czech Republic): shear zone-hosted, late Variscan and post-Variscan hydrothermal mineralization. Miner Deposita 44:99–128

Kushev VG (1972) Alkaline Metasomatites of the Precambrian, Nedra, Leningrad, 191 p, (in Russian)

Lange G, Freyhoff G (1991) Geologie und Bergbau in der Uranlagerstätte Ronneburg/Thüringen. Erzmetal 44:264–269

Langmuir D (1978) Uranium solution-mineral equilibria at low temperatures with applications to sedimentary ore deposits. Geochim Cosmochim Acta 42:547–569

Langmuir D, Herman JS (1980) The mobility of thorium in natural waters at low temperatures. Geochim Cosmochim Acta 44:1753–1766

Lentz D (1996) U, Mo, and REE mineralization in late-tectonic granitic pegmatites, south west Grenville Province, Canada. Ore Geol Rev 11:197–227

Leroy J (1984) Episyénitisation dans le gisement d'uranium du Bernardan (Marche): comparaison avec des gisements similaires du Nord Quest du Massif Central français. Miner Deposita 19:26–35

Leroy J, George-Aniel B (1992) Volcanism and uranium mineralisations: the concept of source rock and concentration mechanism. J Volcanol Geotherm Res 50:247–272

Lindahl I (2007) Thorium resources in Norway. The Geological Survey of Norway internal report

Linnen RL, Cuney M (2005) Granite-related rare-element deposits and experimental constraints on Ta-Nb-W-Sn-Zr-Hf mineralization. In: Linnen RL, Samson IM (eds) Rare-element geochemistry and mineral deposits. Geological association of Canada, short course notes 17. Geological Association of Canada, St. John's, pp 45–67

Lobato LM, Forman JMA, Fuzikawa K, Fyfe WS, Kerrich R (1983) Uranium in overthrust Archaean basement. Bahia. Brazil. Can Mineral 21:647–654

Lorilleux G, Jebrak M, Cuney M, Baudemont D (2002) Polyphase hydrothermal breccias associated with unconformity-related uranium mineralization (Canada): from fractal analysis to structural significance. J Struct Geol 24:323–338

Lottermoser BG (1990) Rare-earth element mineralization within the Mt. Weld carbonatite laterite, Western Australia. Lithos 24:151–167

Ludwig KR, Simmons KR (1992) U-Pb dating of uranium deposits in collapse breccia pipes of the Grand Canyon region. Econ Geol 87:1747–1765

Luo M, Guolong G, Guolin G, Liu C (2010) Occurrence state of thorium in Baotou rare earth concentrate. Chin J Rare Metals 34:292–316 (in Chinese)

Maas R, McCulloch MT, Campbell IH, Page RW (1987) Sm–Nd isotope systematics in uranium–rare earth mineralisation at the Mary Kathleen uranium mine, Queensland. Econ Geol 82:1805–1826

Mann AW, Deutscher RL (1978) Genesis principles for the precipitation of carnotite in calcrete drainage in Western Australia. Econ Geol 73:1724–1737

Mathieu R, Cuney M, Cathelineau M (2000) Geochemistry of palaeofluids circulation in the Franceville basin and around Oklo natural nuclear reaction zones (Gabon). J Geochem Explor 69–70:245–249

Mathieu R, Zetterström L, Cuney M, Gauthier-Lafaye F, Hidaka H (2001) Alteration of monazite and zircon and lead migration as geochemical tracers of fluid paleocirculations around the Oklo-Okélobondo and Bangombé natural nuclear reaction zones (Franceville basin, Gabon). Chem Geol 171:147–171

McGill BD, Marlatt RB, Matthews RB, Sopuck VJ, Homeniuk LA (1993) The P2 North uranium deposit; Saskatchewan: Canada. Explor Min Geol 2:321–331

McKay D, Miezitis Y (2001) Australia's uranium resources, geology and development of deposits. Mineral resource report 1. AGSO Geoscience Australia, Canberra, 200 p

Meneghel L (1979) Uranium occurrence in the Katanga System of north-western Zambia. In: Uranium deposits in Africa, geology and exploration. IAEA, Vienna, pp 97–122

Mernagh TP, Miezitis Y (2008) A review of the geochemical processes controlling the distribution of thorium in the Earth's crust and Australia's thorium resources. Geoscience Australia, 2008/05, Canberra, 48 p

Min MZ, Luo XZ, Mao SL (2002) The Saqisan mine – a paleokarst uranium deposit, South China. Ore Geol Rev 19:79–93

Moine B, Rakotondratsima C, Cuney M (1985) Les pyroxénites à urano-thorianite du Sud Est de Madagascar: conditions physico-chimiques de la métasomatose. Bull Minér 108:325–340

Moine B, Ramambazafy A, Rakotondrazafy M, Ravololomiandrinarivo B, Cuney M, De Parseval P (1998) The role of flour-rich fluids in the formation of the thorianite and sapphire deposits from SE Madagascar. Miner Mag 62:999–1000

Montel JM (1993) A model for monazite/melt equilibrium and application to the generation of granitic magmas. Chem Geol 110:127–146

Morales G, Nadal PJ, Merino JL, Gasos P (1985) Uranium recovery as a by-product from radioactive coal. In: Advances in uranium processing and recovery from

non-conventional sources. IAEA TC 491/16. IAEA, Vienna, pp 275–287
Morton RD, Abut A, Gandhi SS (1978) Fluid inclusion studies of the Rexpar uranium fluorite deposit, Birch Island, British Columbia. In: Current research. Bul Geol Surv Can Pap 78–1B:137–140
Nash JT, Granger HC, Adams SS (1981) Geology and concept of genesis of important types of uranium deposits. Econ Geol (75th Anniv vol):63–116
Netto AM (1983) Le gisement d'uranium d'Itataïa (Brésil). Thesis, Clermont Ferrand University, France
Ngongo-Kashisha (1975) Sur la similitude entre les gisements uranifères (type Shinkolobwe) et les gisements cuprifères (type Kamoto) au Shaba Zaïre. Soc Géol Belg Ann 98:449–462
OECD/NEA (2007) Management of recyclable fissile and fertile materials. OECD, Paris, 84 p
Oliver NHS, Pearson PJ, Holcombe RJ, Ord A (1999) Mary Kathleen metamorphic hydrothermal uranium – rare-earth element deposit: ore genesis and numerical model of coupled deformation and fluid flow. Aust J Earth Sci 46:467–484
Page RW (1983) Chronology of magmatism, skarn formation, and uranium mineralization, Mary Kathleen, Queensland, Australia. Econ Geol 78:838–853
Pagel M, Cavellec S, Forbes P, Gerbaut O, Vergely P, Wagani I, Mathieu R (2005) Uranium deposits in the Arlit area (Niger). In: Proceedings of the 8th SGA meeting, Beijing, China. Mao J, Bierlein FP (eds) Mineral deposit research: meeting the global challenge. Springer, Berlin/Heidelberg/New York, pp 303–305
Pearson JM, Taylor WR, Barley ME (1996) Geology of the alkaline Gifford creek complex, Gascoyne complex, Western Australia. AustJ Earth Sci 43:299–309
Peiffert C, Nugyen-Trung C, Cuney M (1996) Uranium in granitic magmas: Part 2. Experimental determination of uranium solubility and fluid-melt partition coefficients in the uranium oxide-haplogranite-H2O-NaX (X=Cl, F) system at 770°C, 2 kbar. Geochim Cosmochim Acta 60:1515–1529
Petrov VA, Golubev VN, Golovin, VA (2003) An unique uranium mineralization in pillow lavas, Dornot ore field, Mongolia. Proc Inter Conf, M. Cuney Edit., "Uranium Geochemistry 2003", Nancy, pp 289–292
Polito PA, Kyser KT, Stanley C (2007) The Proterozoic, albitite-hosted, Valhalla uranium deposit, Queensland, Australia, a description of the alteration assemblage associated with uranium mineralisation in diamond drill hole. Miner Deposita 44:11–40
Poty B, Leroy J, Cathelineau M, Cuney M, Friedrich M, Lespinasse M, Turpin L (1986) Uranium deposits spatially related to granites in the French part of the Hercynian orogen. In: Vein type uranium deposits, IAEA-TECDOC-361. International Atomic Energy Agency, Vienna, pp 215–246
Pretorius DA (1961) The nature of the Witwatersrand gold-uranium deposits. In: Wolf KH (ed) Handbook of stratabound and stratiform ore deposits. Elsevier, Amsterdam, pp 29–88
Ragheb M, Tsoukalas L (2010) Global and USA thorium and rare earth elements resources. In: Proceedings of the 2nd thorium energy alliance conference, the future thorium economy, Google Campus, Mountain View, California, pp 1–17
Ranstead TW (1994) Brockman project East Kimberley region, Western Australia. Holdings Limited, West Coast Holdings Ltd. Annual Report (1993) Prospecting licences p80/1255–1268, open file company report (unpublished) to the Department of Industry and Resources, Western Australia A 40911 of I 10983
Reynolds LJ (2000) Geology of the Olympic Dam Cu-U-Au-Ag deposit. In: Porter TM (ed) Hydrothermal iron oxide copper-gold & related deposits, 1. Porter Geoconsultancy Pty Ltd, Linden Park, pp 93–104
Richard A, Pettke T, Cathelineau M, Boiron MC, Mercadier J, Cuney M, Derome D (2010) Brine–rock interaction in the Athabasca basement (McArthur River U deposit, Canada): consequences for fluid chemistry and uranium uptake. Terra Nova 22:303–308
Richards MN (1985) Annual report for 1984 on exploration licence 80/113, Cummins Range, Mt. Bannerman, SE 52–13, Western Australia. CRA Exploration Pty. Ltd: Open File Company Report (unpublished) to the Department of Industry and Resources, Western Australia, 246 p
Roy PS (1999) Heavy mineral beach placers in southeastern Australia: their nature and genesis. Econ Geol 94:567–588
Sanguinetti H, Oumarou J, Chantret F (1982) Localisation de l'uranium dans les figures de sédimentation des grès hôtes du gisement d'Akouta, Niger. C R Acad Sci Paris 294:591–594
Scherbak MP, Bobrov OB (2005) Mineral deposits of Ukraine, vol I: Metalliferous mineral deposits, 783 p
Scaillet S, Cuney M, Le Carlier VC, Cheilletz A, Royer JJ (1996) Cooling patterns and mineralization history of the Saint Sylvestre and Western Marche leucogranite plutons, French Massif Central. II – Thermal modelling and implications for the mechanisms of U-mineralization. Geochim Cosmochim Acta 60:4673–4688
Schorscher HD, Shea ME (1992) The regional geology of the Poços de Caldas alkaline complex: mineralogy and geochemistry of selected nepheline syenites and phonolites. J Geochem Explor 45:25–51
Schovsbo NH (2002) Uranium enrichment shorewards in black shales: a case study from the Scandinavian Alum Shale. GFF 124:107–115
Sibbald TII, Munday RJC, Lewry JF (1976) The geological setting of uranium mineralization in northern Saskatchewan; in Uranium in Saskatchewan (Dunn CE Ed). Saskatchewan Geol Soc Spec Publ 3:51–98
Sircombe KN (1997) Detrital mineral SHRIMP geochronology and provenance analysis of sediments in Eastern Australia: Unpublished PhD thesis, Australian National University, 411 p
Smellie JAT, Laurikko J (1984) Skuppesavon, northern Sweden: a uranium mineralisation associated with alkali metasomatism. Miner Deposita 19:183–192

Sørensen H, Rose-Hansen J, Nielsen BL, Løvborg L, Sørensen E, Lundgaard T (1974) The uranium deposit at Kvanefjeld, Ilímaussaq intrusion, South Greenland. Rep Grønlands Geol Unders 60:54

Soudry D, Segal I, Nathan Y, Glenn CR, Halicz L, Lewy Z, VonderHaar DL (2004) 44Ca/42Ca and 143Nd/144Nd isotope variation in Cretaceous–Eocene Tethyan francolites and their bearing on phosphogenesis in the southern Tethys. Geology 32:389–392

Spirakis CS (1996) The roles of organic matter in the formation of uranium deposits in sedimentary rocks. Ore Geol Rev 11:53–69

Staatz MH (1978) I and L uranium and thorium vein system, Bokan Mountain, southeastern Alaska. Econ Geol 73:512–523

Staatz MH (1979) Geology and mineral resources of the Lemhi Pass thorium district, Idaho and Montana. USGS Professional Paper 1049-A, 90 p

Staatz MH, Armbrustmacher TJ, Olson JC, Brownfield IK, Brock MR, Lemons JF, Coppa LV, Clingan BV (1979) Principal thorium resources in the United States. USGS Circ 805:42

Stumpfl EF, Kirikoglu MS (1985) Fluorite-barite-rare earths deposits at Kizilcaoren, Turkey. S Mitt Österr Geol Ges 78:193–200

Syed HS (1999) Comparison studies adsorption of thorium and uranium on pure clay minerals and local Malaysian soil sediments. J Radioanal Nucl Chem 241:11–14

Taylor WR, Page RW, Esslemont G, Rock NMS, Chalmers DI (1995) Geology of the volcanic-hosted Brockman rare-metals deposit, Halls Creek Mobile Zone, northwest Australia. I. Volcanic environment, geochronology and petrography of the Brockman Volcanic. Miner Petrol 52:209–230

Trueman NA, Long JVP, Reed SJB, Chinner GA (1988) The lead-uranium systematics, and rare earth element distributions of some Olympic Dam and Stuart shelf mineralisation. Western Mining Corporation, Adelaide. Unpublished report K/3143

Turpin L, Maruejol P, Cuney M (1988) U-Pb – Rb-Sr and Sm-Nd chronology of granitic basement, hydrothermal albitites and uranium mineralization (Lagoa Real, South-Bahia. Brazil). Contrib Mineral Petrol 98:139–147

Turpin L, Leroy J, Sheppard SMF (1990) Isotopic systematics (O, H, C, Sr, Nd) of superimposed barren and U-bearing hydrothermal systems in a granite, Massif Central, France. Chem Geol 88:85–96

Twenhofel WS, Buck KL (1956) The geology of thorium deposits in the United States. In: Proceedings of the international conference on peaceful uses atomic energy. New York, United Nations: Geology of U and Th, pp 562–567

Upton BGJ, Emeleus, CH (1987) Mid Proterozoic alkaline magmatism is southern Greenland. In: Fitton JG, Upton BGJ (eds) Alkaline igneous rocks. Geological Society of London, Special Publication 30. Blackwells, Oxford, pp 449–471

Van Gosen BS, Lowers HA (2007) Iron Hill (Powderhorn) carbonatite complex, Gunnison county, CO – a potential source of several uncommon mineral resources. Min Eng 59:56–62

Velichkin VI, Kushnerenko VK, Tarasov NN, Andreeva OV, Kiseleva GD, Krylova TL, Donikova OA, Golubev VN, Golovin VA (2005) Geology and formation conditions of the Karku unconformity type deposit in the Northern Ladoga region (Russia). Geol Ore Deposits 47:87–112

Vels B, Fritsche R (1988) Sodium metasomatism in the Kitongo uranium occurrence near Poli, Cameroon. Uranium 4:365–383

Verwoerd WJ (1986) Mineral deposits associated with carbonatites and alkaline rocks. In: Anhaeusser CR, Maske S (eds) Mineral deposits of South Africa, vol I and II. Geol Soc South Africa, Johannesburg, pp 2173–2192

Von Pechman E, Binaconi F (1982) Synmetamorphic uranium mineralization from Tiraun, Graubfinden, Switzerland. Miner Mag 46:173–178

Waber N (1992) The supergene thorium and rare-earth element deposit at Morro do Ferro, Poços de Caldas, Minas Gerais, Brazil. J Geochem Explor 45:113–157

Wallace AR, Whelan JF (1986) The Schwartzwalder uranium deposit; III, Alteration, vein mineralization, light stable isotopes, and genesis of the deposit. Econ Geol 81:872–888

Wang J, Tatsumoto M, Li X, Premo WR, Chao ECT (1994) A precise 232Th/208Pb chronology of fine-grained monazite-age of the Bayan Obo REE-Fe-Nb ore deposit. China Geochim Cosmochim Acta 58:3155–3169

Weber F (1968) Une série précambrienne du Gabon: le Francevillien. Sédimentologie, géochimie et relation avec les gîtes minéraux associés: Mém Serv Carte Géol Alsace Lorraine, 328 p

Wenrich KJ, Titley SR (2008) Uranium exploration for northern Arizona (USA) breccia pipes in the 21st century and consideration of genetic models. Arizona Geol Soc 22:295–309

Woolley AR (1987) The alkaline rocks and carbonatites of the world. Part I: North and South America. British Museum (Natural History) and University of Texas Press, London, 216 p

Woolley AR (2001) Alkaline rocks and carbonatites of the world: Africa. The Geological Society of London/The Natural History Museum, London, 372 p

Yamazaki IM, Geraldo LP (2003) Uranium content in phosphate fertilizers commercially produced in Brazil. Appl Radiat Isot 59:133–136

Yuan Z, Bai G, Wu C, Zhang Z, Ye X (1992) Geological features and genesis of the Bayan Obo REE ore deposit, Inner Mongolia, China. Appl Geochem 7:429–442

Zachariah JK, Bhaskar Rao YJ, Srinivasan R, Gopalan K (1999) Pb, Sr and Nd isotope systematics of uranium mineralised stromatolitic dolomites from the Proterozoic Cuddapah Supergroup, south India: constraints on age and provenance. Chem Geol 162:49–64

The Principal Rare Earth Elements Deposits of the United States: A Summary of Domestic Deposits and a Global Perspective

Keith R. Long, Bradley S. Van Gosen, Nora K. Foley, and Daniel Cordier

Abstract

Demand for the rare earth elements (REE, lanthanide elements) is estimated to be increasing at a rate of about 8% per year due to increasing applications in consumer products, computers, automobiles, aircraft, and other advanced technology products. Much of this demand growth is driven by new technologies that increase energy efficiency and substitute away from fossil fuels. Production of these elements is highly concentrated in China, which is reducing its exports of REE raw materials as part of its industrial policy. The ability of the rest of the world to replace supply from China depends on the quality of known REE resources and the degree to which those resources have been explored and evaluated. A review of United States resources in a global context finds that the United States could make significant contributions to future REE production. Aside from two advanced projects in the United States and Australia, however, there are no REE projects advanced enough to meet short-term demand.

Keywords

Rare earth elements • United States • Reserves • Resources • Lead times • Economic factors

K.R. Long (✉)
U.S. Geological Survey, Tucson, AZ, USA
e-mail: klong@usgs.gov

B.S. Van Gosen
U.S. Geological Survey, Denver, CO, USA

N.K. Foley • D. Cordier
U.S. Geological Survey, Reston, VA, USA

R. Sinding-Larsen and F.-W. Wellmer (eds.), *Non-Renewable Resource Issues: Geoscientific and Societal Challenges*, International Year of Planet Earth,
DOI 10.1007/978-90-481-8679-2_7, © Springer Science+Business Media B.V. 2012 (Outside the USA)

Introduction and Background

The rare earth elements (REE) are 15 elements with atomic numbers 57–71, from lanthanum to lutetium ("lanthanides"), plus yttrium (39), which is chemically similar to the lanthanide elements and typically included with the rare earth elements. Although industrial demand for these elements is relatively small in tonnage terms, they are essential for a diverse and expanding array of high-technology applications. REE-containing magnets, metal alloys for batteries and lightweight structures, and phosphors are essential for many current and emerging alternative energy technologies, such as electric vehicles, energy-efficient lighting, and wind power. REE are also critical for a number of key defense systems and other advanced materials.

The purpose of this chapter is to provide an overview of reserves and resources of REE in the United States and of possibilities for utilizing those resources in lieu of foreign imports. At the present time, the United States imports almost all of the REE it uses from China, which raises serious issues of supply security. This report restates basic geologic facts about REE relevant to assessing security of supply. This is followed by a review of current United States consumption and imports of REE, current knowledge of domestic resources, and possibilities for future domestic production. The necessary steps to develop domestic resources are discussed in a separate section that leads into a review of current domestic exploration and a discussion of the value of a future national mineral resource assessment of REE. This report also includes an overview of known global REE resources (outside China) and discusses the reliability of alternative foreign sources of REE.

The Rare Earth Elements

The rare earth elements (REE) comprise 15 elements that range in atomic numbers from 57 (lanthanum) to 71 (lutetium) on the periodic table (Fig. 1). These elements are also commonly referred to as "lanthanides." Yttrium (atomic number 39) is also included in the REE group because it shares chemical and physical similarities with the lanthanides.

Traditionally, the REE are divided into two groups on the basis of atomic weight: the light rare earth elements are lanthanum through europium (atomic numbers 57 through 64), and the heavy rare earth elements are gadolinium through lutetium (atomic numbers 65 through 71). Yttrium, although light, is included with the heavy REE group because it has common chemical and physical affiliations with the heavy REE in nature.

Most of the REE are not as rare as the group's name suggests. They were named rare earth elements because most were identified during the eighteenth and nineteenth centuries as oxide components within seemingly rare minerals. Cerium is the most abundant REE, and it is actually more common in the Earth's crust than copper or lead. All of the REE, except promethium, are more abundant than silver or mercury (Taylor and McLennan 1985). The rare earth elements are commonly found together in the Earth's crust because they share a trivalent charge ($^{3+}$) and similar ionic radii. Detailed information on the REE is described in Emsley (2001), and an overview of the geology, production, and economics of REE is provided by Castor and Hedrick (2006).

Basic Geology of Rare Earth Elements

Several geologic aspects of the natural occurrence of rare earth elements strongly influence the supply of REE raw materials. These geologic factors are presented as statements of facts followed by a detailed discussion.

> Although rare earth elements are relatively abundant in the Earth's crust, they are rarely concentrated into mineable ore deposits.

In the Earth's crust, the estimated average concentration of the rare earth elements, which ranges from around 150–220 ppm (Table 1), exceeds that of many other metals that are mined

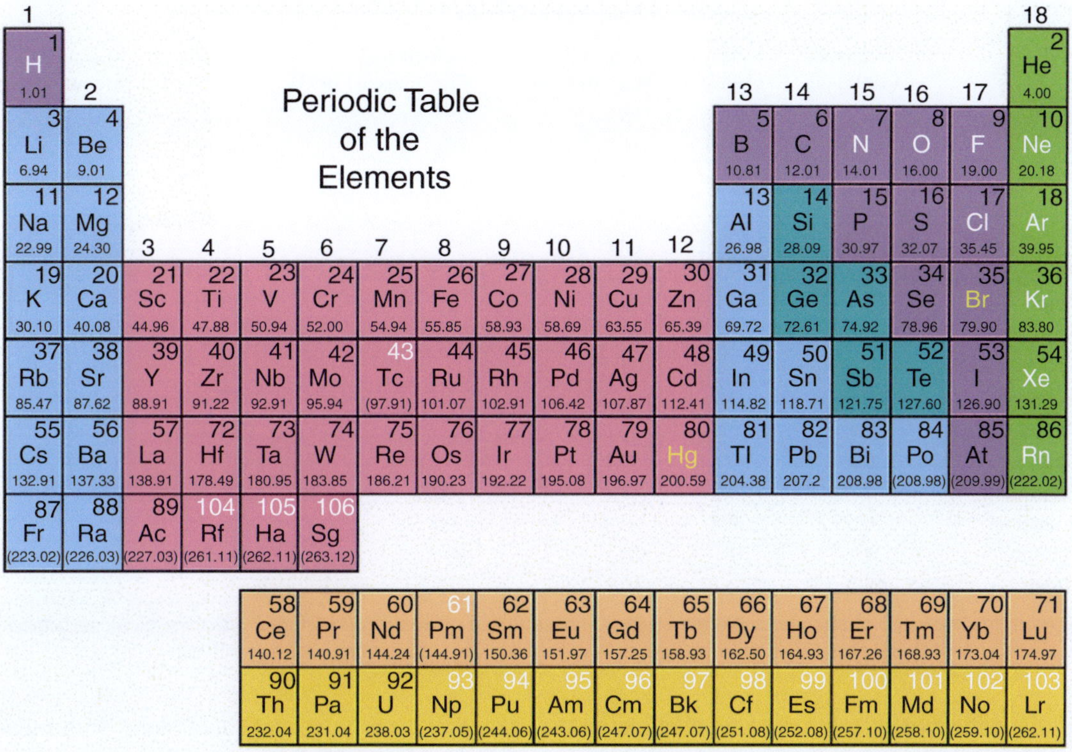

Fig. 1 Periodic table of the elements. The rare earth elements comprise 15 elements, which range in atomic numbers from 57 to 71, including lanthanum (La) to lutetium (Lu). The elements are also commonly referred to as "lanthanides." Yttrium (Y, atomic number 39) is also typically included in the rare earth elements group because it shares chemical, physical, and application properties with the lanthanides

on an industrial scale, such as copper (55 ppm) and zinc (70 ppm). Unlike most commercially mined base and precious metals, however, rare earth elements are rarely concentrated into mineable ore deposits. The principal concentrations of rare earth elements are associated with uncommon varieties of igneous rocks, namely, alkaline rocks and carbonatites. Other potentially useful concentrations of REE-bearing minerals are also found in placer deposits, residual deposits formed from deep weathering of igneous rocks, pegmatites, iron-oxide copper-gold deposits, and marine phosphates (Table 2).

Alkaline igneous rocks form from cooling of magmas derived by small degrees of partial melting of rocks in the Earth's mantle. The formation of alkaline rocks is complex and not fully understood but can be thought of as a geologic process that extracts and concentrates those elements that do not fit into the structure of the common rock-forming minerals. The resulting alkaline magmas are rare and unusually enriched in elements such as zirconium, niobium, strontium, barium, lithium, and the rare earth elements. When these magmas ascend into the Earth's crust, their chemical composition undergoes further changes in response to variations in pressure, temperature, and composition of surrounding rocks. The result is an astonishing diversity of rock types that are variably enriched in economic elements, including the rare earth elements. The mineral deposits associated with these rocks are likewise quite diverse and awkward to classify, in that the distinctive features of these deposits and their rarity can result in classifications that have only one or a few known examples.

Table 1 Estimates of the crustal abundances of rare earth elements

	Wedepohl (1995)	Lide (1997)	Sabot and Maestro (1995)	Jackson and Christiansen (1993)	McGill (1997)	Mason and Moore (1982)
Lanthanum	30	39	18	29	5–18	30
Cerium	60	66.5	46	70	20–46	60
Praseodymium	6.7	9.2	5.5	9	3.5–5.5	8.2
Neodymium	27	41.5	24	37	12–24	28
Samarium	5.3	7.05	6.5	8	4.5–7	6
Europium	1.3	2	0.5	1.3	0.14–1.1	1.2
Gadolinium	4	6.2	6.4	8	4.5–6.4	5.4
Terbium	0.65	1.2	0.9	2.5	0.7–1	0.9
Dysprosium	3.8	5.2	5	5	4.5–7.5	3
Holmium	0.8	1.3	1.2	1.7	0.7–1.2	1.2
Erbium	2.1	3.5	4	3.3	2.5–6.5	2.8
Thulium	0.3	0.52	0.4	0.27	0.2–1	0.5
Ytterbium	2	3.2	2.7	0.33	2.7–8	3.4
Lutetium	0.35	0.8	0.8	0.8	0.8–1.7	0.5
Yttrium	24	33	28	29	28–70	33
Total:	168.3	220.17	149.9	205.2	–	184.1

Rare earth elements are listed in order of increasing atomic number; yttrium (Y) is included with these elements because it shares chemical and physical similarities with the lanthanides. Unit of measure, parts per million

Classification of ores related to alkaline rocks is also controversial. Table 2 presents a relatively simple classification that follows analogous categories for deposits related to nonalkaline igneous rocks. Some of the more unusual alkaline rocks that host, or are related to, REE ores are carbonatite and phoscorite, igneous rocks composed principally of carbonate and phosphate minerals, respectively. Carbonatites, and especially phoscorites, are relatively uncommon, as there are only 527 known carbonatites and phoscorites in the world (Woolley and Kjarsgaard 2008). Economic concentrations of REE-bearing minerals occur as well in some alkaline rocks, skarns and carbonate-replacement deposits associated with alkaline intrusions, veins and dikes cutting alkaline igneous complexes and surrounding rocks, and soils and other weathering products of alkaline rocks.

The weathering of all types of rocks yields sediments that are deposited in a wide variety of environments, such as streams and rivers, shorelines, alluvial fans, and deltas. The process of erosion concentrates denser minerals, most notably gold, into deposits known as placers. Depending on the source of the erosion products, certain REE-bearing minerals, such as monazite and xenotime, can be concentrated along with other heavy minerals. The source need not be an alkaline igneous rock or a related rare earth deposit. Many common igneous, metamorphic, and even older sedimentary rocks contain enough monazite to produce a monazite-bearing placer. As a result, monazite is almost always found in any placer deposit. However, the types of placers with the greatest concentrations of monazite are typically ilmenite-rich heavy-mineral placers, which are primarily mined for titanium oxide pigments, and cassiterite placers, which are mined for tin.

In tropical environments, rocks are deeply weathered to form a characteristic soil profile consisting of laterite, an iron- and aluminum-rich soil, which can be as much as many tens of meters thick. The processes of soil formation commonly concentrate heavy minerals in residual deposits, resulting in an enriched metal layer overlying unweathered bedrock. When a rare earth deposit undergoes such weathering, the resulting concentrations of REE-bearing minerals may be of economic interest.

A particular type of REE deposit, the ion-absorption type, is formed by the leaching of REE from tin-bearing granitic rocks and fixing the elements onto clays in soil (Ren 1985; Wu et al. 1996).

Table 2 Classification of rare earth elements–bearing mineral deposits

Association	Type	Example
Peralkaline igneous rocks	Magmatic—alkali-ultrabasic	Lovozero, Russia
	Pegmatite dikes (alkali-ultrabasic)	Khibina Massif, Russia
	Pegmatite dikes (peralkaline)	Motzfeldt, Greenland
	Hydrothermal veins and stockwork	Lemhi Pass, Idaho
	Volcanic	Brockman, Western Australia
	Metasomatic-albitite	Miask, Russia
Carbonatites	Magmatic	Mountain Pass, California
	Dikes and dialational veins	Kangankunde Hill, Malawi
	Hydrothermal veins and stockwork	Gallinas Mtns., New Mexico
	Skarn	Saima, China
	Carbonate rock replacement	Bayan Obo, China
	Metasomatic-fenite	Magnet Cove, Arkansas
Iron oxide copper gold	Magnetite-apatite replacement	Eagle Mountain, California
	Hematite-magnetite breccia	Olympic Dam, South Australia
Pegmatites	Abyssal-HREE	Aldan, Russia
	Abyssal-LREE	Five Mile, Ontario
	Muscovite-REE	Spruce Pine, North Carolina
	Rare element-REE-allanite-monazite	South Platte, Colorado
	Rare element-REE-euxenite	Topsham, Maine
	Rare element-REE-gadolinite	Ytterby, Sweden
	Miarolitic-REE-topaz-beryl	Mount Antero, Colorado
	Miarolitic-REE-gadolinite-fergusonite	Wausau Complex, Wisconsin
Porphyry molybdenum	Climax type	Climax, Colorado
Metamorphic	Migmatized gneiss	Music Valley, California
	U-REE skarn	Mary Kathleen, Queensland
Stratiform phosphate	Platform phosphorite	Southeast Idaho
Residual	Carbonatite-associated	Mount Weld, Western Australia
	Granite-associated laterite	South China
	Baddeleyite bauxite	Poços de Caldas, Brazil
	Karst bauxite	Montenegro
Paleoplacer	Uraniferous pyritic quartz pebble conglomerate	Elliot Lake, Ontario
	Auriferous pyritic quartz pebble conglomerate	Witwatersrand, South Africa
Placer	Shoreline Ti-heavy mineral placer	Cooljarloo, Western Australia
	Sn stream placer	Malaysia

Sources: Castor and Hedrick (2006), Richardson and Birkett (1996), and Sinclair (1996)

These poorly understood deposits are only known in southern China and Kazakhstan. In China, these deposits account for 14% of Chinese REE production (Wu et al. 1996).

Laterite REE deposits are formed by weathering of carbonatites in tropical or forested warm environments (Richardson and Birkett 1996). Laterites are soils rich in iron and aluminum, which, if derived from carbonatites with REE-bearing minerals such as monazite, can concentrate the REE-bearing minerals. Significant deposits of laterite REE are found in Brazil and southern Africa.

Among pegmatites, a group of very-coarse-grained intrusive igneous rocks, the niobium-yttrium-fluorine family comprises a large number of subtypes formed in different geologic environments (Sinclair 1996). These subtypes are granitic in composition and are usually found peripheral to large granitic intrusions. In general, however, REE-bearing pegmatites are small and usually of economic interest only to mineral collectors.

The iron-oxide copper-gold type of deposit has been recognized as a distinct deposit type only since the discovery of the giant Olympic Dam deposit in South Australia in the 1980s. The Olympic Dam deposit is unusual in that it contains large amounts of rare earth elements and uranium (Oreskes and Einaudi 1990). An economic method for recovering rare earth elements from these deposits has not yet been found. Many other deposits of this type have been identified around the world, but information on their REE content is commonly lacking. Trace amounts of rare earth elements have also been identified in similar magnetite-apatite replacement deposits such as Mineville, New York (McKeown and Klemic 1956).

Karst bauxites, aluminum-rich soils that accumulate in cavernous limestone (underlying karst topography) in Montenegro and elsewhere, are enriched in rare earth elements, but the resulting concentrations are not of economic interest at present (Maksimović and Pantó 1996).

Some marine phosphate deposits can contain as much as 0.1% REE oxides (Altschuler et al. 1966). As a result, recovery of rare earth elements as a by-product of phosphate fertilizer manufacture has been investigated, but no economic process has been found.

> The ores of rare earth elements are mineralogically and chemically complex and commonly radioactive.

The metallurgical recovery of rare earth elements can be better understood given some background on how other metals are extracted. For example, in many base and precious metal deposits, the metals extracted are highly concentrated in a single mineral phase, such as copper in chalcopyrite ($CuFeS_2$) or zinc in sphalerite (ZnS). Separation of a single mineral phase from rock is a relatively easy task. The final product is a concentrate typically sent to a smelter for final extraction and refining of the metals. Zinc, for example, is almost entirely derived from the mineral sphalerite. Thus, the global zinc smelting and refining industry has developed highly specialized methods for extracting zinc from sphalerite. Thus, production of zinc has a pronounced cost advantage over some other metals because a single standard technology is used, and the development of a new zinc mine is a largely conventional process.

Current mineral-processing practice is capable of sequential separation of multiple mineral phases, but this is not always cost-effective. When elements of interest are found in two or more mineral phases, each requiring a different extraction technology, mineral processing is relatively costly. Many REE deposits contain two or more REE-bearing phases. Therefore, REE deposits in which the rare earth elements are largely concentrated in a single mineral phase have a competitive advantage. To date, REE production has largely come from single-mineral-phase deposits, such as Bayan Obo (bastnäsite), Mountain Pass (bastnäsite), and heavy-mineral placers (monazite).

REE-bearing minerals, once separated, contain as many as 14 individual rare earth elements (lanthanides and yttrium) that must be further separated and refined. The complexity of extracting and refining rare earth elements is illustrated by a metallurgical flow sheet for the Mountain Pass mine in California (Fig. 2). Unlike metal sulfides, which are chemically simple compounds, REE-bearing minerals are quite complex (Table 3). Base-metal sulfide ores, such as sphalerite (ZnS), are typically smelted to burn off sulfur and separate impurities from the molten metal. The resulting metal is further refined to near purity by electrolysis. Rare earth elements, on the other hand, are typically extracted and refined through dozens of chemical processes to separate the different rare earth elements and remove impurities.

The principal deleterious impurity in REE-bearing minerals is thorium, which imparts an unwanted radioactivity to the ores. Because radioactive materials are difficult to mine and handle safely, they are heavily regulated. When a radioactive waste product is produced, special disposal methods must be used. The cost of handling and disposing of radioactive material is a serious impediment to the economic extraction of the more radioactive REE-rich minerals, in particular monazite, which typically contains considerable amounts of thorium. In fact, imposition of tighter regulations on the use of radioactive

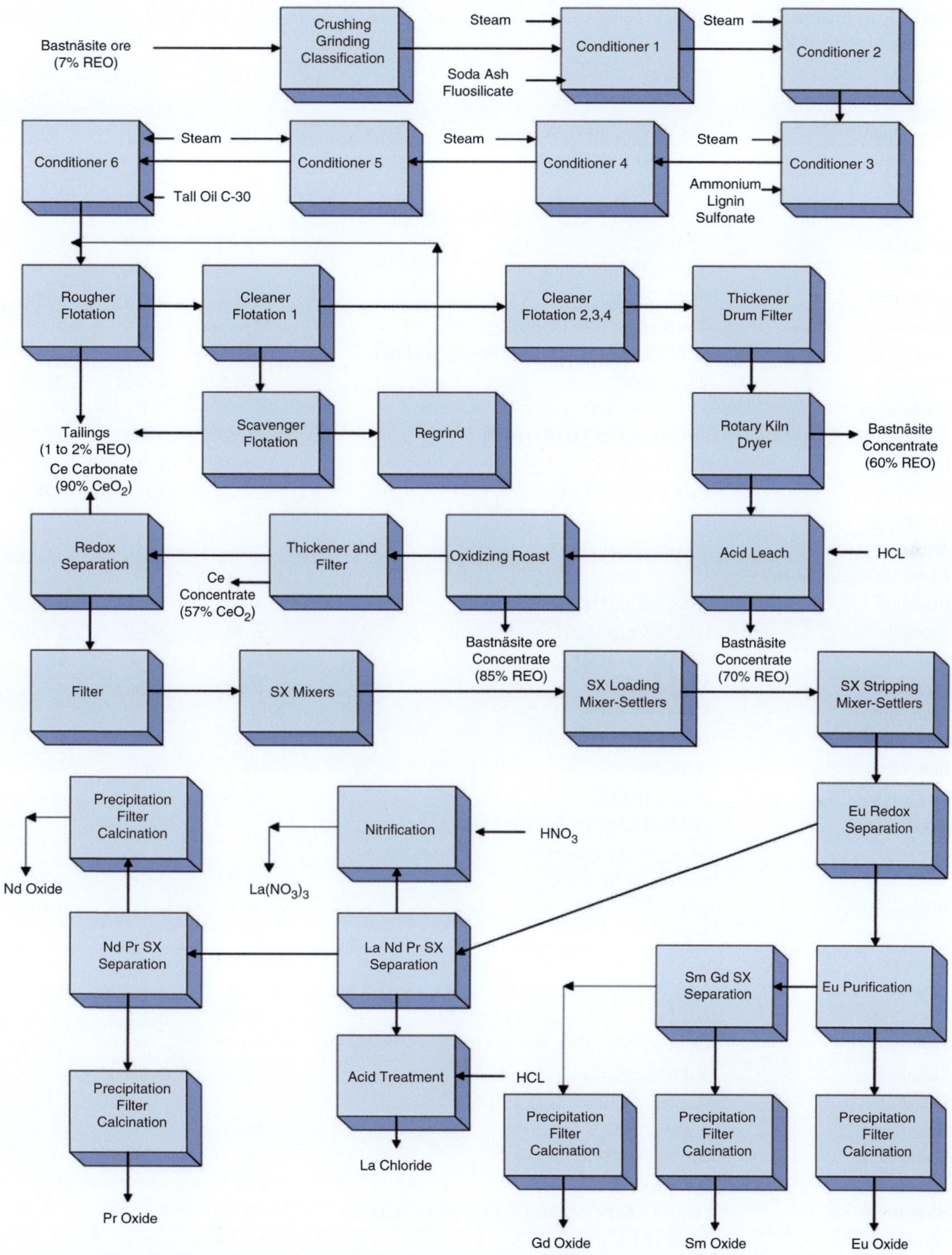

Fig. 2 Rare earth elements mineral-processing flow sheet for the Mountain Pass mine, California, about 1995. From one type of ore, no less than 12 rare earth elements products were obtained. *REO* rare earth oxides, Ce cerium, Eu europium, Gd gadolinium, La lanthanum, Nd neodymium, Pr praseodymium, Sm samarium, HCl hydrochloric acid (Gupta and Krishnamurthy 2005; Castor and Hedrick 2006)

Table 3 Predominant rare earth elements–bearing minerals identified in commercial and potentially commercial deposits of rare earth elements

Mineral	Formula	REO%	ThO_2%	UO_2%
Oxides and hydroxides				
Aeschynite	$(Ce,Th,Ca...)[(Ti,Nb,Ta)_2O_6]$			
Brannerite	$(U,Ca,Y,Ce)(Ti,Fe)_2O_6$	4	4	63
Cerianite	$(Ce^{4+},Th)O_2$	82	5	–
Euxenite	$(Y,Er,Ce,U,Pb,Ca)(Nb,Ta,Ti)_2(O,OH)_6$			
Fergusonite	$YNbO_4$	42–52	0–0.9	1–2.5
Loparite	$(Ce,Na,Ca)(Ti,Nb)O_3$	32–34	0.8	–
Perovskite	$(Ca,REE)TiO_3$	≤37	0–2	≤0.05
Pyrochlore	$(Ca,Na,REE)_2Nb_2O_6(OH,F)$	2.6	0.2	0–10
Samarskite	$(Y,Er,Fe,Mn,Ca,U,Th,Zr)(Nb,Ta)_2(O,OH)_6$			
Uraninite	$(U,Th,Ce)O_2$	0.9–5	0.2–14	70–91
Carbonates				
Ancylite	$Sr(Ce,La)(CO_3)_2(OH)\cdot(H_2O)$	46–53	0–0.4	0.1
Bastnäsite	$(Ce, La,Y)CO_3F$	70–74	0–0.7	–
Parisite	$Ca(Ce,La)_2(CO_3)_3F_2$	59	0–0.5	0–0.3
Synchysite	$Ca(Ce,Nd,Y,La)(CO_3)_2F$	51	–	–
Tengerite	$Y_2(CO_3)_3\cdot n(H_2O)$			
Phosphates and fluorides				
Britholite	$(Na,Ce,Ca)_5(OH)[(P,Si)O_4]_3$	33–61	0.5–21	0.2–1.5
Brockite	$(Ca,Th,Ce)(PO_4)\cdot H_2O$	7–24	24–45	3
Cheralite	$(Ca,Ce,Th)(P,Si)O_4$	27–43	28–32	4
Churchite	$YPO_4\cdot H_2O$	50–53	–	–
Crandallite	$CaAl_3(PO_4)_2(OH)_5\cdot H_2O$			
Florencite	$(La,Ce)Al_3(PO_4)_2(OH)_6$	18–32	1.4	–
Fluocerite	$(La,Ce)F$	83	1.6	–
Gagarinite	$NaCaY(F,Cl)_6$	55–57	–	–
Gorceixite	$(Ba,REE)Al_3[(PO_4)_2(OH)_5]\cdot H_2O$			
Goyazite	$SrAl_3(PO_4)_2(OH)_5\cdot H_2O$			
Monazite	$(Ce,La,Th,Nd,Y)PO_4$	35–71	0–20	0–16
Rhabdophane	$(Ce,La)PO_4\cdot H_2O$	58–69	0.7	0.4
Xenotime	YPO_4	52–57	0.4	0–5
Silicates				
Allanite	$Ca(Ce,La,Y,Ca)Al_2(Fe^{2+},Fe^{3+})(SiO_4)(Si_2O_7)O(OH)$	3–51	0–5	0–3
Chevkinite	$(Ca,Ce,Th)_4(Fe^{2+},Mg)_2(Ti,Fe^{3+})_3Si_4O_{22}$	40–45	0.7–0.8	–
Eudialyte	$(Na,Ca,REE)_5(Fe^{2+}Mn)(Zr,Ti)[(Si_3O_9)_2](OH,Cl)$	0.4–7	–	≤0.09
Gadolinite	$Y_2Fe^{2+}Be_2Si_2O_{10}$	51–55	0–0.9	–
Gerenite	$(Ca,Na)_2(Y,REE)_3Si_6O_{18}\cdot 2H_2O$			
Iimorite	$Y_2(SiO_4)(CO_3)$	69	–	–
Kainosite	$Ca_2(Ce,Y)_2(SiO_4)_3CO_3\cdot H_2O$	38	0.03	–
Sphene	$(Ca,REE)TiSiO_5$	0–4.5	–	0.06
Steenstrupine	$Na_{14}Ce_6Mn_2Fe_2(Zr,Th)(Si_6O_{18})_2(PO_4)_7\cdot 3H_2O$	30–31	2	–
Thalenite	$Y_2[Si_2O_7]$	63–64	–	–
Thorite	$(Th,U)SiO_4$	≤3	72–82	8–16
Zircon	$(Zr,REE)SiO_4$	0–10.5	0–2	0–5

Typical rare earth element (REO), thorium oxide, and uranium oxide contents of minerals reported in weight percent. Mineral names and formulas from Back and Mandarino (2008)

minerals drove many sources of monazite out of the REE market during the 1980s.

The complex metallurgy of rare earth elements is compounded by the fact that no two REE ores are truly alike. As a result, there is no standard process for extracting the REE-bearing minerals and refining them into marketable rare earth compounds. To develop a new rare earth elements mine, the ores must be extensively tested by using a variety of known extraction methods and a unique sequence of optimized processing steps. Compared with a new zinc mine, process development for rare earth elements requires substantially more time and money.

Mineralogy of United States Deposits

The main REE-bearing minerals found in the United States are bastnäsite, euxenite, xenotime, monazite, and allanite (Castor 2008). Samarskite, aeschynite, fergusonite, parisite, synchysite, tengerite, ancylite, florencite, britholite, kainosite, and thalenite have also been identified in United States deposits (Table 3). Euxenite [(Y,Ca,Ce,U,Th)(Nb,Ta,Ti)$_2$O$_6$] is an oxide mineral that forms a series with the mineral polycrase [(Y,Ca,Ce,U,Th)(Ti,Nb,Ta)$_2$O$_6$]. Other REE oxide (REO) minerals, such as fergusonite, aeschynite, and samarskite, have similar properties, making identification in hand sample difficult. Euxenite is black with a tabular to prismatic habit, making it indistinguishable from rutile, a common oxide mineral, when found in massive form. However, euxenite does not have any cleavage planes and, unlike rutile, has a conchoidal fracture. Furthermore, because euxenite is ordinarily found in granite pegmatites, it is commonly associated with quartz, feldspars, columbite (now called ferrocolumbite), tantalite (now called ferrotantalite or manganocolumbite), and monazite.

Bastnäsite (also spelled bastnasite or bastnaesite) is a rare REE-bearing carbonate mineral [(Ce,La,Y)CO$_3$F] that forms a series with the mineral hydroxyl-bastnäsite [(Ce,La)CO$_3$(OH,F)]. Bastnäsite can be pale white, tan, gray, brown, yellow, or pink, with a pearly, vitreous, or greasy to dull luster. Bastnäsite usually forms small rounded hexagonal or short prismatic crystals, though it can also form rosettes and spheres. Both massive and granular varieties have been observed. Bastnäsite is closely related to the mineral parisite [Ca(Ce,La)$_2$(CO$_3$)$_3$F$_2$] and has been known to replace crystals of allanite.

Xenotime is an yttrium-bearing phosphate mineral (YPO_4) and can be yellowish brown to reddish brown with a vitreous to resinous luster. Less common colors include gray, salmon pink, and green. Xenotime is usually an accessory mineral in acidic and alkaline rocks. Although it has been observed in mica schists and quartz-rich gneisses, it may also be a detrital mineral. Xenotime can easily be confused with zircon because of similarities in crystal habit and overall appearance.

Monazite is a REE- and thorium-bearing phosphate mineral [(Ce,La,Nd)PO$_4$] and typically contains 60–62% total rare earth oxides. Monazite's resistance to chemical weathering and its high specific gravity account for its association with other resistant heavy minerals such as ilmenite, magnetite, rutile, and zircon. Because monazite is radioactive, grains have often lost their crystalline structure owing to radioactive decay. Crystals of monazite are yellow to brown or orange brown with a vitreous and resinous or adamantine luster. Monazite grains are usually equant to prismatic with wedge-shaped terminations. Both granular and massive forms exist.

Allanite [CaCeFe^{2+}Al$_2$(Si$_2$O$_7$)(SiO$_4$)O(OH)], which belongs to the epidote mineral group, is one of the more common REE-bearing minerals in igneous rocks but is rarely concentrated enough to form an ore of REE. Allanite grains are tabular and usually black, though dark-brown to brownish-violet varieties also occur. Allanite has a conchoidal fracture and is commonly metamict because of the radioactive decay of thorium. The presence of a halo or dark ring inside the mineral grain is also an effect of its radioactivity. Most commonly, allanite is found as an accessory mineral in igneous rocks, such as granites, syenites, diorites, and associated pegmatites.

The relative abundance of rare earth elements within and among deposits is highly variable, but light rare earth elements are typically more abundant than heavy rare earth elements.

The relative proportion of the different rare earth elements in an ore body is quite variable (Table 4). The chief differences can be seen in the relative proportion of light to heavy rare earth elements. REE ore bodies are typically somewhat enriched in the light REE—lanthanum to europium—compared with average crustal abundances. In comparison, most rare earth ores are notably depleted in the heavy REE—gadolinium to lutetium. A minority of deposits are relatively enriched in heavy REE, most commonly those that contain xenotime as the principal REE mineral.

Given that each individual rare earth element has its own particular uses and market (Table 5), the proportions of the various rare earth elements in a deposit are unlikely to parallel those of demand for rare earth elements. For instance, the most abundant rare earth element, cerium, is available in quantities that exceed demand for traditional uses (Heymann 2010). Most REE deposits currently (2010) considered for development are enriched in light REE and would likely flood the market for cerium if put into production. By contrast, heavy REE are in short supply with limited reserves. Certain rare earth elements, such as lutetium, presently have no market and are not worth recovering at this time.

Rare earth elements are typically obtained as a by-product or coproduct of mining other mineral commodities.

When the economic viability of a mining project is assessed, the potential mineral products are divided into principal products and by-products. The principal product, for example, zinc in a zinc mine, contributes most to the value of the minerals produced. Generally, returns from the principal product are sufficient to pay the costs of mining and processing. All other products are referred to as by-products, whose returns typically bolster the overall profitability of a mine. Where two or more products of essential value are obtained, they are called coproducts. A salient feature of REE mining is that REE-rich minerals may be by-products or coproducts of mining other mineral commodities.

Mine production decisions are driven by demand for principal products, not for by-products. Thus, production of by-product REE will vary subject to changes in demand for principal products and will be relatively unresponsive to demand for REE. China currently (2011) accounts for about 96% of global REE production (Table 6). Of a total production of 120,000 metric tons, about 55,000 metric tons was produced as a coproduct of the Bayan Obo iron mine. This fact means that at least 44% of world REE production is a co product. Of the remaining Chinese production, about 25,000 metric tons is produced in southern China as a primary product from ion-adsorption deposits. The status of remaining Chinese production is unclear. The balance of global REE production is as a by-product. Conceivably, as much as 90% of global REE production is as a by-product or coproduct.

Current Sources and Domestic Reserves

The United States currently imports all of its REE raw materials from foreign sources, principally China (US Geological Survey 2010). This has not always been the case. Prior to 1998, when production from the Mountain Pass mine in California was curtailed, the United States produced most of the light REE consumed domestically and by free-market countries (Castor and Hedrick 2006). Heavy REE were obtained from imported monazite concentrates. That changed in the 1980s after China became the dominant global supplier of light and heavy REE (Papp et al 2008). In 2002, the Mountain Pass mine in California, the sole domestic producer of REE minerals, shut down. Although the mine has continued to produce REE materials from stockpiled raw materials, no new REE ores were mined until the mine reopened toward the end of 2010. Since then, the United States has obtained all of its REE raw materials from imports, principally from China. China accounts for 95% of global REE production despite having only 36% of identified world reserves (Table 7).

Table 4 Distribution of rare earth elements in selected rare earth elements deposits

REE	Average upper crustal abundance (%)	Bastnäsite, Mountain Pass, USA (%)	Bastnäsite, Bayan Obo, China (%)	Monazite, Green Cove Spring, USA (%)	Xenotime, Lahat, Malaysia (%)	High-Y REE laterite, Longnan, China (%)	Low-Y REE laterite, Xunwu, China (%)	Bulk ore, Bear Lodge, USA (%)	Bulk ore, Strange Lake, Canada (%)
Lanthanum	19.3	33.8	23.0	17.5	1.2	1.8	43.4	30.4	4.6
Cerium	39.2	49.6	50.0	43.7	3.1	0.4	2.4	45.5	12.0
Praseodymium	3.8	4.1	6.2	5.0	0.5	0.7	9.0	4.7	1.4
Neodymium	15.5	11.2	18.5	17.5	1.6	3.0	31.7	15.8	4.3
Samarium	2.8	0.9	0.8	4.9	1.1	2.8	3.9	1.8	2.1
Europium	0.6	0.1	0.2	0.2	Trace	0.1	0.5	0.4	0.2
Gadolinium	1.7	0.2	0.7	6.6	3.5	6.9	3.0	0.7	2.5
Terbium	0.3	0.0	0.1	0.3	0.9	1.3	tr	0.1	0.3
Dysprosium	1.7	0.0	0.1	0.9	8.3	6.7	tr	0.2	8.2
Holmium	0.4	0.0	Trace	0.1	2.0	1.6	tr	0.0	1.7
Erbium	1.3	0.0	Trace	tr	6.4	4.9	tr	0.0	4.9
Thulium	0.2	0.0	Trace	tr	1.1	0.7	tr	<0.01	0.7
Ytterbium	0.9	0.0	Trace	0.1	6.8	2.5	0.3	0.5	4.0
Lutetium	0.2	Trace	Trace	Trace	1.0	0.4	0.1	<0.01	0.4
Yttrium	12.3	0.1	Trace	2.5	61.0	65.0	8.0	<0.01	52.8
Data source	(1)	(2)	(3)	(3)	(3)	(3)	(3)	(2)	(2)

Sources: (1) Wedepohl (1995); (2) Castor (2008); and (3) Hedrick (2004)

Table 5 Distribution of rare earth elements usage by application

Application	La	Ce	Pr	Nd	Sm	Eu	Gd	Tb	Dy	Y	Other
Magnets	–	–	23.4	69.4	–	–	2	0.2	5	–	–
Battery alloys	50	33.4	3.3	10	3.3	–	–	–	–	–	–
Metal alloys	26	52	5.5	16.5	–	–	–	–	–	–	–
Auto catalysts	5	90	2	3	–	–	–	–	–	–	–
Petroleum refining	90	10	–	–	–	–	–	–	–	–	–
Polishing compounds	31.5	65	3.5	–	–	–	–	–	–	–	–
Glass additives	24	66	1	3	–	–	–	–	–	2	4
Phosphors	8.5	11	–	–	–	4.9	1.8	4.6	–	69.2	–
Ceramics	17	12	6	12	–	–	–	–	–	53	–
Other	19	39	4	15	2	–	1	–	–	19	–

Source: Lynas Corporation (2010)
This table illustrates how each individual rare earth element has its own applications and market

Table 6 Current production of rare earth elements mines

Country	Mine	2009 Output metric tons TREO	Primary product	By-products
Brazil	Buena Norte	650	Ilmenite concentrate	Monazite concentrate
China	Bayan Obo	55,000	Iron ore	Bastnäsite concentrate
	Sichuan[a]	10,000	Bastnäsite concentrate	–
	South China[a]	45,000	Rare earth elements	–
India	Heavy-mineral sands	2,700	Ilmenite concentrate	Monazite concentrate
Malaysia	Ipoh sand plant	380	Cassiterite concentrate	Xenotime concentrate
Russia	Lovozero	1,900	Loparite concentrate	REE chloride

Source: USGS Mineral Commodity Summaries (US Geological Survey 2010) except for Lovozero mine (Solikamsk Magnesium Works 2010)
TREO total rare earth elements oxide
[a]Many small producers and a few medium-large producers. The Chinese rare earth elements mining industry is currently undergoing government-directed rationalization to reduce the number of producers

Table 7 World production and reserves of rare earth elements (REE) minerals in 2009, and metric tons of contained total REE oxides (total metric tons of REO)

Country	2009 Production metric tons total REO	2009 Production% share	2009 Reserves metric t TREO	2009 Reserves% share	Reference
Australia	0	0	5,400,000	5	(1)
Brazil	650	0.5	48,000	0.05	(1)
China	120,000	95	36,000,000	36	(1)
India	2,700	2	3,100,000	3	(1)
Malaysia	380	0.3	30,000	0.03	(1)
Russia	1,900	2	19,000,000	19	(2)
United States	0	0	13,000,000	13	(1)
Other	0	0	22,000,000	22	(1)
Total	126,230	–	99,000,000	–	–

Sources: (1) US Geological Survey (2010) and (2) Solikamsk Magnesium Works (2010)
China produces 95% of world rare earth elements, although it has only 36% of REE reserves

Table 8 Measures of concentration for selected world metal mining industries

Mineral commodity	Import reliance (%)	US imports CR2	CR3	NHI	RI	World production CR2	CR3	NHI	RI
Antimony	86	90	98	0.42	1.9	91	94	0.77	2.3
Bauxite	100	50	64	0.19	4.6	46	58	0.16	2.8
Bismuth	95	62	80	0.26	0.8	75	90	0.29	2.3
Cobalt	78	43	56	0.13	1.7	52	63	0.20	1.4
Copper	37	75	88	0.32	1.5	44	51	0.16	2.4
Gallium	99	57	73	0.21	1.3	51	65	0.19	1.9
Indium	100	72	81	0.31	1.3	68	76	0.36	1.4
Manganese	100	54	65	0.21	2.9	46	64	0.17	2.8
Nickel	17	59	68	0.23	1.0	32	46	0.10	2.6
Niobium	100	96	97	0.79	2.7	100	100	0.90	2.9
Platinum	94	50	65	0.17	1.5	91	94	0.63	2.9
REE	100	94	96	0.83	1.9	99	100	0.94	2.0
Rhenium	86	95	98	0.81	1.8	59	68	0.26	2.3
Tantalum	100	35	50	0.13	1.6	75	85	0.35	2.0
Tin	79	69	79	0.31	3.2	74	91	0.30	3.3
Titanium	64	85	94	0.39	3.3	55	77	0.23	2.2
Tungsten	70	50	69	0.19	3.3	81	86	0.57	2.3
Vanadium	100	66	74	0.35	1.5	72	97	0.33	1.5
Yttrium	100	96	99	0.78	1.8	100	100	0.98	2.0
Zinc	58	67	82	0.19	1.1	52	66	0.19	1.9

See text for an explanation of indices. The higher the indices, the more concentrated are mineral production and United States imports. Note that CR2 and CR3 are rounded to the nearest percent resulting in some cases in a slight discrepancy between the concentration ratios and the normalized Herfindahl index. *RI* is a country risk index. See text for explanation. Data are for 2007, the latest year for which complete information is available from the US Geological Survey (2010). *CR2* and *CR3* are two-county and three-country concentration ratios, respectively. *NHI* is the normalized Herfindahl index

Concentration of Supply

The high concentration of production of REE in one country is not unusual for a minor metal commodity. For example, a single mine in the United States supplies 86% of world demand for beryllium, and two mines in Brazil account for 92% of world niobium production (US Geological Survey 2010). Such concentration of supply raises concern in regard to price manipulation and issues related to reliability of supply. Given an equal risk of a natural disaster, industrial accident, labor strike, political strife, or anything else that might interrupt production, a single source of supply is inherently more risky than multiple sources of supply. Even though these various risks are not equal among countries, concentration of supply is a key indicator of mineral-supply risk.

Table 8 compares the supply situation of REE with other internationally traded minerals using several measures of concentration. These measures are used by economists to study market concentration and by regulators for antitrust purposes. In Table 8, concentration ratios, abbreviated CR2 and CR3, measure the total percent share in United States imports and world production of the top two or top three supplier countries, respectively. A high percentage, such as the CR2 of 94% and CR3 of 96% shown for REE (excluding yttrium and scandium), indicates that imports and world production are principally derived from one or two countries. A third measure is the Herfindahl index (Stigler 1983), which was originally developed to measure the degree of competition in an industry. It is calculated according to the equation

$$HI = \sum_{i=1}^{n} s_i^2$$

where s_i is the share in global production or United States imports by country i with n countries. The larger this index, the more concentrated are world production and United States imports by country. The Herfindahl index can be normalized

$$NHI = \frac{HI - \frac{1}{n}}{1 - \frac{1}{n}}$$

such that the index ranges from 0 to 1.0, which facilitates comparison between different mineral commodities. A normalized Herfindahl index of 1.0 indicates concentration in a single country; an index of 0 indicates that all countries have exactly the same share in United States imports or world production.

As shown in Table 8, all three of these indices place REE (including yttrium) at the top of all mineral commodities in terms of concentration of United States imports and world production. Antimony and niobium, which are mostly produced in China and Brazil, respectively, have very similar concentration indices. Rhenium is an example of a mineral commodity that the United States largely imports from a single country, Chile, but whose global production is not particularly concentrated.

On the basis of these data, it is no exaggeration to say that China dominates the world REE industry. This dominance is attributable to China's large, high-quality resources of REE, low labor costs, and lack of environmental regulation (Hurst 2010). Referring back to Table 7, China has only about a third of global REE resources; hence, a lower cost of production is a reasonable explanation for China's position as the world's dominant REE producer. Papp et al. (2008) show that REE prices dropped dramatically from 1997 to 2008, consistent with the introduction of significant amounts of lower-priced Chinese REE.

Risk of Supply Interruption

Assessing our nation's vulnerability to mineral-supply disruptions is a classic exercise in risk analysis. The analysis has two components: the nature and probability of threats, and assessment of potential impact. Quantitative measurement of these components would be useful in ranking the relative security of supply of the various mineral commodities used and imported by the United States. A first step was made by a special committee of the National Research Council when it recommended the criticality matrix as a tool for assessing mineral-supply risk (National Research Council 2008). The criticality matrix is a plot that subjectively contrasts supply risk on one axis and the effect of supply restriction on the other. The authors of the study ranked various imported metals, including REE, on a scale of 1–4 (low to high) as shown in Fig. 3. Rare earth elements were ranked 4 (high) for supply risk and 3 (moderately high) for effect of supply restriction.

An analysis of the effect of supply restrictions requires a level of economic analysis that is outside the traditional responsibilities of the US Geological Survey and beyond the scope of this report. Long (2009) proposed combining the quantitative measures of concentration discussed above with measures of country risk to obtain a relative ranking of minerals by supply risk. A similar approach was independently adopted by the Raw Materials Supply Group of the European Union in a recent study of European mineral security (Raw Materials Supply Group 2010). The European Union study does include a rough measure of the economic effect of a mineral-supply disruption.

There are many measures of country risk from which to choose. Long (2009) used the Country Risk Classification that is published annually by the Organization for Economic Co-operation and Development (Organization for Economic Co-operation and Development 2008). This classification is a measure of a country's credit risk or likelihood that a country will service its external

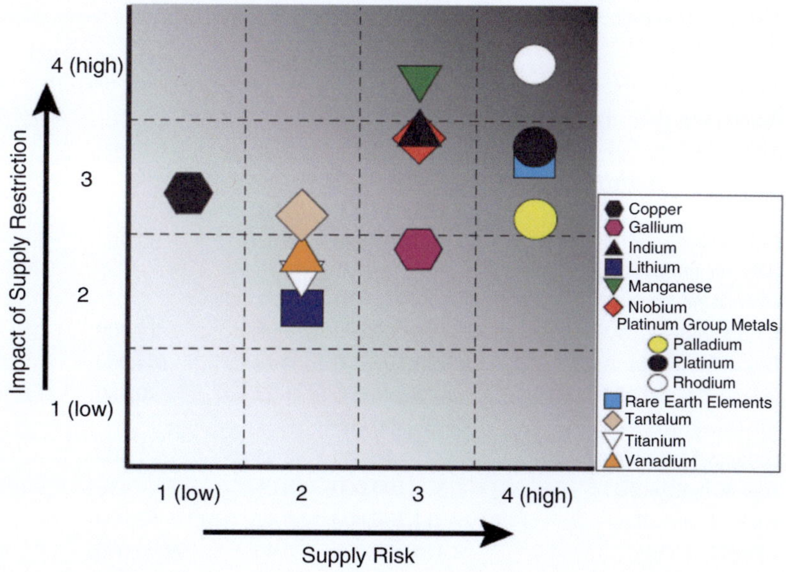

Fig. 3 Criticality matrix for selected imported metals (National Research Council 2008)

debt. Countries are subjectively ranked on a scale from 0 to 7, where 0 is the lowest degree of risk and 7 the highest. An aggregate country risk index for a commodity is obtained as the sum of individual country risk indices weighted by share in United States imports or world production (Table 8). This aggregate country risk index likewise falls on the scale of 0–7. Table 8 illustrates how these indices and ratios can be used. Comparison of concentration indices for United States imports and for world production shows whether imports are more concentrated than global production, thus indicating opportunities for further diversification of supply. A high import concentration index with a low country risk index suggests that imports are obtained from stable trading partners such as Canada and Australia. High indices across the board are cause for greatest concern and indicate those commodities that are of greatest risk.

The European Union study (Raw Materials Supply Group 2010) used the World Bank's World Governance Indicators as a measure of political risk. The World Bank estimates six governance indicators: voice and accountability, political stability, government effectiveness, regulatory quality, rule of law, and control of corruption (World Bank 2010). The European Union study unfortunately does not specify which indicator was used or, if all indicators were used, how they were combined. The US Geological Survey has identified other indicators of country risk, such as the Economic Freedom Score (Heritage Foundation 2010), the Corruption Perceptions Index (Transparency International 2009), and rankings of countries for mining investment (Behre Dolbear 2010; McMahon and Cervantes 2010).

Aside from an indicator of country risk, the European Union study also used a subjective ranking of the degree to which other minerals can substitute for the mineral in question, measures of recycling rates and environmental policy risk, and a rescaled Herfindahl index to measure mineral-supply concentration. Using these indicators, some 14 metals and minerals were short-listed as critical raw materials for European Union member nations, "critical" in this case signifying a high degree of both supply risk and economic importance. The short-listed metals and minerals are antimony, beryllium, cobalt, fluorspar, gallium, germanium, graphite, indium, magnesium, niobium, platinum group metals, REE, tantalum, and tungsten (Raw Materials Supply Group 2010). Long (2009) listed 15 metals and minerals: antimony, barite, chromite, cobalt, fluorspar, gallium, graphite, indium, niobium, platinum group metals, REE,

Table 9 Domestic reserves and resources of rare earth elements, excluding heavy-mineral placer and phosphate deposits

Deposit	Tonnage (metric tons)	Grade (TREO%)	Contained TREO (metric tons)	Reference
Reserves (proven and probable)				
Mountain Pass, CA	13,588,000	8.24	1,120,000	Molycorp, Inc. (2010)
Resources (inferred)				
Bear Lodge, WY	25,038,000	2.64	659,500	Rare Element Resources (2010)
Bokan Mountain (Dotson and I&L Zones), AK	6,702,000	0.580	38,900	McKenzie (2011)
Unclassified resources				
Bald Mountain, WY	18,000,000	0.08	14,400	Osterwald et al. (1966)
Bokan Mountain, AK	34,100,000	0.48	164,000	Keyser and Kennedy (2007)
Diamond Creek, ID	5,800,000	1.22	70,800	Staatz et al. (1979)
Elk Creek, NE	39,400,000			Molycorp, Inc. (1986)
Gallinas Mountains, NM	46,000	2.95	1,400	Jackson and Christiansen (1993)
Hall Mountain, ID	100,000	0.05	50	Staatz et al. (1979)
Hick's Dome, IL	14,700,000	0.42	62,000	Jackson and Christiansen (1993)
Iron Hill, CO	2,424,000,000	0.40	9,696,000	Staatz et al. (1979)
Lemhi Pass, ID	500,000	0.33	1,650	Staatz et al. (1979)
Mineville, NY	9,000,000	0.9	80,000	McKeown and Klemic (1956)
Music Valley, CA	50,000	8.6	4,300	Jackson and Christiansen (1993)
Pajarito, NM	2,400,000	0.18	4,000	Jackson and Christiansen (1993)
Pea Ridge, MO	600,000	12	72,000	Grauch et al. (2010)
Scrub Oaks, NJ	10,000,000	0.38	38,000	Klemic et al. (1959)
Wet Mountains, CO	13,957,000	0.42	59,000	Jackson and Christiansen (1993)

Reserves are proven and probable reserves following definitions and standards of the Securities and Exchange Commission. Resources are divided into inferred resources and unclassified resources. Inferred resources are classified according to the standards of Canadian National Instrument 43–101. The unclassified resources are based on little or no drilling. For data on resources in heavy-mineral placer and phosphate deposits, which are not of economic interest, see Jackson and Christiansen (1993)
TREO total REE oxides

rhenium, tantalum, titanium, and tungsten as those minerals with the greatest supply risk to the United States economy. The two lists are very similar. The differences are beryllium, which is imported by Europe from the United States; barite, critical to the US oil and gas industry; and chromite, rhenium, and titanium, whose supply was rated as less risky in the European Union report.

In both the USGS and European Union studies of mineral-supply risk, REE rank highest as mineral raw materials of critical concern, given uncertain future supplies and their importance to advanced industrial economies. Neither of these studies addressed measures to mitigate these risks, but each did recommend further study, including examination of mineral policy options. There are geologic factors, however, that should be considered in future studies of REE supply, such as the extent and quality of domestic REE resources, undeveloped resources in other low-risk countries, and the time it takes to develop these resources into producing mines.

Domestic and World Resources

The principal domestic resources of REE are summarized in Table 9. It is important to recognize that resource estimates are of differing accuracy and reliability, depending on the degree of exploration undertaken to date. Many of the estimates in Table 9 are obtained by inference from surface exposures of mineralization, a small number of samples, and inferences or assumptions about how deep mineralization extends. Some deposits have been explored at depth by core drilling. A very few have been drilled on a narrowly spaced grid sufficient for an estimate of how much mineralized material may be economic to mine. The potential economic viability of any of these resources can be reliably assessed only with sufficient drilling, pilot plant–scale metallurgical testing, and definitive economic analysis. Only one domestic deposit, Mountain Pass, California, meets those criteria and can be reported to contain a sizable reserve of REE-bearing ore.

Table 10 reports reserves and resources in REE deposits worldwide, divided into three categories. The first category comprises deposits sufficiently explored to estimate a mine plan resource. Although a mine has been designed or already exists for each of these deposits, they have not been demonstrated to be economically viable by means of a definitive feasibility study; hence, they are classed as resources. The second category comprises measured, indicated, and inferred resources for well-explored deposits that have not yet been subject to a feasibility study that includes a mine design. The third category, unclassified resources, is a mixed bag of known resources that are unlikely to be exploited, such as Pilanesberg, South Africa, which is now within a national park, and the Olympic Dam mine, Australia, where extensive study has found that REE are not economic to recover even as a by-product. Other deposits in this category have been little explored, and the resources are inferred from surface exposures and limited sampling. No reliable data are available for mines and deposits in China, Russia, and North Korea.

The first two categories of resources are the only short- and medium-term sources of additional REE that might contribute to the global supply. These categories will likely be augmented through further exploration at existing mines and development projects. It is possible that long-term supply can be met through exploration of known deposits that have had little or no drilling and by the discovery of new deposits. The projects listed in the first two of the categories shown in Tables 9 and 10 put an upper limit on a near-term potential for production of REE mineral supplies. This limit can be put at about 14 million tons of contained total REE oxides (TREO), with a country risk index of near zero because almost all of that production potential is in the United States, Australia, and Canada.

Long-term prospects for the discovery of new reserves and resources depend on sufficient exploration. About 150 projects worldwide that are prospecting and exploring for rare earth elements are known to the US Geological Survey (US Geological Survey 2010). Most of this activity began in the last 2 years, and it will be some time before results are known. Whether this exploration further diversifies REE mineral supplies depends on discovering deposits of adequate size and quality in the right countries. Probability of discovery requires a quantitative mineral resource assessment, which has never been done for REE minerals in any country. The last REE deposit discovered and developed into a mine in the United States was the Mountain Pass mine in California, discovered in 1949 and put into production in 1953. That was more than 50 years ago and is not indicative of the time required to discover and develop REE deposits in today's regulatory climate. During the past 50 years outside of China, there has been little REE exploration and almost no mine development; hence, we have little in the way of a REE exploration and development record to draw upon for assessing the future pace of discovery and development.

Developing Rare Earth Elements Resources

Rare earth elements resources are distributed between many mineral deposits, but only a proportion will be economic to develop and mine. By convention, that portion of resources that is economic to mine is classified as a "reserve." That a rare earth deposit contains reserves does not mean that it will be developed and mined—it means only that it is probably economic to do so. Among the many rare earth reserves available, mining companies will select the most profitable

Table 10 Reserves and resources of rare earth elements outside of the United States, excluding heavy-mineral placer deposits and phosphate deposits

Deposit	Tonnage (metric tons)	Grade (TREO%)	Contained TREO (metric tons)	References
Measured in-pit resources				
Brockman, Australia	4,290,000	0.2	8,600	Chalmers (1990)
Mount Weld, Australia	2,100,000	15.5	326,000	Lynas Corporation (2010)
Thor Lake (Nechalacho), Canada	12,010,000	1.70	204,000	Paul and Stubens (2009)
Steenkampskraal, South Africa	249,500	11.80	29,500	Great Western Minerals Group Ltd. (2009)
Resources (measured, indicated, and inferred)				
Brockman, Australia	50,000,000	0.23	115,000	Chalmers (1990)
Cummins Range, Australia	4,170,000	1.72	72,000	Navigator Resources Ltd. (2009)
Dubbo, Australia	73,200,000	0.89	651,500	Alkane Resources (2010)
Mount Weld, Australia	17,490,000	8.2	1,434,000	Lynas Corporation (2010)
Narraburra, Australia	55,000,000	0.03	16,500	Capital Mining Ltd. (2009)
Nolans Bore, Australia	30,300,000	2.80	849,000	Arafura Resources Ltd. (2010)
Eco Ridge (Elliot Lake), Canada	47,433,000	0.141	66,900	Pele Mountain Resources (2011)
Hoidas Lake, Canada	2,847,000	2.00	57,000	Dunn (2009)
Strange Lake, Canada	137,639,000	0.97	1,335,000	Daigle and Maunula (2010)
Thor Lake (Nechelacho), Canada	311,700,000	1.37	4,270,000	Bakker et al. (2011)
Thor Lake (North T), Canada	1,136,000	0.71	8,000	Palmer and Broad (2007)
Zeus (Kipawa), Canada	39,060,000	0.28	109,400	Camus and Laferrière (2010)
Kvanefjeld, Greenland	619,000,000	1.07	6,600,000	Greenland Minerals and Energy Ltd. (2011)
Sarfartoq, Greenland	15,216,000	1.445	220,000	Druecker and Simpson (2011)
Kangankunde Hill, Malawi	2,530,000	4.24	107,000	Lynas Corporation Ltd. (2007)
Songwe, Malawi	1,781,000	1.6	28,500	Scott and Wells (2010)
Zandkopsdrift, South Africa	43,730,000	2.17	948,900	Venter et al. (2010)
Norra Kärr, Sweden	99,300,000	0.45	447,000	Reed (2011)
Unclassified resources				
John Galt, Australia	382,000	7.96	30,400	Northern Uranium Ltd. (2010)
Olympic Dam, Australia	>2,000,000,000	0.50	>10,000,000	Oreskes and Einaudi (1990)
Yangibana, Australia	3,500,000	1.70	59,500	Jackson and Christiansen (1993)
Araxá, Brazil	450,000,000	1.80	8,100,000	Filho et al. (2005)
Catalão I, Brazil	10,000,000	0.90	90,000	Hirano et al. (1990)
Pitinga, Brazil	164,000,000	0.15	246,000	Bastos Neto and Pereira (2009)
Poços de Caldas, Brazil			115,000	Wedow (1967)
Seis Lagos, Brazil	2,900,000,000	1.50	43,500,000	De Sousa (1996)
Tapira, Brazil	5,200,000	10.5	546,000	Hirano et al. (1990)
Kasagwe, Burundi	67,000	1.50	1,000	Jackson and Christiansen (1993)
Oka, Canada	210,000,000	0.127	267,000	Orris and Grauch (2002)
Mrima Hill, Kenya	50,800,000	0.59	300,000	Mossop and Carter (2010)
Ak-Tyuz, Kyrgyzstan	15,000,000	1.00	150,000	Malyukova et al. (2005)
Karajilga, Kyrgyzstan	957,000	0.70	6,700	Bogdetsky et al. (2001)
Kutessai II, Kyrgyzstan	20,228,000	0.22–0.3	<60,000	Stans Energy Corp. (2010)
Sarysai, Kyrgyzstan	7,000,000	0.20	14,000	Bogdetsky et al. (2001)
Mushgai Khudag, Mongolia	8,700,000	1.1	95,700	Tagaki pers. Comm. (2010)
Lovozero (Alluaiv), Russia	178,840,000	1.25	2,235,000	FODD (2010)

(continued)

Table 10 (continued)

Deposit	Tonnage (metric tons)	Grade (TREO%)	Contained TREO (metric tons)	References
Lovozero (Karnasurt), Russia	23,759,000	1.33	316,000	FODD (2010)
Lovozero (Kedykvypakhk), Russia	9,000	1.52	137	FODD (2010)
Lovozero (Partomchorr), Russia	877,400,000	0.20	1,755,000	FODD (2010)
Lovozero (Umbozero), Russia	180,469,000	0.95	1,714,000	FODD (2010)
Pilanesberg, South Africa	13,500,000	0.70	94,500	Lurie (1986)
Kizilcaören, Turkey	4,695,000	2.78	130,500	Morteani and Satir (1989)
Dong Pao, Vietnam	500,000,000	1.40	7,000,000	Kušnir (2000)
Mau Xe North, Vietnam	557,000,000	1.40	7,800,000	Kušnir (2000)

Heavy-mineral placers are mined for REE in only a few places, including India and Malaysia, and reserve information is unavailable. Reserves and resources are classified according to one of several national standards, including Canadian National Instrument 43–101, JORC, and SAMREC codes. Unclassified resources are based on little or no drilling. Reliable data on rare earth elements reserves and resources in China, North Korea, and Russia are not available. Where reserves and resources are given for the same deposit, the resources include the reserves. For data on other resources in heavy-mineral placers and in phosphate deposits, see Jackson and Christiansen (1993)
TREO total REE oxides

to develop, potentially leaving less profitable reserves undeveloped. Reserves may also be undeveloped because of adverse land use restrictions, civil strife, and a host of other political and social factors.

Developing a new mine requires a prolonged effort of prospecting, exploration, process development, feasibility studies, permitting, construction, and commissioning. These efforts are broadly sequential but commonly overlap. The time required to complete all steps is variable but appreciable, particularly compared with the time typically required by nonextractive industries. Studies of the time required to complete this process typically examine the interval between recognition of a potentially economic deposit and the commencement of commercial production. Peters (1966) examined the development history of a large number of mines and divided them into four classes. The first class requires a preproduction period of 2 years or less. This class of mine is characterized by simple ores of high unit value, the applicability of conventional mining and mineral-processing methods, the absence of need for much additional transportation or power infrastructure, an assured market, and the requirement for only short-term financing. A modern example in this class would be a small- to medium-sized open pit-heap leach gold deposit in Nevada. Gold has a high unit value and ready market. The pertinent mining and mineral processing technology is also simple and well known, with excellent local infrastructure, and regulatory authorities have much experience with this type of mining.

The other three categories are characterized by increasing complexity of ores, size of operations, infrastructure and financing needs, and lower unit values. The last category has a preproduction period of 7 years or more. A modern example would be a nickel laterite mine, the metallurgy of which is quite troublesome; many such nickel mines have required in excess of 10 years of process development plus delays because of market timing. Developing a new nickel laterite mine is best timed to open during a period of high nickel prices. A new REE mine would almost certainly fall into this last category for many of the same reasons—complex metallurgy and restricted opportunities for market entry.

The time to develop a mine in each category is likely to take longer today than it did when Peters (1966) did his study. Most modern mines are developed under a comprehensive regulatory environment where, in the United States at

Table 11 Time that was required to obtain permits, construct, and commission recent metal mines in the United States

Mine	Commodity	Permitting begun	Permitting completed	Production begun	Commercial operations begun	Litigation reported
Alta Mesa, TX	U	1999	2004	10/2005	1/2006	–
Arizona 1, AZ	U	mid-2007	2009	NYA	NYA	Yes
Ashdown, NV	Mo Au	2/2004	11/2006	12/2006	NYA	–
Buckhorn, WA	Au	1992	9/2006	10/2008	11/2008	Yes
Carlota, AZ	Cu	2/1992	6/2007	12/2008	1/2009	Yes
Eagle, MI	Ni Cu Co PGE	4/2004	1/2010	NYA	NYA	Yes
East Boulder, MT	PGE	1995	1998	6/2001	1/2002	–
Kensington, AK	Au	3/1988	6/2005	9/2010	NYA	Yes
Leeville, NV	Au	7/1997	8/2002	10/2006	4Q/2006	–
Lisbon Valley, UT	Cu	2/1996	7/2004	1Q/2006	NYA	Yes
Pend Oreille, WA	Zn	1992	9/2000	1/2004	8/2004	–
Phoenix, NV	Au	1/1999	1/2004	10/2006	4Q/2006	–
Pogo, AK	Au	12/1997	4/2004	2/2006	4/2007	–
Rock Creek, AK	Au	2003	8/2006	9/2008	NYA	Yes
Rossi (Storm), NV	Au	1990	3Q/2006	3/2007	12/2007	–
Safford, AZ	Cu	4/1998	7/2006	4Q/2007	2 H/2008	Yes
Turquoise Ridge, NV	Au	9/1995	5/2003	2004	NYA	

Data are from unpublished historical research by the lead author (Keith R. Long)
NYA, or Not Yet Achieved, indicates that production has not yet begun or that commercial operations were not achieved.
Yes, under the "Litigation Report" column, indicates significant permitting and development delays because of litigation by government agencies and nongovernmental organizations

least, environmental studies, due diligence studies by financing sources, permitting, public participation, and due process require substantial amounts of time. Further delays may be caused by public controversy and litigation against a proposed mine. These delays are illustrated in Table 11, which tallies the time it took to develop metal mines that opened in the United States since 2000. The time to obtain a permit has required as many as 17 years, and one mine, the Pogo, Alaska gold mine, was developed under an expedited permitting schedule that still took 7 years. For a small gold mine in Nevada, once permits were obtained, the time to construct and put a mine into operation took 1 month. For the Kensington, Alaska, gold mine, however, whose operating permits were contested in the courts, the process lasted 63 months. Ramp-up times for new mines took from 2 to 12 months; the longest was for Pogo, Alaska, which had unexpected metallurgical difficulties.

Developing a Rare Earth Elements Mine

The first step in developing a new REE mine is to locate a suitable REE deposit. Exploration for new deposits may be in either of two contexts: greenfield exploration to find new deposits in areas not previously mined and brownfield exploration in and adjacent to current and past mining activity. Greenfield exploration may target frontier areas with no previous exploration or may follow up on past exploration results. Brownfield exploration includes searching for extensions to known reserves and resources within or near a mine as well as new deposits in the vicinity of existing operations. The latter may extend the life of an existing mine or result in the complete redevelopment of a former mine.

Exploration is an uncertain process conducted with limited capital. An explorer's objective is to find a deposit of the targeted type, size, and quality with the least amount of expense. Thus, any particular location of merit is rarely exhaustively explored. Exploration will proceed on the basis of favorable indicators so long as objectives are met within budget. An exploration project will be curtailed if evidence is found that contraindicates the prospectivity of the target or if results are insufficient to justify further work. Exploration commonly runs in cycles, prompted by short- to medium-term increases in mineral prices. The low side of a price cycle may prompt little or no exploration activity. Any particular prospect may undergo many episodes of exploration by different parties during many decades; the exploration episodes are motivated by price cycles, new cost-saving technology, improved exploration concepts and methods, and the vagaries of land access and exploration management.

In the case of REE, very little exploration has ever been done, and the most intense period ever of direct exploration for REE is currently underway (2010). Many important past discoveries have been serendipitous—the world's largest REE mine, Bayan Obo, was first mined for iron ore. The "funny" steel produced from this ore was investigated and found to be contaminated with REE (Laznicka 2006). Mountain Pass, California, was found in the course of a US Geological Survey radioactivity reconnaissance project that expected to find uranium. Many REE-bearing carbonatite deposits were initially explored as a source of niobium or phosphates.

The discovery of an REE deposit must be proved by extensive trenching, drilling, and sampling. Drilling will initially be on a widely spaced grid to evaluate the extent and richness of mineralization. If results are favorable, the deposit will be drilled on progressively narrower spacing until a measured resource of adequate size is established. Concurrently, baseline environmental studies will be completed and bulk samples obtained for metallurgical testing. A conceptual mine plan and metallurgical plan will be engineered, and a preliminary economic feasibility study will be undertaken. If the results of studies are favorable, further work will develop an optimized mine plan, validate metallurgical processing on a pilot plant scale, begin application for permits, and conduct a definitive economic feasibility study. If the results of the feasibility study are positive, financing will be sought and permitting further advanced along with detailed plant engineering. Permitting will typically require an approved plan of operations, a positive environmental impact study, and some kind of final permission by a government agency. If external financing is required, an independent due diligence study will verify the results of the feasibility study.

Once financing and regulatory approval are in place, a construction decision will be announced. Ongoing detailed engineering design will be completed, long lead-time items will be ordered, construction work will be bid on, and contracts will be awarded. Construction begins as soon as possible and is executed quickly to minimize interest and opportunity costs. As mine components are completed, they will be tested and placed in operation with a ramp-up period until full commercial production at the planned output rate is achieved. The ramp-up process may be completed smoothly in a matter of months or may encounter unanticipated difficulties that require an extended period of time to solve. Some mines have failed at this stage because of ore grades that are lower than expected or metallurgical processes that do not work as planned.

Up to this point, a mining company has been spending large amounts of money without any revenue from mineral sales. Capital must be supplied internally, from net revenues of a mining company's other operations, or externally from banks or investors. The cost of borrowing these funds is considerable, and repayment cannot even start until a mine goes into production. The largest of currently (2010) proposed REE mining operations, including Mountain Pass, California, have reported pre-mining capital requirements of a half a billion dollars or more.

A mine will operate until reserves are exhausted. Today, mine closure and reclamation are planned before a mine is built; reclamation is undertaken concurrent with operations wherever possible. Mobile equipment and structures are removed. Roads, waste stockpiles, tailings, and surface facilities are also reclaimed in an attempt to restore the land to its previous use and appearance. Large open pits are not typically backfilled because of

Table 12 Time that was required to develop selected mines outside of the United States

Mine	Country	Commodity	Discovery date	Permitting completed	Production begun	Commercial operations begun
Ambatovy	Madagascar	Ni Co	1960	3/2007	Late 2010	NYA
Araxa	Brazil	Nb	1955	1960	1Q/1961	1966
Barro Alto	Brazil	Ni	1981	12/2006	3/2010	2011
Bulong	Australia	Ni	1971	9/1996	3/1999	9/1999
Catalão I	Brazil	Nb	1970	1974	1976	1977
Cawse	Australia	Ni Co	1994	9/1996	1/1999	6/2000
Goro	New Caledonia	Ni Co	1982	10/2004	Late 2010	NYA
Murrin Murrin	Australia	Ni Co	1984	5/1996	5/1999	12/2003
Niobec	Canada	Nb	6/1967	11/1973	1/1976	3/1976
Onça Puma	Brazil	Ni	1970s	8/2005	3Q/2010	Early 2011
Ravensthorpe	Australia	Ni Co	Late 1960s	3/2004	10/2007	NYA
Urumu Utsumi	Brazil	U	1971	1975	7/1981	1982
Vermelho	Brazil	Ni	1966	7/2005	4Q/2012	NYA

Data from unpublished historical research by the lead author (Keith R. Long)
These mines were selected as close analogies to the mining and metallurgical methods required for a rare earth elements mine. *NYA* indicates that commercial production has not yet been achieved

the large energy and CO_2 emission costs of doing so. Instead, to minimize environmental impact, they are reclaimed as landscape features.

Only one REE mine, at Mountain Pass, California, has ever been developed in the United States. That occurred in the early 1950s and is not illustrative of what is required to develop such a mine today. It is worth noting that the Mountain Pass mine was closed in 1998 and that permitting for reopening the mine began at about that time. The last required permit was obtained in November 2010, and stripping of waste for reactivating the open pit mine began shortly after. The mine is scheduled for commercial production late in 2012 after completion and commissioning of a new mineral processing plant. Nickel laterite deposits were suggested above as a useful analog for the development of a new REE mine. This comparison is particularly apt for a REE laterite deposit, such as Mount Weld, Australia. Niobium-bearing carbonatites are another close analog. These two deposit types share complex and difficult metallurgy and typically require extended periods of time to develop. The time that was required to bring into production recently developed nickel laterite mines and most niobium carbonatite mines is shown in Table 12. All of these mines were developed outside of the United States, and the permitting delays noted in Table 11 do not pertain in these cases.

The time from discovery to initial production for the mines listed in Table 12 ranges from 5 to 50 years, permitting to initial production 1–7 years, and ramp-up times were 3–42 months, except in the case of Araxá, Brazil, which suffered some unusual political delays. These development times are in line with the results of Peters (1966) and indicate that, in some instances, a relatively rapid pace of development can be obtained. It should be noted, however, that the most advanced REE projects other than Mountain Pass, California, namely, Mount Weld and Dubbo in Australia, have required years of metallurgical testing and development and fall into the fourth 7-year-or-more category of Peters (1966).

Summary

United States domestic reserves and inferred resources of REE are approximately 1.8 million metric tons, which are large compared with peak domestic consumption of REE of 10,200 tons in 2007 (US Geological Survey 2010). How much of that reserve and resource will be economically available, when, and at what rate cannot be addressed with the data at hand. It can be said that the reserves and inferred resources reported in Table 9 are of light REE and that these potential mines may not be able to meet domestic needs for heavy REE with the production plans currently (2010) proposed. The pipeline of new REE projects within the United States is rather thin, with 10 out of 150 REE exploration projects identified worldwide. If we extend our analysis to reliable trading partners, such as Australia and Canada, prospects for diversifying supply and meeting future demand are considerably improved. Unfortunately, the time required for development of new REE mines is on the order of at least a decade, perhaps much longer in the United States, and forecasting future supply that far into the future is problematic.

The lack of mining industry exploration of REE deposits in the last few decades is paralleled by a low level of geological research. The US Geological Survey has demonstrated in related studies that the first step in improving our understanding of REE resources and prospects for further discoveries is to conduct national and global mineral resource assessments. Rare earth elements are one of the commodities under consideration for the next National Resource Assessment, scheduled to begin in 2012.

References Cited

Alkane Resources Ltd (2010) The Dubbo zirconia project. Alkane Resources Ltd, Perth, 6 p

Altschuler ZS, Berman S, Cuttitta F (1966) Rare earths in phosphorite—geochemistry and potential recovery. U.S. Geol Surv Open-File Rep 66–3, 35 p

Arafura Resources Ltd (2010) Annual operations report 2009 (year ended June 30). Arafura Resources Ltd, Perth, 32 p

Back ME, Mandarino JA (2008) Fleischer's glossary of mineral species 2008. The Mineralogical Record, Tucson, 345 p

Bakker F, Delaney B, Mercer B, Qi D (2011) Technical report on the Nechalacho deposit, Thor Lake Project, Northwest Territories, Canada. Avalon Rare Metals Inc., Toronto, 289 p

Bastos Neto AC, Pereira VP (2009) The world-class Sn, Nb, Ta, F (Y, REE, Li) deposit and the massive cryolite associated with the albite-enriched facies of the Madeira A-type granite, Pitinga mining district, Amazonas State, Brazil. Can Miner 47:1329–1357

Behre Dolbear (2010) 2010 ranking of countries for mining investment. Behre Dolbear Group Inc, Denver, 18 p

Bogdetsky V, Stavinskiy V, Shukurov E, Suyunbaev M (2001) Mining industry and sustainable development in Kyrgyzstan. International Institute for Environment and Development, London, 95 p

Camus Y, Laferrière A (2010) NI 43–101 technical report, mineral resource estimation, Kipawa deposit, Zeus Project, Quebec, Matamec Explorations Inc. SGS Canada Inc., Blainville, 95 p

Capital Mining Ltd (2009) Annual report 2009 (year ended June 30). Capital Mining Ltd, Philip, 68 p

Castor SB (2008) Rare earth deposits of North America. Resour Geol 58:337–347

Castor SB, Hedrick JB (2006) Rare earth elements. In: Kogel JE, Trivedi NC, Barker JM, Krukowski ST (eds) Industrial minerals & rocks—commodities, markets, and uses, 7th edn. Society for Mining, Metallurgy, and Exploration, Littleton

Chalmers DI (1990) Brockman multi-metal and rare-earth deposit. In: Hughes FE (ed) Geology of the mineral deposits of Australia and Papua New Guinea, vol 1, Australasian institute of mining and metallurgy monograph 14., pp 707–709

Daigle P, Maunula T (2010) Strange lake project B zone deposit, Québec, national instrument 43–101 resource estimate for quest uranium corporation. Wardrop, Toronto, 188 p

De Sousa MM (1996) The niobium deposit in Morro dos Seis Lagos. North-Brazil Int Geol Congr Abstr 30:784

Druecker MD, Simpson RG (2011) Amended technical report on the Sarfartoq project, west Greenland. GeoSim Services, Vancouver, 108 p

Dunn B (2009) Form 43–101 mineral resource report on the Hoidas Lake rare earth project, Northern Mining district, Saskatchewan, prepared for Great Western Minerals Group Ltd. Barr, Minneapolis, 730 p

Emsley J (2001) Nature's building blocks—an A–Z guide to the elements. Oxford University Press, Oxford, 538 p

Filho AI, Riffel BF, de Faria Sousa CA (2005) Some aspects of the mineralogy of CBMM niobium deposit and mining and pyrochlore ore processing, Araxá, MG, Brazil. Companhia Brasileira de Metalurgia e Mineração, Araxá, 12 p

FODD (2010) Fennoscandian Ore Deposit Database, http://en.gtk.fi/ExplorationFinland/fodd/

Grauch RI, Verplanck PK, Seeger CM, Budahn JR, Van Gosen BS (2010) Chemistry of selected core samples, concentrate, tailings, and tailings pond waters—Pea Ridge iron (-lanthanide-gold) deposit, Washington County, Missouri. U.S. Geol Surv Open-File Rep 2010–1080, 15 p

Great Western Minerals Group Ltd (2009) Great Western minerals signs option agreement to recommission rare earth mine in South Africa. News release dated 13 Jan 2009. Accessed 15 July 2010 at http://www.gwmg.ca/html/ newsmedia-releases/index.cfm?ReportID=203280

Greenland Minerals and Energy Ltd (2011) Greenland minerals releases an upgraded resource estimate for the Kvanefjeld multi-element project (Rees, U, Zn). Greenland Minerals and Energy Ltd, West Perth, 12 p

Gupta CK, Krishnamurthy N (2005) Extractive metallurgy of rare earths. CRC Press, Boca Raton, 484 p
Hedrick JB (2004) Rare earths. In: Minerals yearbook, vol 1, metals and minerals. U.S. Geological Survey, Washington, DC, pp 60.1–60.15
Heritage Foundation (2010) 2010 index of economic freedom. Heritage Foundation, Washington, DC
Heymann L (2010) The valuation of rare earth deposits. Paper presented at the society for mining, metallurgy, and exploration annual meeting, Phoenix, Arizona, March, 2010
Hirano H, Kamitani M, Sato T, Sudo S (1990) Niobium mineralization of Catalão I carbonatite complex, Goias, Brazil. Bull Geol Surv Jpn 41:577–594
Hurst C (2010) China's rare earth elements industry—what can the West learn? Institute for the Analysis of Global Security, Washington, DC, 42 p
Jackson WD, Christiansen G (1993) International Strategic Minerals Inventory summary report—Rare earth oxides. US Geol Surv Circ 930–N:68
Keyser HJ, Kennedy J (2007) Geological report on the Bokan Mountain property, Prince of Wales Island, Alaska. Landmark Minerals Inc, Vancouver, 48 p
Klemic H, Heyl AV Jr, Taylor AR, Stone J (1959) Radioactive rare-earth deposit at Scrub Oaks mine, Morris County, New Jersey. US Geol Surv Bull 1082–B:34
Kušnir I (2000) Mineral resources of Vietnam. Acta Montan Slovaca 5:165–172
Laznicka P (2006) Giant metallic deposits—future sources of industrial minerals. Springer, Berlin, 732 p
Lide DR (1997) Abundance of elements in the earth's crust and sea. In: CRC handbook of physics and chemistry, 78th edn. CRC Press, Boca Raton, p 14
Long KR (2009) Quantitative measures of mineral supply risk. Poster presented at the American geophysical union, fall meeting, San Francisco, December, 2009
Lurie J (1986) Mineralization of the Pilanesberg alkaline complex. In: Anhaeusser CR, Maske S (eds) Mineral deposits of Southern Africa, vol 2. Geological Society of South Africa, Johannesburg, pp 2215–2228
Lynas Corporation Ltd (2007) Lynas acquires new rare earths resource in Malawi. Press release dated 6 Sept 2007. Accessed 15 July at http://www.lynascorp.com/content/upload/files/Announcements/Acquisition_of_New_Rare_Earths_Resource_in_Malawi_060907_(2).pdf
Lynas Corporation Ltd (2010) Increase in resource estimate for deposit with elevated heavy rare earths. Lynas Corporation Ltd., Sydney, 7 p
Maksimović ZJ, Pantó G (1996) Authigenic rare earth minerals in karst-bauxite and karstic nickel deposits. In: Jones AP, Wall F, Williams T (eds) Rare earth minerals—chemistry, origin and ore deposits. Chapman & Hall, London, pp 257–279
Malyukova N, Kim V, Tulyaev R (2005) Zonation of polymetallic, rare-earth, molybdenum, zirconium, beryllium and tantalum-niobium mineralization in the Ak-Tyuz ore deposits (northern Tien Shan). In: Jingwen M, Bierlein FP (eds) Mineral deposit research; meeting the global challenge; proceedings of the 8th biennial SGA society for geology applied to mineral deposits meeting, vol 2. Springer, Berlin, pp 1323–1326
Mason B, Moore CB (1982) Principles of geochemistry, 4th edn. Wiley, New York, 344 p
McGill I (1997) Rare earth metals. In: Habashi F (ed) Handbook of extractive metallurgy, vol 3. Wiley-VCH, Weinheim, pp 1695–1741
McKenzie J (2011) Ucore releases NI 43–101 compliant resource for Bokan. Ucore Rare Metals Inc., Halifax, 4 p
McKeown FA, Klemic H (1956) Rare-earth-bearing apatite at Mineville, Essex County, New York. US Geol Surv Bull 1046–B:23
McMahon F, Cervantes M (2010) Survey of mining companies 2009/2010. Fraser Institute, Toronto, 88 p
Molycorp Inc (1986) Internal memorandum dated 5 Feb 1986, quoted by Quantum Rare Earth Developments Corp. Accessed 14 July 2010 at http://www.quantumrareearth.com/elk-creek.html
Molycorp Inc (2010) Prospectus dated 16 Apr 2010, Form S-1, Registration Statement under the Securities Act of 1933, 108 p
Morteani G, Satir M (1989) The bastnaesite-fluorite-barite deposit of the Kizilcaören District, Eskisehhir, Turkey. In: Möller P, Černý P, Saupé F (eds) Lanthanides, tantalum and niobium—mineralogy, geochemistry, characteristics of primary ore deposits, prospecting, processing and applications. Springer, Berlin, pp 189–194
Mossop A, Carter RC (2010) National Instrument 43–101 technical report for the Mrima Hill rare earth/niobium project in southern Kenya. Sound Mining Solution (Pty) Ltd, Johannesburg, 145 p
National Research Council (2008) Minerals, critical minerals, and the U.S. Economy. National Academies Press, Washington, DC, 264 p
Navigator Resources Ltd (2009) 2009 annual report (year ending June 30). Navigator Resources Ltd, West Perth, 72 p
Northern Minerals Ltd (2010) West Perth, Australia, Website. www.northernminerals.com.au/ news/ index.php?options=com_content&view=article&id=57&Itemid=116/rare_earths/12. Accessed 15 July 2010
Oreskes N, Einaudi MT (1990) Origin of rare earth element-enriched hematite breccias at the Olympic Dam Cu-U-Au-Ag deposit, Roxby Downs, South Australia. Econ Geol 85:1–28
Organization for Economic Co-Operation and Development (2008) Country risk classification for participants to the arrangement on officially supported export credits, 1999 to 2009. Downloaded 13 July 2009 from www.oecd.org/dataoecd/9/12/35483246.pdf
Orris GJ, Grauch RI (2002) Rare earth element mines, deposits, and occurrences. U.S. Geol Surv Open-File Rep 2002–0189, 174 p
Osterwald FW, Osterwald DB, Long JS, Wilson WH (1966) Mineral resources of Wyoming. Wyo Geol Surv Bull 50:287

Palmer K, Broad P (2007) Technical report on the Thor Lake rare metals project, NT, for Avalon Ventures Ltd. Wardrop, Toronto, 197 p

Papp JF, Bray EL, Edelstein DL, Fenton MD, Guberman DE, Hedrick JB, Jorgenson JD, Kuck PH, Shedd KB, Tolcin AC (2008) Factors that influence the price of Al, Cd, Co, Cu, Fe, Ni, Pb, rare earth elements, and Zn. U.S. Geol Surv Open-File Rep 2008–1356, 61 p

Paul J, Stubens TC (2009) Thor Lake-Lake Zone mineral resource update for Avalon Rare Metals Inc. Wardrop, Toronto, 194 p

Pele Mountain Resources (2011) Pele Mountain announces total rare earth oxide resources of 52-million pounds indicated plus 96-million pounds inferred. Pele Mountain Resources Inc., Toronto, 3 p

Peters WC (1966) The pre-production interval of mines—summary of an industry-wide survey. Miner Eng 18:63–64

Rare Element Resources (2010) Rare element reports 50% increase in rare-earth mineral reserves at Bear Lodge Project. Rare Element Resources, Vancouver, 2 p

Raw Materials Supply Group (2010) Critical raw materials for the EU—report of the Ad-hoc Working Group on defining critical raw materials. Raw Materials Supply Group, European Commission, Enterprise and Industry, 220 p

Reed GC (2011) NI 43–101 technical report, Norra Kärr REE-zirconium deposit, Gränna, Sweden. Pincock, Allen & Holt, Lakewood, 128 p

Ren X (1985) A brief account of rare earth mineralization in China. In: Xu Q (ed) New frontiers in REE science and applications. Science Press, Beijing, pp 39–41

Richardson DG, Birkett TC (1996) Residual carbonatite-associated deposits. In: Eckstrand OR, Sinclair WD, Thorpe RI (eds.) Geology of Canadian mineral deposit types. Geology of Canada no 8, Geological Survey of Canada, Ottawa, pp 108–119

Sabot JL, Maestro P (1995) Lanthanides. In: Kroschwitz JI, Grant MH (eds) Kirk-Othmer encyclopedia of chemical technology, 4th edn. Wiley, New York, pp 1091–1115

Scott N, Wells A (2010) Technical report on the Songwe rare earth element (REE) project, Republic of Malawi. Saint Barbara LLP, London, 255 p

Sinclair WD (1996) Granitic pegmatites. In: Eckstrand OR, Sinclair WD, Thorpe RI (eds.) Geology of Canadian mineral deposit types. Geology of Canada no 8, Geological Survey of Canada, Ottawa, Rivonia, South Africa, pp 503–512

Solikamsk Magnesium Works (2010) Annual report for 2009. JSC Solikamsk Magnsium Works, Solikamsk

Staatz MH, Armbrustmacher TJ, Olson JC, Brownfield IK, Brock MR, Lemons JF Jr, Coppa LV, Clingan BV (1979) Principal thorium resources in the United States. US Geol Surv Circ 805:42

Stans Energy Corporation (2010) Toronto, Canada, Website. www.stansenergy.com. Accessed 15 July 2010

Stigler G (1983) The organization of industry. University of Chicago Press, Chicago, 356 p

Taylor SR, McLennan SM (1985) The continental crust, its composition and evolution—an examination of the geochemical record preserved in sedimentary rocks. Blackwell Scientific, Oxford, 312 p

Transparency International (2009) Global corruption report 2009—corruption and the private sector. Cambridge University Press, Cambridge, 462 p

U.S. Geological Survey (2010) Mineral commodity summaries 2010. U.S. Geological Survey, Reston, 193 p

Venter M, Hall M, Siegfried P, Brown J (2010) Amended NI 43–101 resource estimate and technical report on the Zandkopsdrift rare earth element (REE) project, located in the Republic of South Africa. MSA Group, Cape Town, 189 p

Wedepohl KH (1995) The composition of the continental crust. Geochim Cosmochim Acta 59:1217–1232

Wedow H Jr (1967) The Morro de Ferro thorium and rare-earth ore deposit, Pocos de Caldas District, Brazil. US Geol Surv Bull 1185-D:34

Wooley AR, Kjarsgaard BA (2008) Paragenetic types of carbonatite as indicated by the diversity and relative abundances of associated silicate rocks—evidence from a global database. Can Miner 46:741–752

World Bank (2010) Worldwide governance indicators, 1996–2008. Accessed 8 Sept 2010, at http://info.worldbank.org/governance/wgi/sc_chart.asp

Wu C, Yuan Z, Bai G (1996) Rare earth deposits in China. In: Jones AP, Wall F, Williams CT (eds) Rare earth minerals—chemistry, origin, and ore deposits. Chapman & Hall, New York, pp 281–310

Evaluating Supply Risk Patterns and Supply and Demand Trends for Mineral Raw Materials: Assessment of the Zinc Market

Peter Buchholz, Maren Liedtke, and Marko Gernuks

Abstract

The economic growth of emerging nations will inevitably require an increasing supply of primary mineral raw materials and metals. Fears about the availability of mineral raw materials have been raised during resource booms in the past and also during the last decade. This chapter refers to a recently developed method for identifying and assessing long-term supply risks and applies it to the case of zinc. By analysing boom-and-bust cycles over the past 50 years, we have quantified indicators and defined benchmarks for identifying problematic market situations for zinc supply. This method, which is based on a systematic and comprehensive analytical approach and a numerical evaluation model, makes it possible to identify supply shortage risks at an early stage. On the basis of this market assessment, companies are in a better position to make informed decisions and can use suitable risk mitigation instruments to counteract problematic developments.

Keywords

Indicators • Market analysis • Mineral resources • Mining • Raw material • Riskt mitigation • Supply risk • Supply and demand • Zinc

Introduction

Concerns about the security of raw materials supply have repeatedly been raised in the past. After World War II, the economic growth and increasing demand for mineral raw materials in North America and Western Europe led to the appointment of the "Paley Commission" in the 1950s by President Truman, known as the President's Minerals Policy Commission (Paley Commission 1952), which raised these concerns. In the 1960s and 1970s, this discussion was taken up by the

P. Buchholz (✉) • M. Liedtke
Economic Geology of Mineral Resources,
Federal Institute for Geosciences and Natural Resources (BGR), Stilleweg 2, Hannover 30655, Germany
e-mail: peter.buchholz@bgr.de; maren.liedtke@bgr.de

M. Gernuks
Volkswagen AG, Group Research, Environmental Affairs Product, Letter Box 1774, Wolfsburg 38436, Germany
e-mail: marko.gernuks@volkswagen.de

R. Sinding-Larsen and F.-W. Wellmer (eds.), *Non-Renewable Resource Issues: Geoscientific and Societal Challenges*, International Year of Planet Earth,
DOI 10.1007/978-90-481-8679-2_8, © Springer Science+Business Media B.V. 2012

report to the Club of Rome on "The limits of growth" (Meadows et al. 1972). As a result, new mineral policies to securing supply were developed in most industrialised nations.

Since the rise of the Chinese economy in the early twenty-first century and sharp increases in raw material prices, the discussion about possible supply bottlenecks for mineral raw materials has been back on the table – despite the crash on the financial markets in 2008. Over the last two decades, China's demand for mineral raw materials has grown substantially. In 1997, China consumed about 10% of global iron ore and zinc production. Ten years later, in 2007, the country's consumption had increased to about 45% of global iron ore production and about 40% of zinc production. China's significant role in the commodity markets is related not only to its industrial sector's heavy demand for raw materials but also to its ability to influence raw material markets and its growing interest in acquiring shares in deposits, mines or mining companies abroad.

As most industrialised and emerging nations are largely import dependent with regard to metals and industrial minerals, new raw material policies are being developed to improve the supply situation and reduce ecological impacts along the supply chain. In 2005, the Federation of German Industry initiated the presidential working group "international raw materials issues" (BDI 2005, 2007) accompanied by new raw materials policy reports from the German government (BMWi 2005, 2007). Recently, the Committee on Critical Mineral Impacts on the U.S. Economy has published an expert report on critical minerals to the USA with the aim of helping decision makers to take steps to avoid mineral supply constraints (Committee on Critical Mineral Impacts on the U.S. Economy 2008). In November 2008, the European Commission announced its intention to define critical raw materials for the European Union in its new Raw Materials Initiative (European Commission 2008). New raw material policies are being developed in Japan (Meti 2009), India (India Ministry of Mines 2008) and China (MLR 2009).

Against this background, Volkswagen AG (VW) and the Federal Institute for Geosciences and Natural Resources (BGR) – the Federal German Geological Survey – have developed a method of identifying and assessing long-term supply risks for mineral raw materials (Rosenau-Tornow et al. 2009). By using the method, we define thresholds (benchmarks) and quantify indicators for supply risks to help recognise problematic market situations. This method can help to ensure better assessment of future developments in raw material markets and thus help companies to make better informed decisions, react in good time and implement measures ensuring security of supply. Understanding future supply and demand trends is especially useful when selecting new technologies for products which require intensive use of raw materials. The automotive sector in particular is facing challenging times with its task of developing consumer and eco-friendly cars. The availability of mineral raw materials is fundamental to this sector in view of forthcoming changes in technology as well.

Summary of the BGR-VW Method

For the simple reason that we have barely explored its subsurface in our efforts to find mineral raw materials, our planet still hides many mineral deposits yet to be discovered. Thus, it is not so much the discussion about the finite nature of mineral raw materials, but rather the crucial requirement of secure and sustainable supply which is an important issue for our economies.

Supply analyses may consider current production levels and price, production and investment costs, freight capacities, capacity increases due to new exploration, mining and refinery projects or political and social issues (Citigroup 2005; Deutsche Bank 2008; World Bank 2008; Radetzki 2006).

The BGR-VW method combines market indicators of this kind with demand estimates and presents a systematic approach to evaluating mineral raw material markets, including time series analyses and a numerical evaluation model. Integrating such market indicators into a systematic set of manageable analytical instruments is a challenge and a prerequisite for balanced judgement of future market trends. The strength of this analysis is that it looks back on the past 50 years,

Rating scale[#]: relaxed (1 2 3) | moderate (4 5 6) | problematic (7 8 9)

Zinc	2007	Trend 2013
1. Current supply and demand (average)	7	
Market balance		
Stock keeping		
Refinery**/mine utilization*		
2. Production costs	7	moderate
"Cash operating costs" 2008	7	falling
3. Geostrat. Risiks (country conc. + country risk)	4	moderate
4. Market power/ Company concentration	1	relaxed
5. Supply and demand trends (average)	5	problematic
Market balance in 5 years (2013) (at 2.8 / at 3.2 % growth in demand per year)	6.5 / 7	problematic
Degree of exploration (2007)	5***	moderate
Investments (2008/2009)	4	moderate
Recommendations (not shown)		

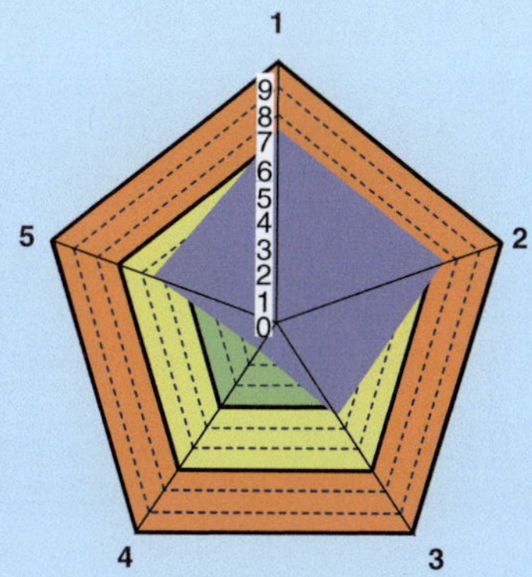

according to supply risk
* Mine production + direct scrap and totel metal consumption
** refined production, -consumption, change in stocks
*** only Australia

Fig. 1 Summary of the evaluated and rated market situation for zinc

sets benchmarks for problematic market situations and uses scenarios to evaluate the extent to which supply problems may arise in future. The main aim is to quantify whether a market will be balanced within a foreseeable future of 5–15 years under certain preconditions of growth in supply and demand, and whether other parameters such as political risk in mining countries may affect supply.

For this approach we have suggested to evaluate technical aspects of supply and demand (Fig. 1): (1) the current market balance including stock keeping and mining/refining capacity utilisation (and freight capacity, e.g. for iron ore); (2) production costs for mining; (3) geostrategic risks including country risk and country concentration of production; (4) exercise of market power through company concentration and, the final step, (5) evaluating supply and demand trends. In the final step, we look at future supplies originating from new exploration and mining projects, at exploration and mining investments and at the reserves and resources situation. We then compare the supply scenarios with possible future demand scenarios.

For each indicator, we review and evaluate past market trends and derive maximum and minimum values for each indicator corresponding to problematic or relaxed market situations of the past (benchmarks). We then compare these benchmarks with current market trends, thus enabling us to conclude whether the supply risk is likely to increase or decrease. For each raw material under consideration, the benchmark values are different. To permit better comparison of the supply risk for various raw materials, the results for each indicator are normalised on a scale from 1 to 9, subdivided into "relaxed" (values 1–3), "moderate" (values 4–6) or "problematic" (values 7–9) market situations. Finally, weighting of the indicators indicates the major risks – labelled 7–9 – that affect security of supply (Table 1). This method puts analysts in a position to use quantifiable data for their market outlook based on profound knowledge of the past. For more details of the methodology, see Rosenau-Tornow et al. (2009).

Table 1 Indicators for supply risk analysis, risk scale and summary of rated results for the zinc market

Indicator	Situation in 2007/2008 and trend	Rating scale according to supply risk 1 2 3 4 5 6 7 8 9
Indicators for market assessment		
1. Current supply and demand		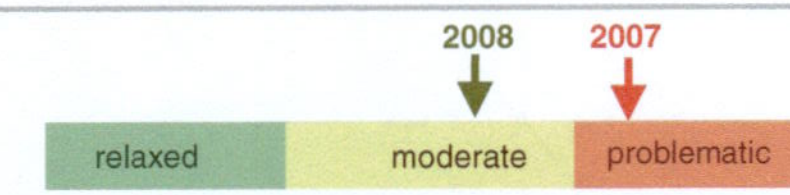
Current supply and demand Current market balance (Mb) (without stocks) $Mb_{i\Delta t} = P_{it} - Con_{it}$ $Mb_{i\Delta t} = \frac{P_{it} - Con_{it}}{P_{it}} \times 100$	Mb = 2007: 256,000 t, 1.9% share of production; 2008: 860,000 t, 6.3% share of production; trend moderate Difference between production (*P*) (mine production + direct scrap) and consumption (Con) (total metal: refinery + direct scrap)	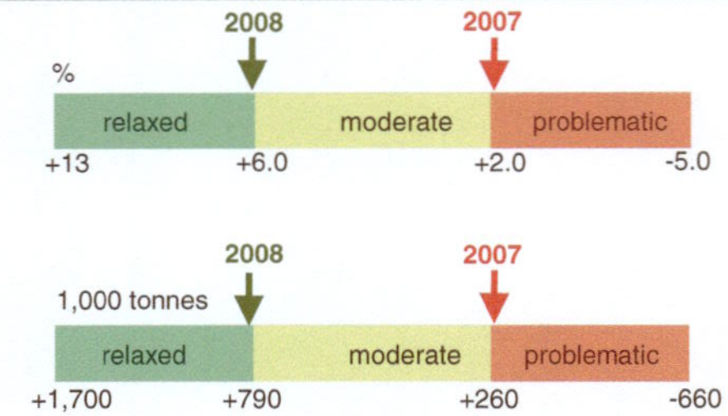
Current supply and demand Current market balance (Mb) (with change in stocks) $Mb_{i\Delta t} = P_{it} - \Delta S_{it} - Con_{it}$ $Mb_{i\Delta t} = \frac{P_{it} - \Delta S_{it} - Con_{it}}{P_{it}} \times 100$	Mb = 2007: −92,800 t, −0.8% share of production; 2008: 13,000 t, 0.1% share of production; trend moderate Difference between production (*P*) (mine production + direct scrap), change in stocks (Δ*S*) and consumption (Con) (total metal: refinery + direct scrap)	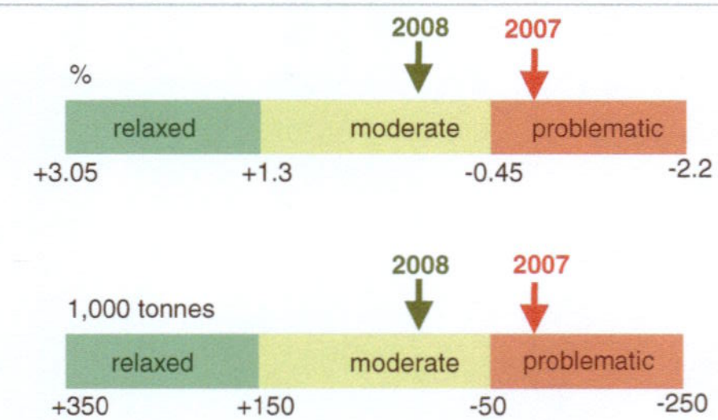
Stock keeping (Sk) $Sk_{i\Delta t} = \frac{S_i \times 100}{RP - \Delta S_i}$	Sk = 2007: 5.2%; 2008: 6.7%; trend problematic Ratio of global stocks (*S*) to refinery production (RP)	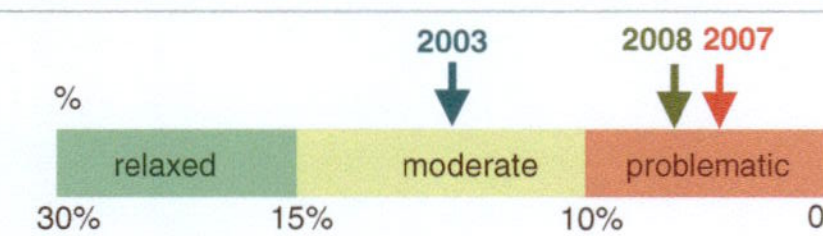
Mine (Mcu) or refinery (Rcu) capacity utilisation $Mcu_{i\Delta t} = \frac{MP_i \times 100}{Mc_i}$ $Rcu_{i\Delta t} = \frac{RCon_i \times 100}{Rc_i}$	not available Ratio of mine production (MP) to mine capacity (Mc) or refined metal consumption (RCon) to refinery capacity (Rc)	not available

2. Production costs "Cash costs" (CC) $CC_{i\Delta t} = \frac{US\$}{lb_i}$	CC = 0.53 US$/lb (1,170 US$/t); 2008, problematic; trend unchanged (2001: 0.35 US$/lb, 770 US$/t) Cash costs (CC) per pound (lb) of raw material produced	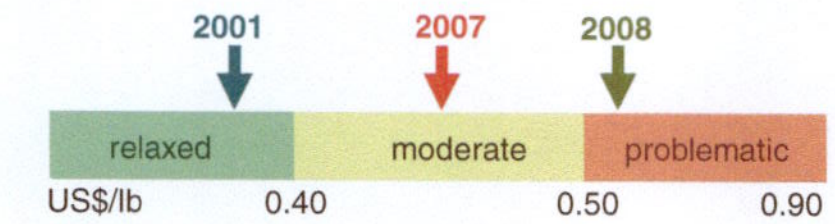
3. Geostrategic risks Country concentration (HHI_{Cc}) $HHI_{Cc} = \sum_{n=1}^{x} aiCP^2$	HHI_{Cc} Mine production = 1,274 Refinery production = 1,320 Herfindahl–Hirschman Index (HHI): sum of squared values of raw materials production (in %) in each country (aiCP)	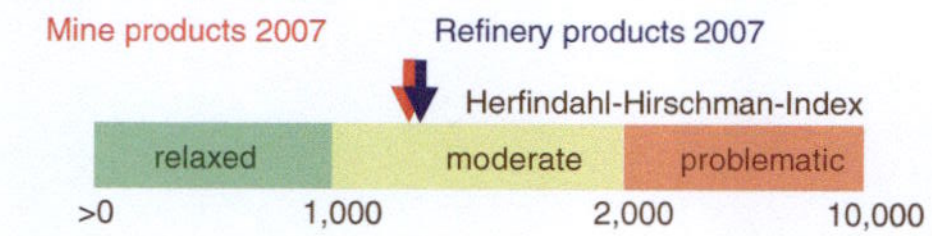
Country related risk (Cr) $Cr = \sum_{n=1}^{x} P_{countries} \times G_{WB}$	Cr Mine production = 5.5 Refinery production = 5.6 Sum of global raw materials production (*P*, in % for each country) weighted with the governance index (*G*) of the World Bank (WB) for each country (based on six indicators)	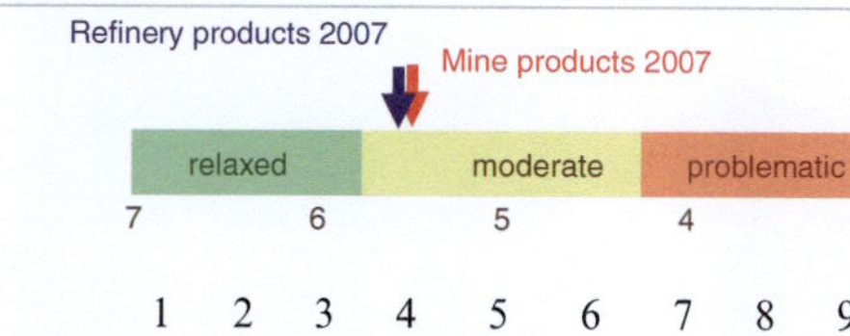
4. Market power Company concentration (HHI_{Compc}) $HHI_{Compc} = \sum_{n=1}^{x} aiComP^2$	HHI_{Compc} Mining companies = 226 Refineries = 320 Herfindahl–Hirschman Index (HHI): sum of squared values of raw materials production (in %) of each company (aiComP)	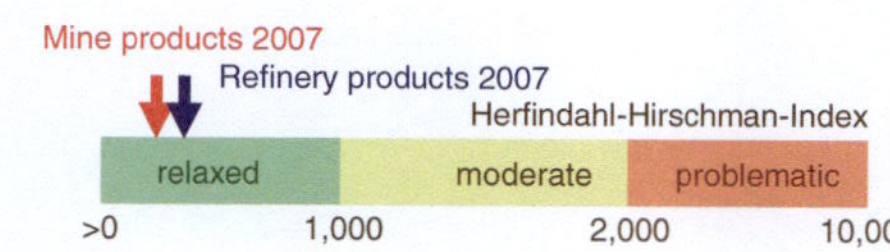
5. Supply and demand trends Future market capacity (McF) in 2013	McF = 2.0% (at 2.8% growth of demand)	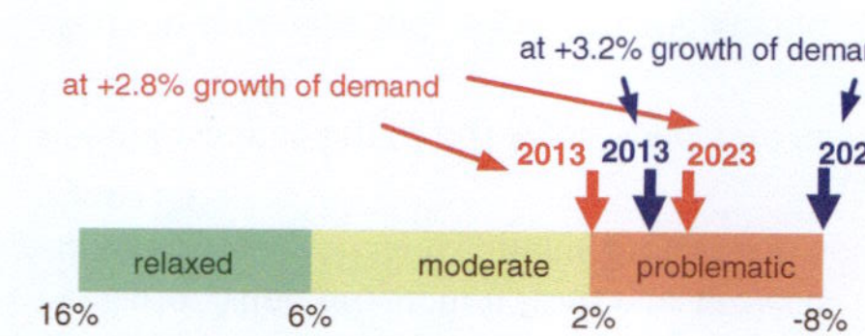

(continued)

Table 1 (continued)

Indicators for market assessment		
Indicator	Situation in 2007/2008 and trend	Rating scale according to supply risk 1 2 3 4 5 6 7 8 9
$McF_{it} =$	McF = –0.3% (at 3.2% growth of demand)	Scenario: Mine production from new projects without mine extensions, minus projects on hold, minus 25%
$\frac{(Dem_{it\ at\ x\%\ growth\ of\ demand/year})\times 100}{(MP_{it} + APc_{i\Delta t}F + Rec_{i\Delta t})} - 100$	Ratio of expected future demand (Dem) at a forecasted growth in demand for mining products of x%/year to mine production plus additional annual production capacity of future projects plus recycling material (MP + APcF + Rec)	Possible demand until 2013, just moderate market balance = 2.8% Medium-term demand, expected problematic market balance = 3.2%
Degree of exploration (E) $Lt_{i\Delta t} = \frac{R_i}{MP_i}$; $IEx_{i\Delta t} = \frac{ExB_i}{MP_i}$	Lt = 16.3 years (reserves), IEx Australia = 39 US$/t (2003–2008) Weighted values of static life time (Lt) and investments in exploration (IE) Lt is the ratio of reserves (*R*) to mine production (MP) IE is the exploration budget (E × B) for each tonne of mine production	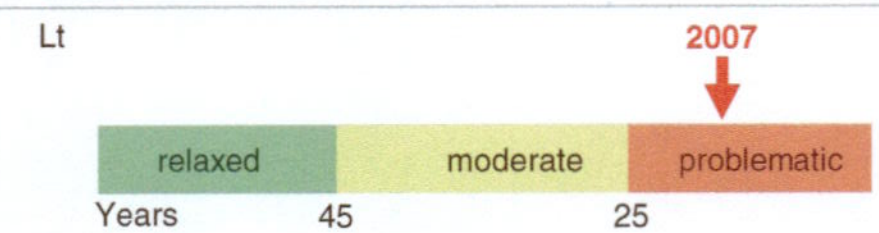 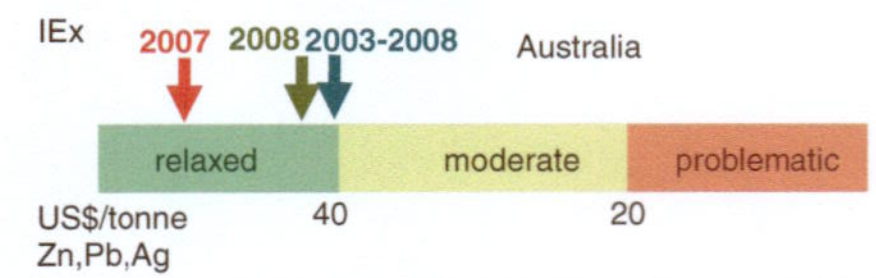
Investment in mining (IM) (2004–2009) $IM_{i\Delta t} = \frac{CapC_i}{MP_i}$	IM = 150 US$/t extracted Zn, 2008–2009 IM = 89 US$/t extracted Zn, 2004–2009 Ratio of capital costs (CapC) for new mining projects to mine production (MP)	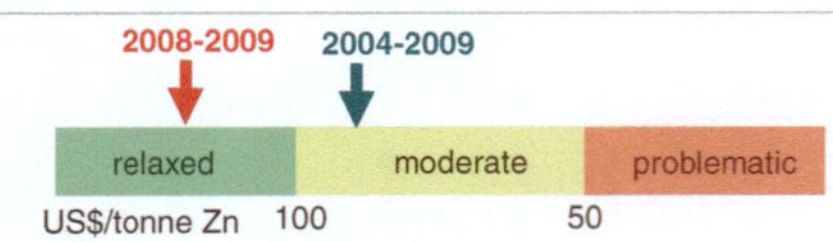

Analysis for Zinc

Data Situation

The statistical database for analysing the zinc market is relatively good. This analysis used the following data sources, dating from 2008 to 2009:

- International Lead and Zinc Study Group (ILZSG: Recycling, stocks, production, consumption, price)
- Initiative Zinc (www.initiative-zink.de: consumption, recycling).
- World Bureau of Metal Statistics (WBMS: production, consumption, stocks, prices)
- World Bank (country risk based on six indicators, economic data)
- Raw Materials Group (mines and projects: annual production capacity, production, mergers and acquisitions, reserves, resources, ore content; capital expenditures)
- United States Geological Survey (USGS: production, reserves, reserve base)
- Metals Economics Group (MEG: exploration budgets/expenses, new projects)
- Australia Bureau of Statistics (ABS: exploration budgets/expenses)
- Business reports from mining companies, market analysts and banks.

The BGR database combines statistics on production, consumption, recycling, the annual production capacity of new mining and exploration projects, prices and a digital archive of mining and raw material reports and publications. The statistical data on global supply and demand for the period from 1960 to 2008 is largely complete. However, worldwide data on exploration and capital expenditures as well as cash costs have only been available from the late 1990s onwards and are to some extent incomplete.

Price Trends

Zinc prices are closely related to global industrial growth and the steel market. The price of zinc more than tripled between 2002 and 2006 during a global economic boom and – in real terms – exceeded the peak of 1989 (Fig. 2). The annual average price rose from 800 to 3,500 USD per tonne and reached record levels of over 4,400

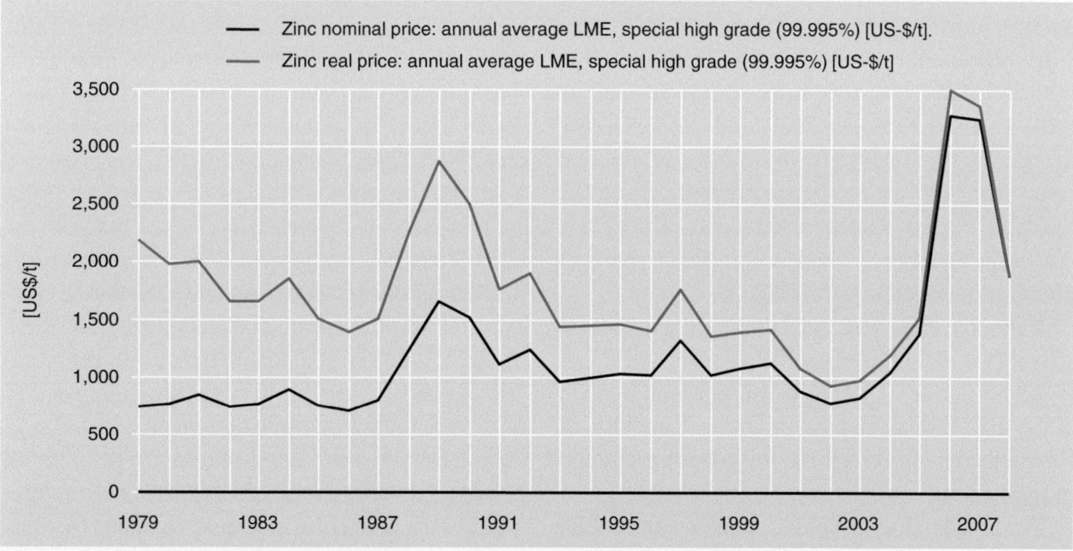

Fig. 2 Historic price trend for zinc, nominal and real (Data source: LME, BGR database)

USD per tonne in 2006. Unexpectedly high demand and the inability of supply to meet this demand in the short run led to both high prices and low stocks. Declining stocks between 2002 and 2006 combined with expectations of high demand from Asia kept the market nervous and prices high. From early 2007 to mid 2008, demand for zinc and other base metals shrank because of a slowdown in global industrial growth. As a result, zinc prices fell just below 2,000 USD by mid-2008 and crashed to 2002 levels at the beginning of the financial crisis. From early 2009 to early 2010, prices have climbed to 2,500 USD due to resuming industrial production.

Risk Assessment According to the Five Indicators

Supply and Demand

Market balance (rating: 7/2007, problematic): Worldwide mine production for zinc increased by 235%, from 3.33 million tonnes in 1960 to 11.14 million tonnes in 2007 at an average annual growth rate of 2.6% (Fig. 3). During this period, India and PR China had the highest growth rates of 10.4% and 8.4%, respectively. Between 1960 and 1974, the average annual growth of global mine production was significantly higher at 4.4%. Canada made massive increases in its mine production during that time. Between 1974 and 1991, the annual growth in global mine production was down to 1.0%, after which, it climbed back to 3.4% between 2000 and 2007. During that period, the major growth in mine production came from PR China, Peru and India. In 2007, the three biggest mining producers – PR China, Australia and Peru – had a combined share of global production of 54% (Fig. 3).

Between 1960 and 2007, refinery production showed an annual average increase of 2.8% and followed a similar trend to mine production. In 2007, PR China, Canada and the Republic of Korea accounted for 46.1% of global production of refined zinc.

Worldwide zinc recycling increased from 1.26 million tonnes in 1980 to 2.04 million tonnes in 2007, a rise of 61% at an average annual growth rate of 1.8%. Since 1995, secondary zinc as a percentage of total production has fallen from 20% to 15% due to a stronger increase in mine production (Fig. 4).

Between 1960 and 2007, the average growth rate for the consumption of refined zinc was 2.8%. The three biggest consumers were PR China, USA and Germany. In PR China and India, the growth rates were 8.7% and 4.4%, respectively, but in industrialised countries such as the USA, Japan and Germany, they were only between 0.5% and 2.4% (Fig. 5). From 2000 to 2007, the average annual growth in global refined zinc consumption was 3.6%. PR China held first place with an average annual growth rate of 14.4%. By contrast, the growth of refined zinc consumption in the industrialised countries was low or – as in the case of the USA and Japan – even negative.

Zinc is primarily used as an anti-corrosion coating in the iron and steel industry. Around 50% is used for galvanising steel, about 17% is processed mainly for zinc die-casting alloys and another 17% for brass. In the chemical industry, zinc oxide (ZnO) and zinc dust are used. Zinc oxide, for example, is used in pharmaceutical and cosmetic products, pigments, paints and varnishes, rubber (tyres), ceramics, animal feeds and fertilisers. The main applications for zinc products are the construction and transportation sector, consumer goods and machinery (Fig. 6).

The calculated difference between refined zinc production and consumption shows that there were supply deficits in the mid-1960s, early 1970s, late 1980s, mid-1990s and between 2004 and 2007. These – combined with strong demand – resulted in price increases of more than 100% (e.g. in the late 1980s) (Fig. 7, see marked circles). These shortfalls were offset by using stocks held by producers, suppliers and exchanges, however not all of these stock changes are published. In 2007, the overall market balance was problematic. It was not possible to cover the supply deficit of 67,800 tonnes by utilising registered stocks. In 2008, the situation eased slightly, with a small positive market balance of 13,000 tonnes of zinc.

Looking at the historic market balance (including movements in stocks) over time, we have set benchmarks between +150,000 and −50,000

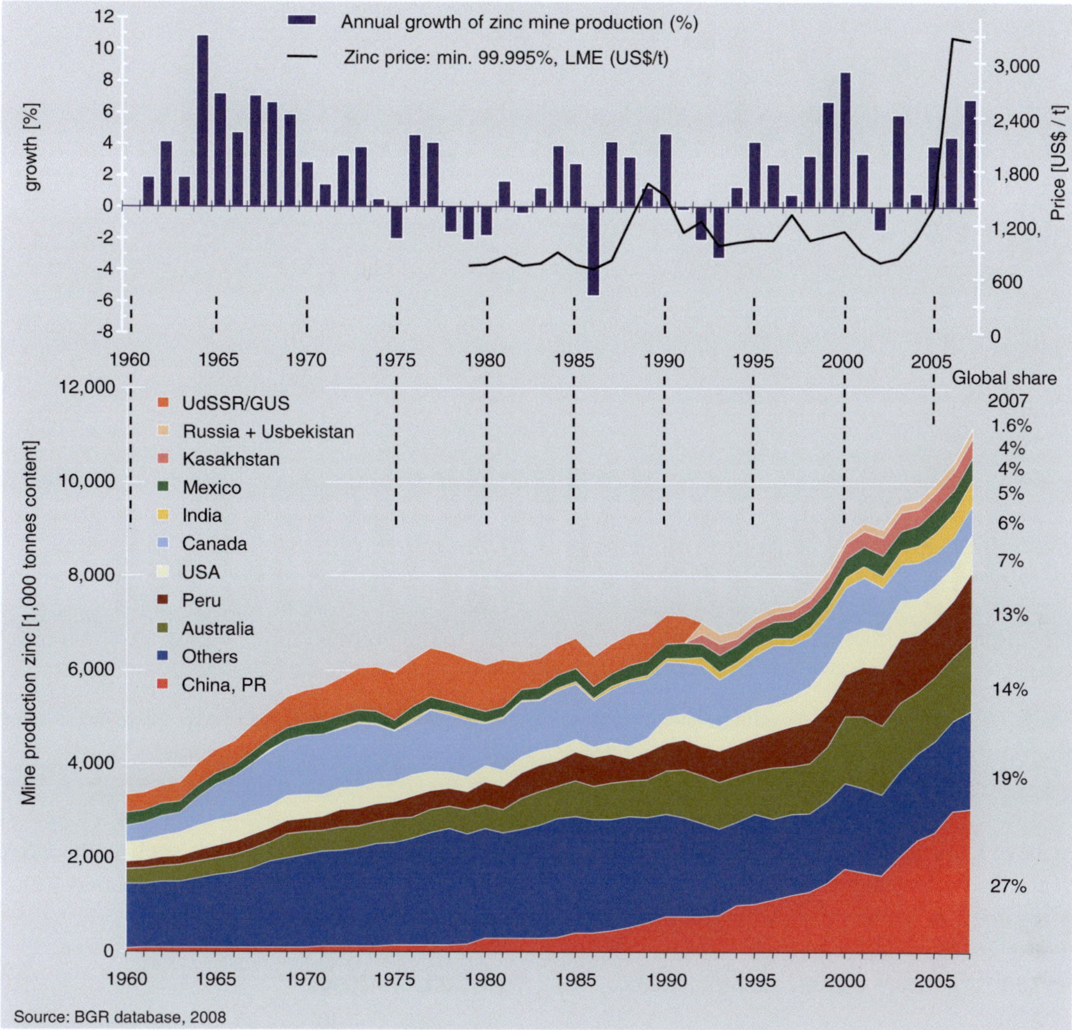

Growth rate %						
Period:	**1960-2007**	**1960-1974**	**1974-1991**	**1992-2000**	**1992-2007**	**2000-2007**
China, PR	8.4	4.5	10.9	11.3	9.7	8.0
Australia	3.3	2.6	4.8	4.2	2.7	0.9
Peru	4.8	6.5	3.1	5.2	6.0	6.8
USA	1.5	1.0	1.1	5.6	2.5	-0.8
Canada	1.1	9.0	-0.8	-2.2	-4.3	-6.6
India	10.4	9.5	10.4	4.0	9.0	15.1
Mexico	1.1	-0.2	1.1	4.2	3.0	1.7
Kasakhstan				5.6	5.2	4.8
Russia + Usbekistan				-6.5	-1.8	3.8
UdSSR/GUS		6.9	-2.9			
others	0,9	3,4	-0,2	-1,2	0,3	2,0
World	**2,6**	**4,4**	**1,0**	**3,0**	**3,1**	**3,4**

Fig. 3 Development of mine production of zinc by country, growth rates and zinc price (Data sources: World Bureau of Metal Statistics (1960–2009) LME)

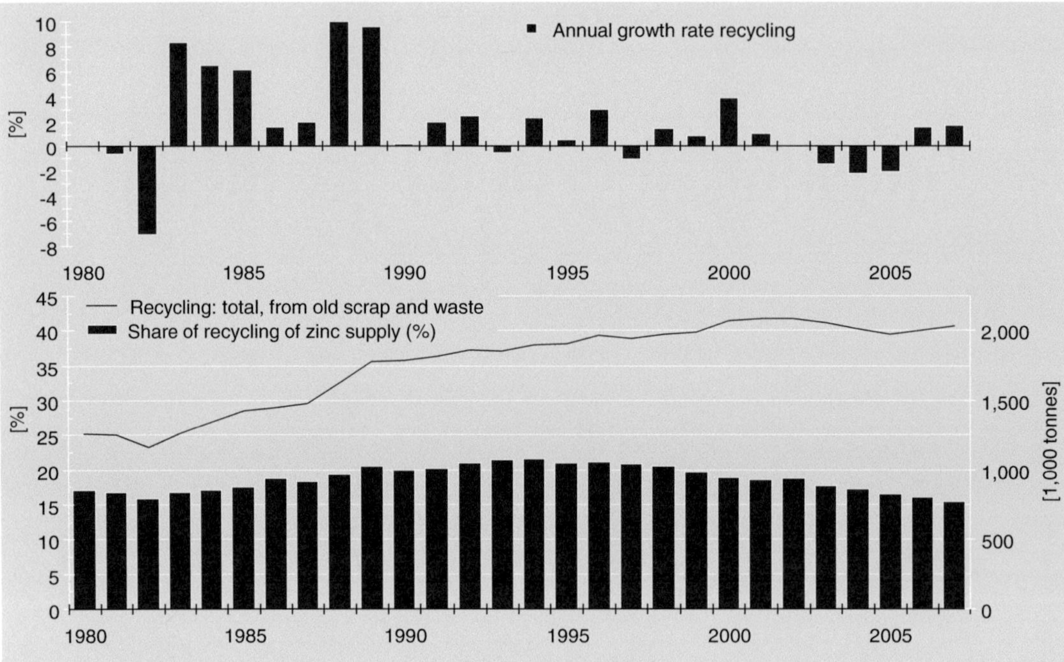

Fig. 4 Development of zinc recycling, share of zinc supply and annual growth rate (Data source: International Lead and Zinc Study Group 2009)

tonnes (1.3% to –0.45% of refined production; Table 1). Within that range, the market situation is regarded as moderate. Levels above +150,000 tonnes are considered relaxed, and levels below –50,000 tonnes are considered problematic.

Stock keeping (rating: 8/2008, problematic): The stock-keeping situation is evaluated as problematic. Between 1973 and 2007, average annual stocks were 1.17 million tonnes. In 1994, the ratio of stocks to refined production peaked at 30% (2.13 million tonnes). After that, it fell to a problematic 5.2% in 2007 (Fig. 8). In 2008, it showed a slight rise of 1.5% to reach 6.7%. In order to quantify the supply risk, historic stock keeping is considered moderate here when stocks amount to around 10–15% of refinery production (Table 1). If stocks are below 10% of refinery production for a period of more than 1 year, there would seem to be serious problems maintaining the supply. Although this indicator is more appropriate for understanding the current situation and describing short-term risks, strategic stock planning by governments such as the USA and more recently by China and Japan may influence the market in the longer term.

Production Costs

Cash costs (rating: 7/2008, problematic, tendency unchanged): Cash costs (C1) for zinc were around 0.53 USD/lb (1,170 USD/t) in 2008 and around 0.55 USD/lb (1,200 USD/t) in early 2009 (Fig. 9). Compared to around 0.35 USD/lb (772 USD/t) in 2001, they increased by over 50%. High material costs, the opportunity to bring more high-cost deposits into production, and the weak USD were the main factors responsible for this increase.

In 2008, zinc was produced by individual companies in a cash cost range from below 0.20 USD/lb (<450 USD/t) to about 0.90 USD/lb (>2,000 USD/t). After the financial crisis, zinc prices dropped to around 0.54 USD/lb (1,200 USD/t; approx. 50th percentile of cash cost curve), which means that almost half of the capacity was produced with a negative cash flow.

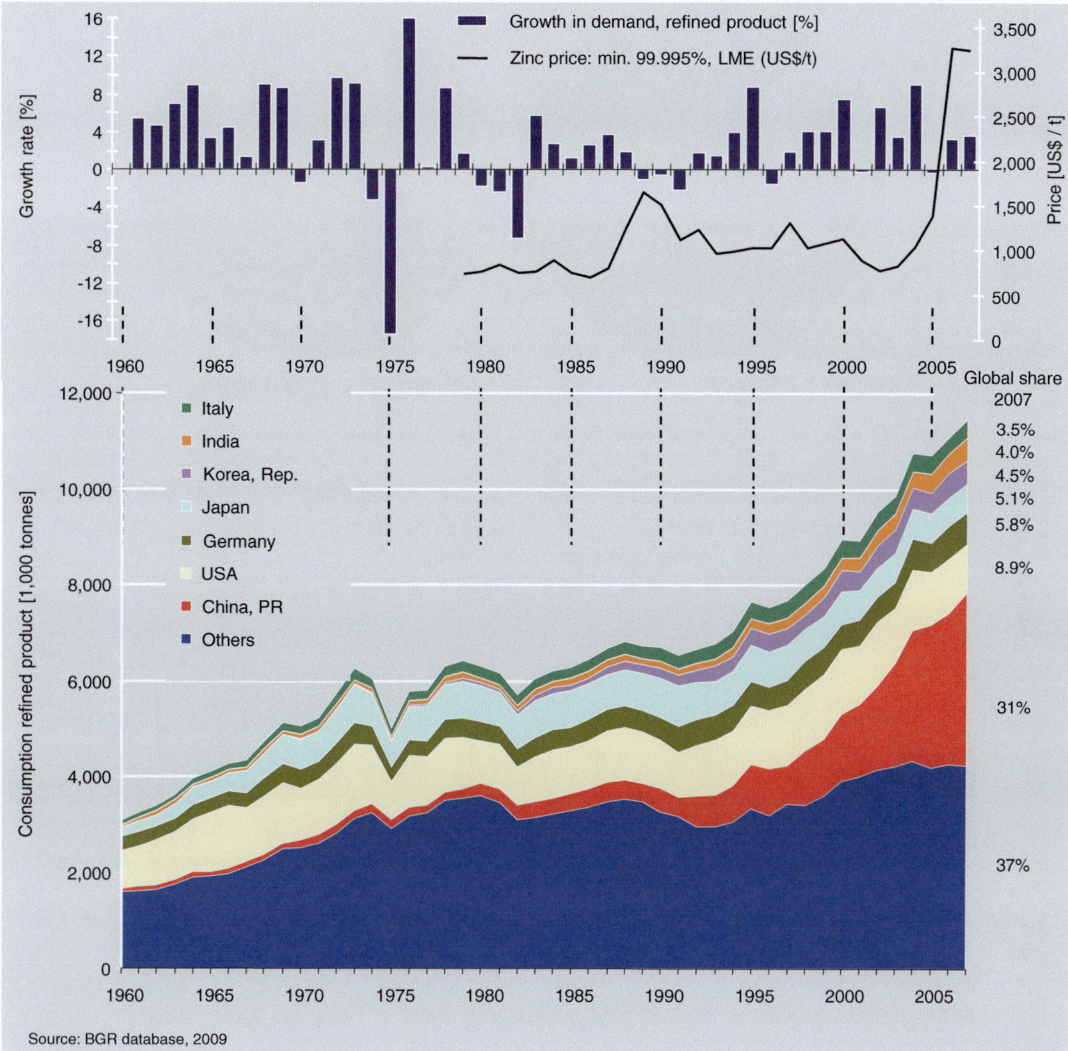

Growth rate%

Period:	**1960-2007**	**1960-1974**	**1974-1991**	**1992-2000**	**1992-2007**	**2000-2007**
China, PR	8.7	7.0	4.7	10.2	12.1	14.4
USA	0.5	3.1	-1.5	3.4	-0.1	-4.0
Germany	1.7	2.0	1.9	0.0	1.5	3.3
Japan	2.4	9.7	1.2	-1.8	-1.9	-2.0
Korea, Rep.			14.3	6.4	4.5	2.3
India	4.4	2.6	2.4	8.6	8.2	7.7
Italy	3.3	6.4	2.0	2.9	1.9	0.8
Others	2,1	5,3	-0,1	3,5	2,4	1,2
World	**2,8**	**5,0**	**0,5**	**3,7**	**3,7**	**3,6**

Fig. 5 Development of refinery consumption of zinc by country, growth rates and zinc price (Data sources: World Bureau of Metal Statistics, (1960–2009) LME)

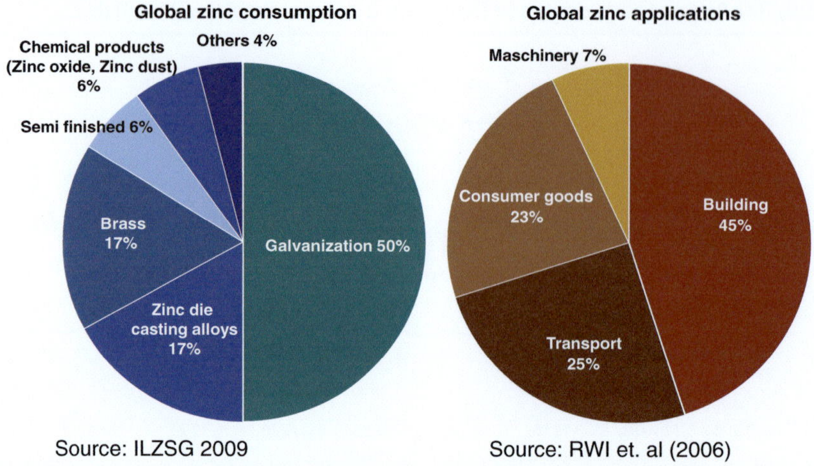

Fig. 6 Global zinc consumption by application and by sector (International Lead and Zinc Study Group 2009)

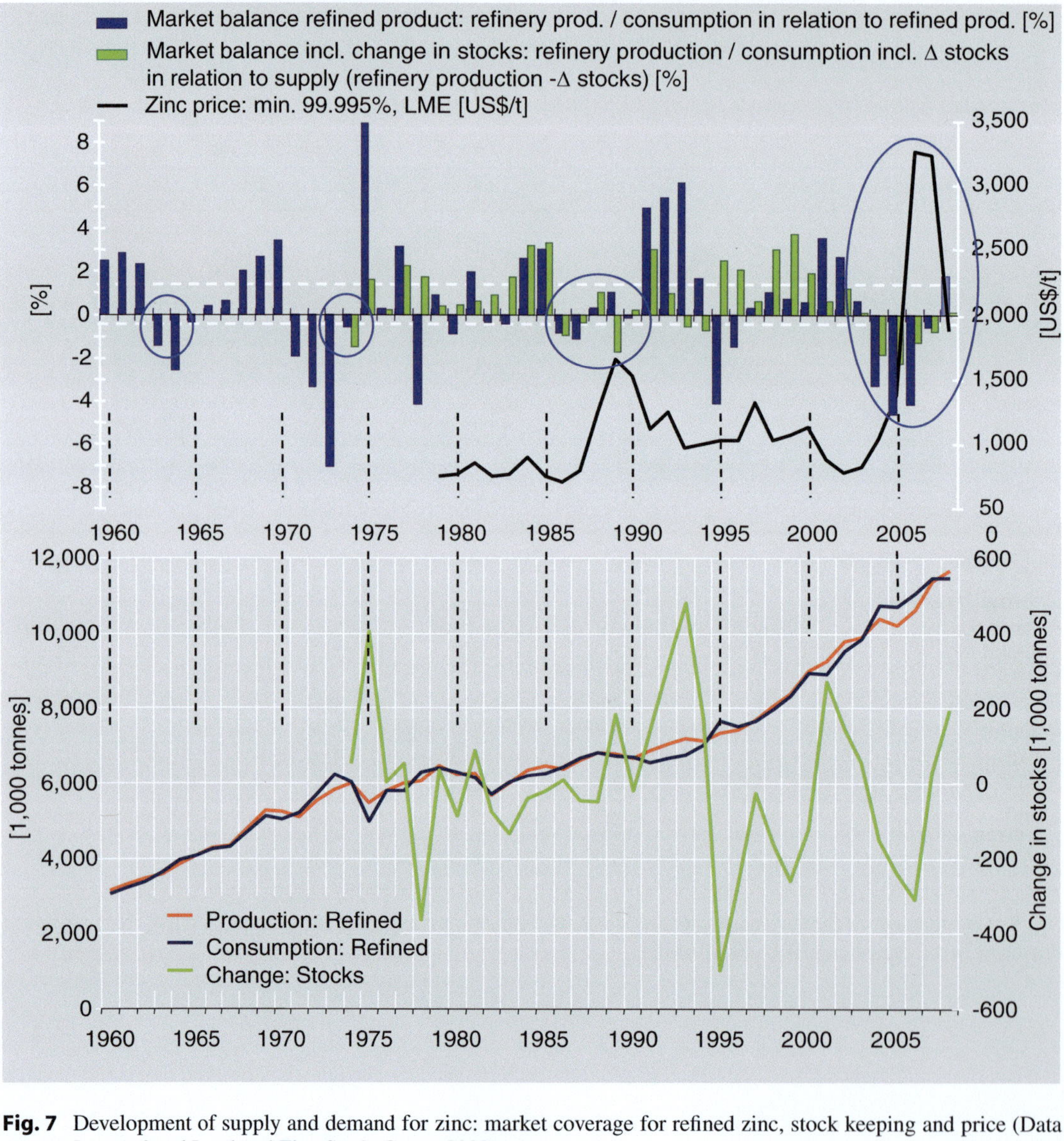

Fig. 7 Development of supply and demand for zinc: market coverage for refined zinc, stock keeping and price (Data source: International Lead and Zinc Study Group 2009)

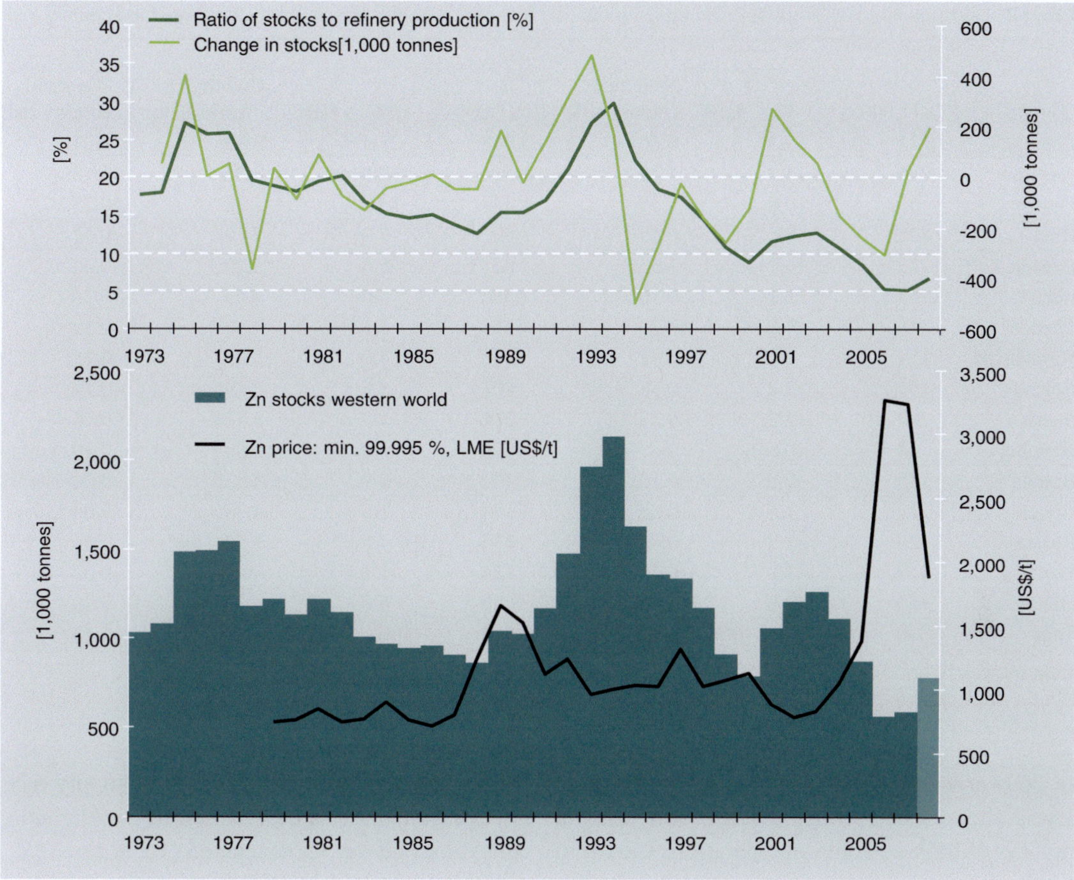

Fig. 8 Development of stocks, ratio of stocks to refinery production and zinc price (Data sources: International Lead and Zinc Study Group 2009, LME)

Fig. 9 Development of C1 cash costs for zinc mines and mining projects (Data source: BGR database; C1 cash operating costs including mining, processing, site administration and net of by-product credits)

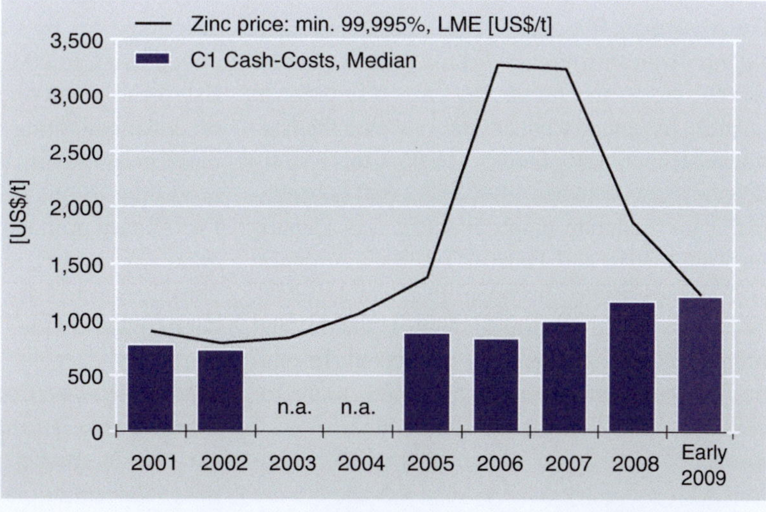

Table 2 Geostrategic risk for global zinc mine production including country concentration and weighted country risk

	1996				2007				
Country	Zn [1,000 t]	%	HHI	Weighted country risk	Zn [1,000 t]	%	HHI	Country risk (World Bank)[a]	Weighted country risk
China, PR	1,121	15.3		0.66	3,048	27.4		3.93	1.07
Australia	1,076	14.7		1.15	1,514	13.6		8.23	1.12
Peru	761	10.4		0.45	1,444	13.0		4.28	0.55
USA	628	8.6		0.69	803	7.2		7.49	0.54
Canada	1,222	16.7		1.36	623	5.6		8.29	0.46
India	154	2.1		0.10	558	5.0		4.63	0.23
Mexico	348	4.8		0.22	452	4.1		4.66	0.19
Kazakhstan	159	2.2		0.08	446	4.0		3.85	0.15
Ireland	165	2.2		0.18	401	3.6		8.19	0.29
Sweden	160	2.2		0.18	214	1.9		8.57	0.16
Namibia	34	0.5		0.03	196	1.8		5.66	0.10
Bolivia	145	2.0		0.10	194	1.7		3.52	0.06
Brazil	117	1.6		0.08	194	1.7		4.78	0.08
Russia	126	1.7		0.06	177	1.6		3.50	0.06
Poland	159	2.2		0.14	124	1.1		5.97	0.07
Others	953	13.0		0.66	756	6.8		–	0.30
Total	7,330	100	**976**	**6.12**	11,144	100	**1,274**	–	**5.45**

Data sources: World Bank (2009) and World Bureau of Metal Statistics (2007–2009)
[a]World Bank scale (–2.5 to +2.5) converted by BGR to values between 1 and 10; 1 = highest risk

In 2009 and early 2010, zinc prices increased to levels above 2,000 USD/t, which will bring a number of high-cost mining operations back into production.

Against this historic background, if real average cash costs increase to a level above 1,100 USD/t (0.50 USD/lb) over a couple of years, the situation may be termed problematic with respect to possible long-term price escalation (Table 1).

Geostrategic Risks

Country concentration and country risks (rating: 4/2007, moderate; trend: moderate): In order to quantify the country concentrations with the Herfindahl-Hirschman Index (HHI), this study defines scores between 1,000 and 2,000 as benchmarks for moderate supply risk (the U.S. Department of Justice and Federal Trade Commission (1997) uses values between 1,000 and 1,800). Scores above 2,000 are rated problematic, and scores below 1,000 are seen as relaxed. In relation to the standardised World Bank scale for country risks (World Bank 2009, values of –2.5 to +2.5 converted to a scale from 1 to 10), countries with a country risk between 4.5 and 5.5 are defined as benchmarks for moderate supply risk. Values below 4.5 are classified as problematic, and values above 5.5 as relaxed (Table 1).

With a Herfindahl-Hirschman Index (HHI) of 1,274 for zinc, the country concentration for mine production is moderate but has increased compared with 1996 (Table 2, rating: 4.5; see also Fig. 10 for global distribution of production). Although the increased country concentration for mine production is largely due to the share accounted for by China, which had an elevated country risk in 2007 (3.9, rating: 7), the country risk for mine production is still moderate. Based on equal weighting of these indicators, the combined geostrategic risk is thus moderate. Because, according to this principle, the geostrategic risk for refinery production is evaluated at 4.5 (moderate, not shown), the overall result is a moderate geostrategic risk for zinc mine and refinery production worldwide.

Market Power/Company Concentration

Company concentration (rating: 1, relaxed): The exercise of market power through global company concentration for zinc mine and refinery

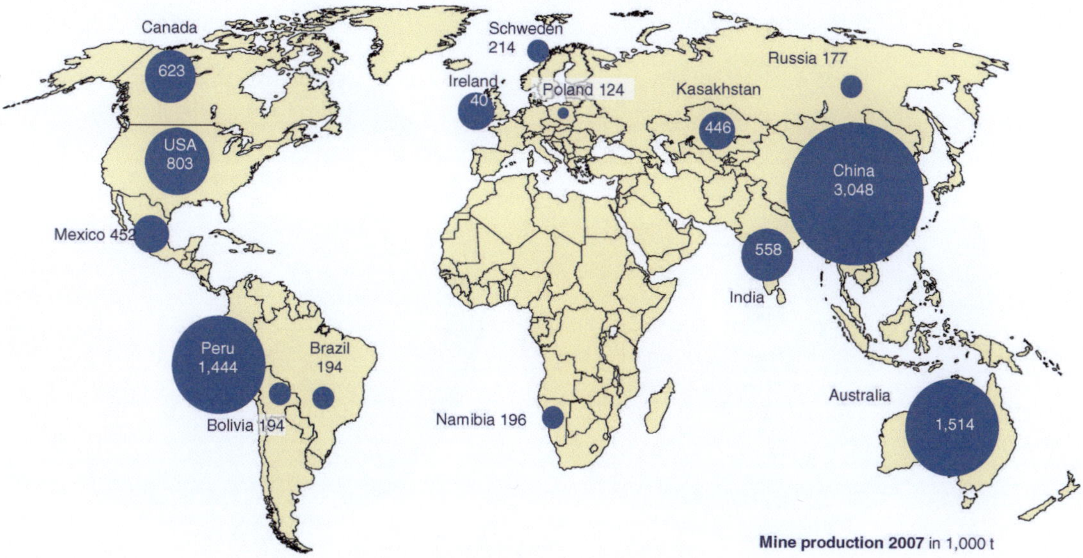

Fig. 10 Global zinc mine production by country (Data source: World Bureau of Metal Statistics)

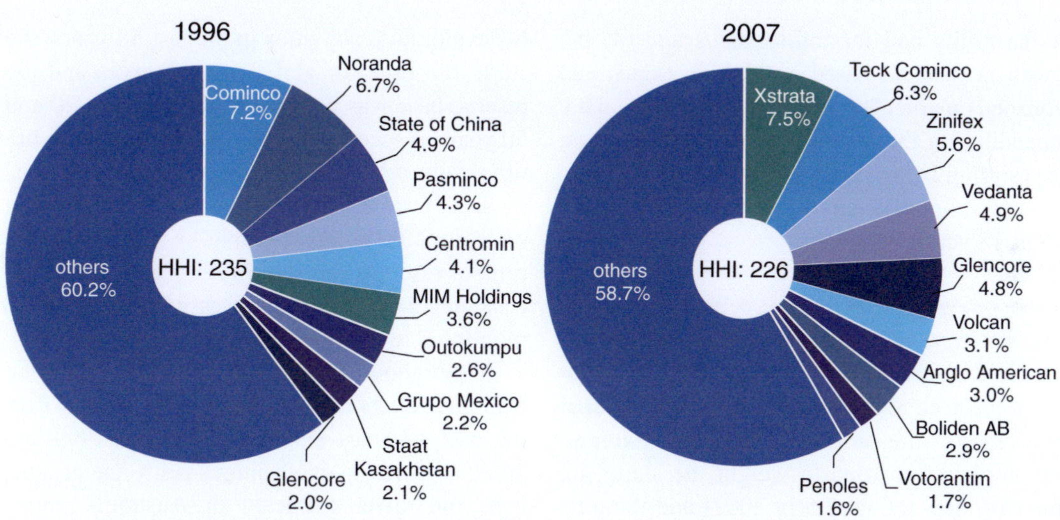

Fig. 11 Company concentration for zinc mining, comparison between 1996 and 2007 (*HHI* Herfindahl-Hirschman Index, Source: Raw Materials Group 2009)

production is considered relaxed in overall terms, since in each case the HHI lies well below 1,000 (Fig. 11). As far as competition is concerned, the supplier base can thus be classified as diversified. In order to quantify the company concentration using the HHI, the scale is applied in a similar way to country concentration.

Supply and Demand Trends

The forecast for future supply and demand is based on the analysis of: (1) degree of exploration; (2) development of mining investment; (3) production capacity of mines currently under construction/expansion and production forecasts for exploration projects with grassroots,

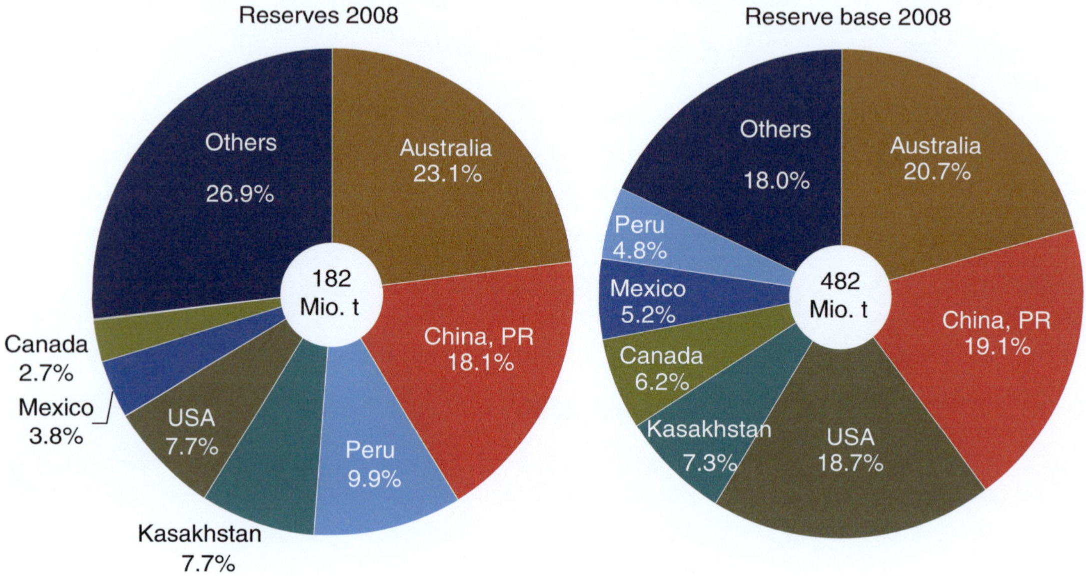

Fig. 12 Distribution of worldwide reserves and reserve base for zinc (Data source: USGS 2009)

pre-feasibility and feasibility status and (4) the growth in demand based on historic trends and published market sector and economic data. Depending on the supply and demand forecast, the resulting balance can mean that supply may or may not be sufficient to meet a given demand in 5 or 15 years.

Degree of exploration (rating: 5, moderate): The static life time of reserves and the reserve base alone is not a measure of shortage. It is an indication of the need to invest in exploration (Wellmer 1998, 2008). The indicator "degree of exploration" is therefore used to weight the static life time (for zinc, rating: 7.5 in 2007) and the total amount of the exploration expenditures or budgets of mining and exploration companies in relation to annual global mine production (for zinc in Australia, rating 2–3: relaxed).

The largest zinc reserves and reserve base in the world lie in Australia, with shares of 20.7% and 23.1%, respectively (Fig. 12). Considering current mines and planned mining projects, total zinc reserves are 182 million tonnes, resources including reserves are 508 million tonnes according to Raw Materials Group (reserve base USGS, 482 million tonnes, Table 3). Thanks to successful exploration activities in the last 50 years, the static life time for global zinc reserves and the reserve base was maintained for between 20 and 50 years, respectively (Fig. 13). For the static life time, values between 25 and 45 years may thus be used as a benchmark for a moderate situation in exploration progress (Table 1). If the static life time of reserves falls below 25 years, as was the case for 2007, exploration activities should be monitored regularly.

Exploration expenditures for zinc (including lead and silver) have increased significantly over the past 10 years (Fig. 14). Between 1999 and 2004, exploration expenditures per tonne of zinc, lead and silver extracted in Australia ranged between 9 and 15 USD/tonne. This was way below the average of 25 USD/tonne for the period 1999–2008. In the light of the past 10 years' experience, we have set a range between 20 and 40 USD/exploration expenditure tonne zinc, lead and silver mine production as a moderate value (Table 1). Although data on exploration expenditures were only available for Australia, we assume that global exploration budgets follow a similar trend.

Because companies' exploration expenditures have been rising again since 2005, a positive trend is emerging regarding the possible discovery of

Table 3 Summary of worldwide zinc reserves and resources in operating mines, extensions, mines under construction and deposits under exploration

	Raw Materials Group				USGS	
Status	Zn resources[a] [mill. t]	Number	Zn reserves [mill. t]	Number	Reserve base[a] [mill. t]	Reserves [mill. t]
Under construction	29	28	16	18		
Restart	18	22	7	6		
Feasibility	59	36	24	19		
Pre-feasibility	30	41	3	9		
Grassroots/conceptional	87	183	0.5	6		
Subtotal	**223**	**310**	**51**	**58**		
Operating mines	231	154	111	138		
Mine extensions	76	27	33	24		
Subtotal	**308**	**201**	**144**	**148**		
Total	**508**	**491**	**182**	**206**	**482**	**182**

Sources: BGR database, Raw Materials Group (2009), and USGS (2009)
[a]Reserves included in resources

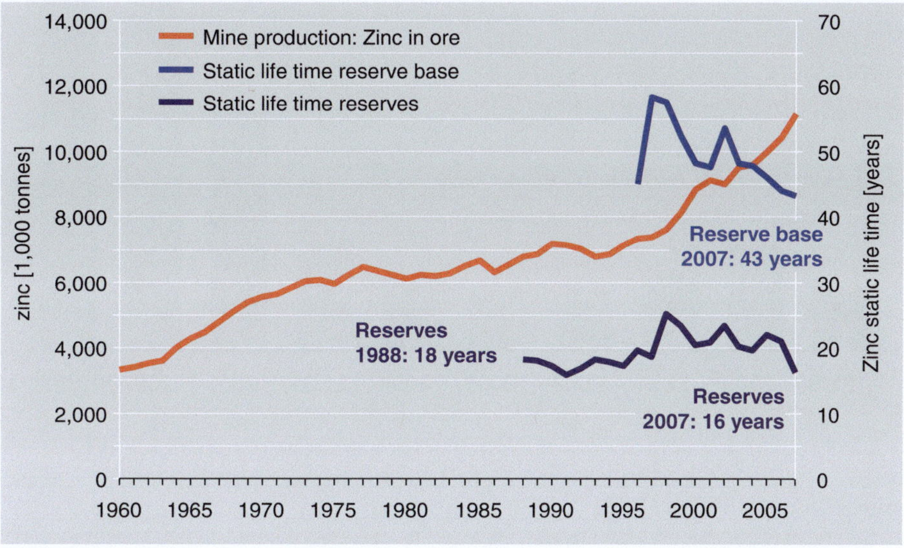

Fig. 13 Development of static life time for zinc reserves and the reserve base (Data sources: World Bureau of Metal Statistics (2007–2009) USGS)

new mine locations. In the long term, this would lead to an increase in reserves and the reserve base. For the past 5 year period 2003–2008, the rating of 4.5 for zinc (equal weighting of static life time and exploration expenditures) indicates that the degree of exploration is moderate.

Investment (for 2004–2009, rating: 4, moderate): Between 1998 and 2009, the trend for capital expenditure on new zinc mines was similar to the development of exploration costs (Fig. 15). Per tonne of zinc extracted worldwide, investment fell from 95 USD/tonne in 1998/1999 to the problematic level of 24 USD/tonne in 2002/2003. Between 2000 and 2005, it was significantly below the average of 71 USD/tonne for the period 1998–2009. Capital expenditure on new zinc mines has increased significantly since 2003.

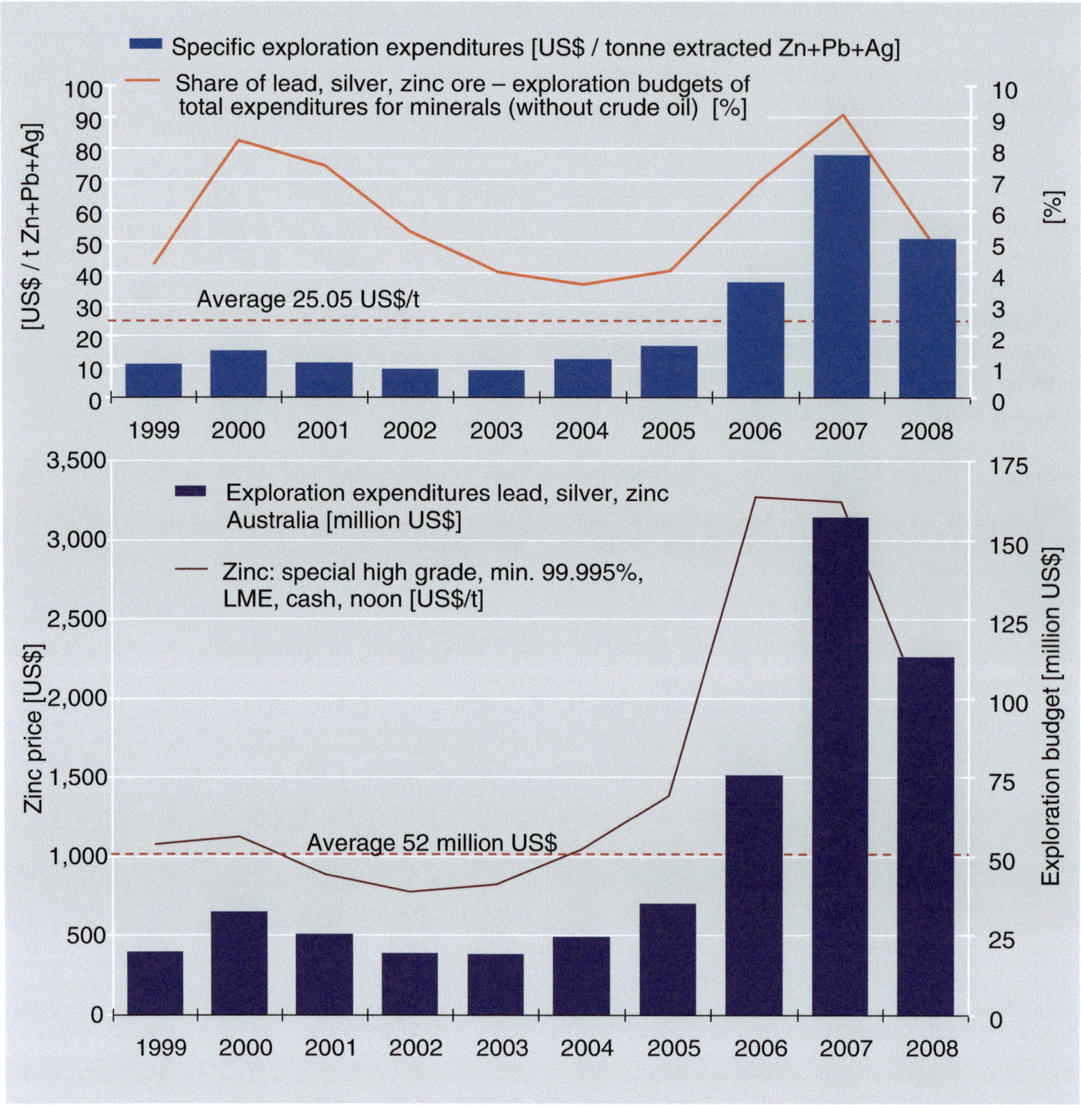

Fig. 14 Development of exploration expenditures for zinc, lead and silver and specific exploration expenditures (Data source: Australia Bureau of Statistics 2009)

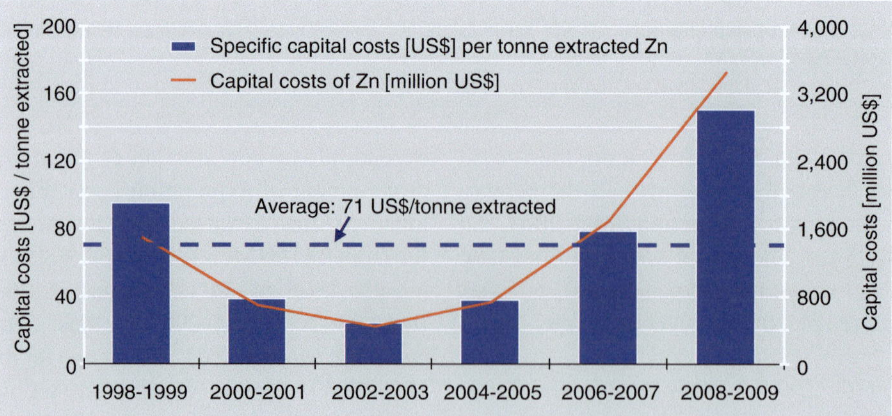

Fig. 15 Development of capital costs for zinc mines and projects and specific capital costs (Data source: Raw Materials Group 2009)

Table 4 Summary of potential additional annual production capacity for zinc from mine extensions, mines under construction and new exploration projects available to the market up to and after 2013

Status	Number	Current mine extensions [1,000 t/year Zn]	Additional up to 2013 [1,000 t/year Zn]	On hold [1,000 t/year Zn]	Additional after 2013 [1,000 t/year Zn]
Mine extensions	23	646		14	
Under construction	19		1,183	273	55
Feasibility	41		1,814	80	799
Pre-feasibility	17		107		897
Grassroots/ conceptional	19		48	45	1,890
Restart	15		212	84	98
Subtotal	**134**	**646**	**3,364**	496	**3,739**
Projects on hold	−11	−14	−482		
Total	**123**	**632**	**2,882**		**3,739**

Data sources: BGR database, Raw Materials Group (2009), and USGS (2009)

In 2006/2007, it rose to 79 USD/tonne, and in 2008/2009, it reached about 150 USD per tonne of zinc extracted worldwide.

Market balance in 5 years (rating: 6.5–7.0, problematic): Total additional annual zinc production capacity expected from mines under construction and projects with pre-feasibility and feasibility status is 2.8 million tonnes for the period 2008–2013. This assumes that 40 zinc mining projects for which production is scheduled to start by 2013 or earlier will actually come into production (Table 4, see also Table 5, projects "on ice" and mine extensions not considered). About 24% of the additional production capacity is being developed in Australia and 12–8% in Russia, Mexico, South Africa, Iran, Canada and Bolivia.

Total zinc production in 2013 could be 14.02 million tonnes, on the basis of current mine production (11.14 million tonnes, 2007) and new mine projects (Table 4). This would correspond to annual production growth of 3.9% until 2013, which is just above the average of the recent boom years (2000–2007: 3.4%) and above the long-term trend (1960–2007: 2.6%). Zinc recycling could boost production by 2.27 million tonnes by 2013, assuming average annual growth of 1.8% on the basis of growth rates between 1980 and 2007. Recycling material, current mine production and production from new projects could add up to 16.28 million tonnes zinc available for 2013.

However, several mine projects are currently on hold (about 482,000 tonnes). Furthermore, we estimate in our scenario that around 25% of the remaining mine projects which were scheduled for 2013 will not come into production, partly because of the credit crunch and partly because of other difficulties. This would reduce new mine production capacity to 2.16 million tonnes and total zinc production including recycling material to 15.57 million tonnes. In that case, the annual growth of zinc production until 2013 would average around 3.1%.

Developments in demand are difficult to predict. Demand growth for zinc (refined zinc consumption and old scrap recycling) was 2.1% per year between 1980 and 2007 and 3.2% between 1992 and 2007. If China maintains a zinc demand growth of around 10% per year until 2013, the average annual growth of global zinc demand could reach almost 4%. As the major future markets for zinc are zinc coating, die-casting alloys and brass in building, transport, consumer goods and machinery, these sectors will also dominate zinc demand in the decade ahead. Future technologies such as zinc tin oxide layers in displays or zinc-based batteries may stimulate demand to a certain extent, but in overall terms, the impact will only be slight.

Table 5 Summary of the 25 largest zinc projects with pre-feasibility and feasibility status with an expected annual production capacity of over 50,000 tonnes, project costs and expected start of production

Project	Company	Country	Type	Status	Project costs (mill. US$)	Expected annual capacity (1,000 t Zn)	Expected start of production
Gamsberg Zinc Mine	Anglo-American, Exxaro	Rep. South Africa	OP	Feasibility	860	300	2007
Mehdiabad Lead/Zinc Mine	State of Iran, Union Resources	Iran	OP	Feasibility	1,300	300	2010
Ozernoye Lead/Zinc Mine	IFC Metropol, Lundin Mining	Russia	OP	Feasibility	430	300	2008
San Cristobal Polymetallic Mine	Apex Silver, Sumitomo Corp	Bolivia	OP	Under construction	n.a.	225	2007
Penasquito Gold/Silver Deposit	Goldcorp	Mexico	OP	Under construction	1,490	204	2008
Dugald River Lead/Zinc Deposit	OZ Minerals	Australia	UG	Feasibility	417.5	200	2011
Red Zinc Manto Zinc Deposit	Metalline	Mexico	UG	Pre-feasibility	250	182	After 2013
Citronen Zinc/Lead Deposit	Ironbark Gold	Greenland	UG	Pre-feasibility	205	180	After 2013
Trident Gold Deposit	Avoca Resources	Australia	UG	Feasibility	n.a.	175	2007
Izok Lake Base Metal Deposit	OZ Minerals	Canada	OP	Feasibility	539	150	After 2013
Cerro Lindo Copper Mine	Milpo	Peru	UG	Under construction	n.a.	146	2007
Lady Loretta Zinc Mine	Xstrata, Min Securities	Australia	UG	Feasibility	167.4	125	2008
Dairi Zinc/Lead Deposit	Herald	Indonesia	UG	Feasibility	227	120	After 2013
Perseverance Zinc Mine	Xstrata	Canada	UG	Under construction	135.1	115	2008
Oued Amizour Zinc Deposits	Terramin Aust, State of Algeria	Algeria	UG	Pre-feasibility	n.a.	109	After 2013
San Gregorio Zinc Deposit	Buenaventura	Peru	OP	Pre-feasibility	200	105	After 2013
San Nicolas Copper/Zinc Deposit	Teck Cominco, Goldcorp	Mexico	OP	Feasibility	245.6	95	After 2013
Accha Lead/Zinc Deposit	Southwestern	Peru	OP	Feasibility	65	75	After 2013
Terrazas Copper/Zinc Deposit	Constellation Cu	Mexico	OP	Pre-feasibility	500	68	2009
Aguas Tenidas Base Metals Mine	Iberian Minerals	Spain	UG	Under construction	168.2	67	2009
Prairie Creek Lead/Zinc Deposit	Canadian Zinc	Canada	UG	Pre-feasibility	37.7	60	After 2013
Jabali Zinc Deposit	ZincOx, Ansan Wikfs	Yemen	OP	Feasibility	216	56	2009
Bisha Gold/Silver/Zinc/Copper Deposit	Nevsun, State of Eritrea	Eritrea	OP	Under construction	246	55	2010
Wolverine Zinc Deposit	Yukon Zinc	Canada	UG	Feasibility	193.3	53	After 2013
Sulphur Springs Zinc/Copper Deposit	CBH Resources	Australia	n.a.	Feasibility	187	50	2009

Sources: BGR database, Raw Materials Group (2009)
OP open pit, *UG* underground, *n.a.* not available

For our scenario, we consider an annual growth of 3.2% (1992–2007) in global zinc demand and an expected growth of 3.1% in zinc production. This would lead to a small deficit of –0.04 million tonnes of zinc in 2013 (Figs. 16a and b). The deficit could probably be compensated by production from mine expansions which have not been considered in the estimate.

An additional 3.7 million tonnes of zinc per year are already scheduled to come on stream after 2013. Together with recycling material, this would – from what we know today – add up to 19.75 million tonnes of zinc supply in 2023. This would correspond to an average annual growth rate of 2.7% from 2007 to 2023.

Because of the financial and economic crisis in 2008/2009, we do not expect average annual growth in demand to exceed 3.2% until 2013. Greater demand would seem to be possible for the period 2013–2023, but this will depend on developments in emerging markets, especially China and India. Against the background of past industrialisation phases in Western countries (e.g. the period of high zinc demand growth 1960–1974, see Fig. 5), a global growth rate above 4% is possible. Investment in exploration and mine construction would have to expand accordingly.

Overall rating: Due to strained market coverage and low stock keeping, the overall supply conditions for 2007 were problematic (rating: 7). The situation improved in 2008 although stock keeping was still low. After the financial crisis, zinc price increases in 2009 and early 2010 to levels above 2,000 USD/t will result in a number of mines and mining projects that were formerly "put on ice" being brought back into production and development. This will improve the supply situation. The development of geostrategic risks and the possibility of mining companies and refinery operators exerting their market power to control prices are evaluated as moderate to relaxed due to a high degree of diversification. In contrast, supply and demand trends indicate that the market situation for zinc is problematic until 2013 and could continue to be tense until 2023. Exploration and mining projects should therefore be monitored carefully in order to evaluate future supply.

Discussion and Conclusions

One major constraint in the assessment of supply and demand is the availability of statistical data for mineral raw materials, which are largely based on worldwide surveys. Our analyses for zinc are based on the evaluation of historic trends and developments compared with current and future market trends. This makes it essential to have access to data series covering long periods of time. For this purpose, BGR not only maintains its own statistical database and archives, but also uses selected commercial databases on zinc and other mineral raw materials. For zinc, the database is relatively comprehensive, for other raw materials, the situation may be different.

Data sources which are not necessarily related to the mining industry "maybe problematic". For example, the World Bank Group's Governance index was used to estimate the country risks (World Bank 2009). As for most country risk classifications developed by banks and investment companies the main focus is on overall investment but not on supply risk. This means that the Governance index can, at best, only approximate a potential supply risk for raw materials from a particular country. Another example would be the analysis of the freight market (not used for zinc), where the data available are not specific to the mining industry. Here again, as for all other data, personal experience is necessary to assess data and information quality.

Estimating supply trends is no easy undertaking. Commencement dates and data on annual production capacity from new mine projects are only target figures published by exploration and mining companies. By their very nature, planned mining projects may be postponed or discontinued. Information on reserves and resources also changes during the course of exploration. This means that the status of projects must be updated regularly and new projects have to be comprehensively recorded. Evaluating data on annual production capacity from mine expansions is always difficult. It is unclear to what extent they will offset decreasing production from current

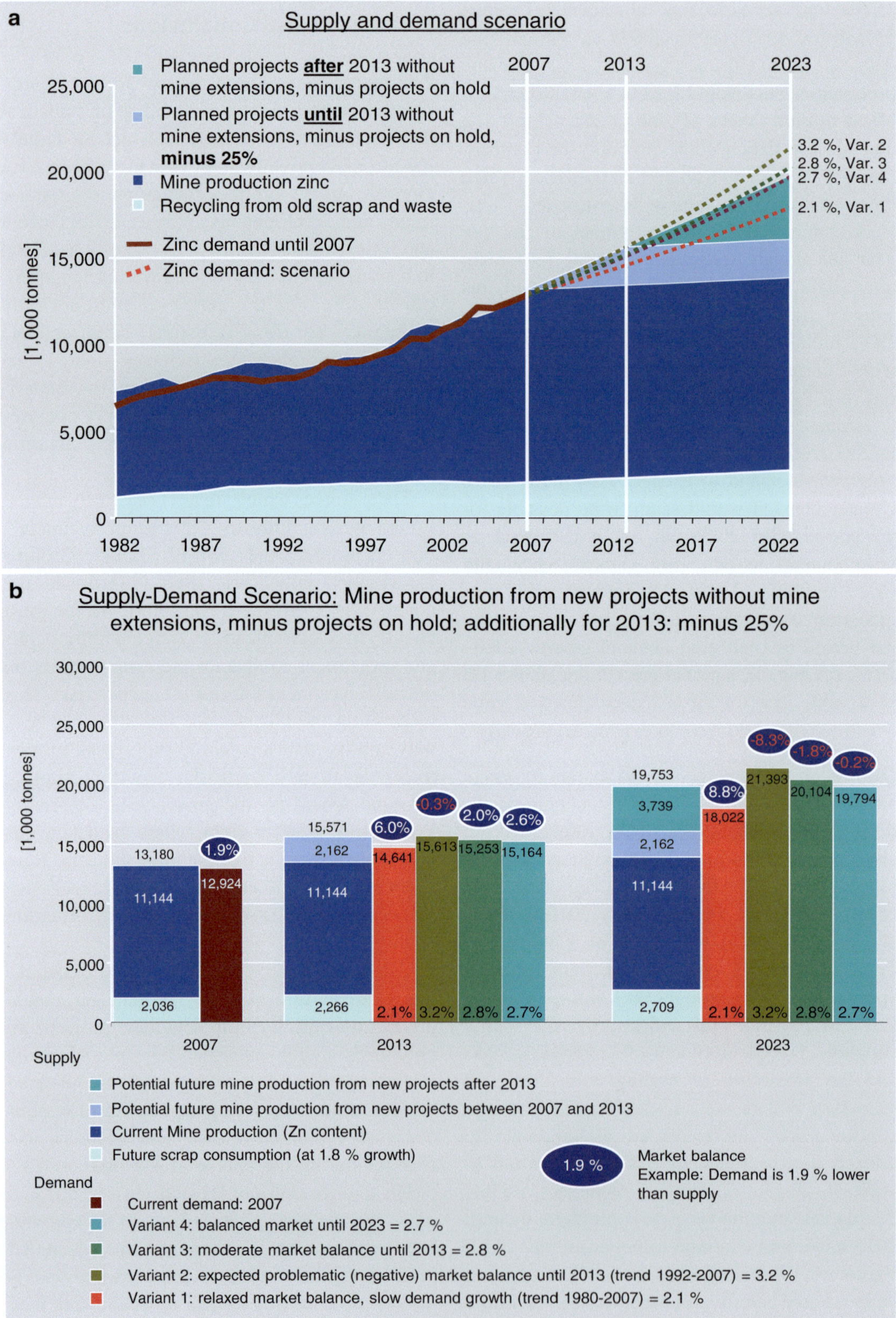
a
Supply and demand scenario
Planned projects after 2013 without mine extensions, minus projects on hold
Planned projects until 2013 without mine extensions, minus projects on hold, minus 25%
Mine production zinc
Recycling from old scrap and waste
Zinc demand until 2007
Zinc demand: scenario
2007
2013
2023
3.2 %, Var. 2
2.8 %, Var. 3
2.7 %, Var. 4
2.1 %, Var. 1
[1,000 tonnes]
25,000
20,000
15,000
10,000
5,000
0
1982
1987
1992
1997
2002
2007
2012
2017
2022
b
Supply-Demand Scenario: Mine production from new projects without mine extensions, minus projects on hold; additionally for 2013: minus 25%
[1,000 tonnes]
30,000
25,000
20,000
15,000
10,000
5,000
0
13,180
11,144
2,036
1.9%
12,924
15,571
2,162
11,144
2,266
6.0%
14,641
-0.3%
15,613
2.0%
15,253
2.6%
15,164
2.1%
3.2%
2.8%
2.7%
19,753
3,739
2,162
11,144
2,709
8.8%
18,022
-8.3%
21,393
-1.8%
20,104
-0.2%
19,794
2.1%
3.2%
2.8%
2.7%
2007
2013
2023
Supply
Potential future mine production from new projects after 2013
Potential future mine production from new projects between 2007 and 2013
Current Mine production (Zn content)
Future scrap consumption (at 1.8 % growth)
1.9 %
Market balance
Example: Demand is 1.9 % lower than supply
Demand
Current demand: 2007
Variant 4: balanced market until 2023 = 2.7 %
Variant 3: moderate market balance until 2013 = 2.8 %
Variant 2: expected problematic (negative) market balance until 2013 (trend 1992-2007) = 3.2 %
Variant 1: relaxed market balance, slow demand growth (trend 1980-2007) = 2.1 %

mine sites or – in absolute terms – add to new supply. For this reason, production information for mine extensions is initially not included in the future supply and demand balance, but it is looked at for final evaluation of the market situation.

Whereas potential future production can at least to some extend be predicted reasonably well within a time period of 5–10 years (lead time for zinc exploration and mining projects), the demand side cannot. However, it would seem to be a legitimate approach to take different demand scenarios and check whether supply is able to meet such demand. Historically, global demand for base metals has not changed rapidly within a 10–15 years time horizon (Frondel et al. 2007). Over such periods, the change in global demand growth has been 1% or 2% annually up or down. As long as there is no prospect of sudden technological innovations, global demand for zinc will not depart from this growth pattern. This is the basis for our evaluation of the demand side.

Further estimates of possible future growth in demand should include detailed studies of the development of emerging technologies, substitution potentials, forecasts of GDP, industrial production, urbanisation and regulatory or other public policy changes (see methods in Barney 1980; Sohn 2006; IMF 2006). Recently, Angerer et al. (2009) have introduced a method of estimating materials demand based on fundamental research into emerging technologies and the need for raw materials. More research in this field would permit better prediction of future demand. As innovation cycles are speeding up, revolutionary new technologies will emerge in exploration and mining as well. Such technologies need to be integrated into scenarios of supply and demand.

Apart from these uncertainties, the method described has some major advantages: a systematic market evaluation and assessment of supply risks using time series analysis ideally for a long period of time, which allows comparison of the current market situation with past market cycles; clear scaling of a range of risk factors, which helps to quantify the supply risk; comparability of results even among different mineral raw material sectors and estimates of future market trends. The latter is essentially based on supply and demand scenarios for a period of 5–15 years and accompanied by monitoring the degree of exploration, investment in mining projects, trends in production costs, geostrategic risks or changes in market power.

The method can be used to help companies make better informed decisions on whether to take action to secure the supply of raw materials. This method – among others – is used by Volkswagen AG and BGR to assess long-term trends in the mineral raw material markets. The automotive industry is particularly strongly affected by the availability and price of raw materials in its pre-production value-added chains. By carefully selecting materials and improving the efficiency of material use, as well as by making targeted use of hedging or long-term contracts, it is possible to counteract negative market trends at an early stage. Especially in the context of new technologies, transparency about risks in raw material supply is crucial.

Acknowledgements The results of this study are based on a joint research project by the research department of Volkswagen AG and Germany's Federal Institute for Geosciences and Natural Resources (BGR). This project, entitled "Criteria and Indicators for Risk Evaluation in the

Fig. 16 (**a**) Scenario of future supply and demand for zinc until 2013 and 2023. Supply up to 2013: includes production capacity of mines under construction and exploration projects in feasibility and pre-feasibility status announced as coming into production by 2013 or earlier. Not considered: mine extensions, projects on hold. Additionally: subtraction of 25% of planned production capacity. Supply up to 2023: includes all projects considered up to 2013 plus feasibility, pre-feasibility and grassroots exploration projects announced as coming into production after 2013 or with indication of planned production capacity without year of commencement. Not considered: mine extensions, projects on hold; demand calculated at growth rates between 2.1% and 3.2%. (**b**) Scenario of future supply and demand for zinc up to 2013 and 2023 including details for variants 1–4 (see text for further description)

Supply of Mineral Raw Materials," was carried out between September 2005 and May 2006. It was followed by the research project "Risk Assessment for Metals" and carried out between 2008 and 2010, which also included the analysis for zinc.

The BGR is the central geoscientific authority providing geo-related advice to the German Federal Government and the economy. This also applies to the activities of the BGR in the field of mineral economics. We would like to thank D. Homberg-Heumann and E. Westphale for the comprehensive preparation of data, tables and graphs. Our thanks also go to Dr. D. Huy and Dr. J. Vasters for critical discussions and revisions.

The research department of Volkswagen AG examines issues such as sustainable material procurement and development. Its findings are used in the evaluation of raw material markets, future technological trends and in corporate planning. We would like to thank Dr. M. Wagner, Dr. U. Elsner, Dr. D. Rosenau-Tornow, A. Riemann and S. Konietzko for their valuable contributions and ideas during the research project.

Finally, we would like to thank the reviewers at Springer for their valuable contributions and advice on improving the manuscript.

References

Angerer G, Erdmann L, Marscheider-Weidemann F, Lüllmann A, Scharp M, Handke V, Marwede M (2009) Rohstoffbedarf für Zukunftstechnologien – Einfluss des branchenspezifischen Rohstoffverbrauchs in rohstoffintensiven Zukunftstechnologien auf die zukünftige Rohstoffnachfrage. Fraunhofer Institut für System- und Innovationsforschung ISI, Institut für Zukunftsstudien und Technologiebewertung IZT GmbH, 384 p

Australia Bureau of Statistics (2009) Mining, exploration, mineral and petroleum exploration, Australia (cat no. 8412.0), Table 5; mineral exploration (other than for petroleum) – expenditure by mineral sought. http://www.abs.gov.au/AUSSTATS/abs@.nsf/DetailsPage/8412.0Dec%202009?. Accessed April 2009

Barney O (ed) (1980) Global 2000 – entering the twenty-first century. The report to the president, vol II. Council on Environmental Quality and the Department of State, Washington, DC, 815 p

BDI Bundesverband der Deutschen Industrie e.V. (2005) BDI-Kongress Rohstoffsicherheit – Herausforderung für die Industrie. BDI-Drucksache Nr. 365, Berlin, 110 p

BDI Bundesverband der Deutschen Industrie e.V. (2007) Rohstoffsicherheit – Anforderungen an Industrie und Politik – 2. BDI-Rohstoffkongress am 20. März 2007 in Berlin, Ergebnisbericht der BDI-Präsidialgruppe "Internationale Rohstofffragen." BDI-Drucksache Nr. 395, Berlin, 102 p

BMWi, Projektgruppe Rohstoffe, Bundesministerium für Wirtschaft und Arbeit (2005) Bericht zur aktuellen rohstoffwirtschaftlichen Situation und zu möglichen rohstoffpolitischen Handlungsoptionen. BMWi, Berlin, 45 p

BMWi (2007) Elemente einer Rohstoffstrategie der Bundesregierung. Die Bundesregierung, Berlin, 17 p

Citigroup (2005) China – the engine of a commodities super cycle. Commodity update, global equity research – Commodities, Smith Barney Division of Citigroup Global Markets Inc., March 2005, 24 p

Committee on Critical Mineral Impacts of the U.S. Economy (2008) Minerals, critical minerals, and the U.S. economy. The National Academy of Sciences, National Academies Press, Washington, DC, 264 p

Deutsche Bank (2008) Commodities outlook, global markets research. Deutsche Bank AG, London, 102 p

European Commission (2008) The raw materials initiative – meeting our critical needs for growth and jobs in Europe. Communication from the Commission to the European Parliament and the Council COM(2008)699, (SEC(2008)2741), 13 p

Frondel M, Grösche P, Huchtemann D, Oberheitmann A, Peters J, Angerer G, Sartorius C, Buchholz P, Röhling S, Wagner M (2007) Trends der Angebots- und Nachfragesituation bei mineralischen Rohstoffen. Rheinisch-Westfälisches Institut für Wirtschaftsforschung (RWI Essen), Fraunhofer-Institut für System- und Innovationsforschung (ISI), Bundesanstalt für Geowissenschaften und Rohstoffe (BGR), 350 p

IMF (2006) The boom in non-fuel commodity prices. Can it last? In: International Monetary Fund, chapter 5. IMF, Washington, DC, 31 p. http://www.imf.org/external/pubs/ft/weo/2006/02/pdf/c5.pdf. Accessed April 2009

India Ministry of Mines (2008) National mineral policy 2008. Government of India, Ministry of Mines, New Delhi, 15 p. http://mines.nic.in/NMP2008.pdf. Accessed April 2010

International Lead and Zinc Study Group (1960–2009) Lead and zinc statistics – Monthly Bulletin

Meadows DH, Meadows DL, Randers J, Behrens WW (1972) The limits of growth. Universe Books, New York, 205 p

Meti Japan Ministry of Economy, Trade and Industry (2009) Announcement of strategy for ensuring stable supplies of rare metals. http://www.meti.go.jp/english/press/data/20090728_01.html. Accessed July 2009

Paley Commission (1952) The President's Materials Policy Commission (Paley Commission). Resources for freedom, 5 vols. Government Printing Office, Washington, DC

Radetzki M (2006) The anatomy of three commodity booms. Resources Policy 36:56–64

Raw Materials Group (2009) Raw materials data, February 2009

Rosenau-Tornow D, Buchholz P, Riemann A, Wagner M (2009) Assessing long-term supply risks for mineral raw materials – a combined evaluation of past and future trends. Resources Policy 34:161–175

Selwyn Resources Ltd. (2008) Zinc – the storm continues to brew. Company report, 26 May 2008

Sohn I (2006) Long-term projections of non-fuel minerals: we were wrong, but why? Resources Policy 30:259–284

U.S. Department of Justice and Federal Trade Commission (1997) Horizontal merger guidelines. U.S. Department of Justice and Federal Trade Commission, Washington, DC. http://www.usdoj.gov/atr/public/guidelines/horiz_book/toc.html. Accessed March 2010

USGS (2009) Commodity statistics and information. Minerals information statistics, United states geological survey. http://minerals.usgs.gov/minerals. Accessed March 2009

Wellmer F-W (1998) Lebensdauer und Verfügbarkeit energetischer und mineralischer Rohstoffe. Erzmetall 51:663–675

Wellmer F-W (2008) Reserves and resources of the geosphere. Z dt Gesellschaft für Geowissenschaften 159(4):575–590

World Bank (2008) The commodity boom: longer-term prospects. In: World Bank (ed) Global economic prospects 2009. Commodity markets at the crossroads. World Bank and the International Bank for Reconstruction and Development, Washington, DC, pp 51–93

World Bank (2009) World development indicators. The World Bank, Washington, DC, CD-ROM

World Bureau of Metal Statistics (1960–2009) Metallstatistik/metal Statistics - various Editions, Ware, England

Stretching the Availability of Non-renewable Resources

Sigurd Heiberg and Friedrich-W. Wellmer

Abstract

Stretching the availability of non-renewable resources at reasonable prices is critical. A number of principles for achieving this are explained by reference to examples. They point to the importance of efficiency and to the need for governments, industry and financiers to act in concert to achieve it.

A classification of resources that is both complete and coherent with other management information is reviewed for its potential both to allow existing resource inventories to be compared and for its potential to serve key stakeholders as a global standard facilitating efficient operations. This is the United Nations Framework Classification for Fossil Energy and Mineral Reserves and Resources (UNFC 2009). It provides a framework for communicating information on reserves and resources by seeing recoverable quantities in light of the projects that recover them. It is reflective of economic and social conditions, industrial and technical conditions and, of course, the natural endowment – the geological conditions.

An account is given of how government and industry have been able to act in concert over nearly four decades to stretch the petroleum resources of Norway while reducing emissions and improving the return on capital employed through staging the developments.

Finally, the need for improved efficiency, demanding common standards for common markets is reiterated. The introduction of international financial reporting standards for extractive activities, now in preparation, is seen as a golden opportunity to implement this.

S. Heiberg (✉)
Chairperson of the UNECE Committee on Sustainable Energy, Geneva, Switzerland

Corporate Strategy Analysis, Statoil ASA, Stavanger N4035, Norway
e-mail: shei@statoil.com

F.-W. Wellmer
Formerly Bundesanstalt für Geowissenschaften und Rohstoffe (BGR), Neue Sachlichkeit 32, Hannover D-30655, Germany
e-mail: fwellmer@t-online.de

R. Sinding-Larsen and F.-W. Wellmer (eds.), *Non-Renewable Resource Issues: Geoscientific and Societal Challenges*, International Year of Planet Earth,
DOI 10.1007/978-90-481-8679-2_9, © Springer Science+Business Media B.V. 2012

Keywords
Canada • Case examples • Circum-pacific region • Crirsco • Efficiency • Finance • Fossil energy • Gas • Global standard • Government • Industry • Mineral resources • Norway • Oil • Policy • Resource classification • Resource managment • SPE • UNECE • UNFC

Introduction

Normally, a mining law states that natural resources have to be exploited wisely to the benefit of a country. Ruthless mining like high grading of a deposit for a short-term gain contradicts "good miner's practice" everywhere in the world. In the oil industry flaring gas, early depressurisation of oil and condensate reservoirs and dilution of oil by inefficient displacement fluids are equivalent examples of this.

To illustrate how to stretch the availability of natural resources, the term "reserves" (in the meaning commercially recoverable quantities) contrary to "resources" has to be understood. Those concentrations of a known mineral or substance which can be recovered by a commercial development or mining operation equate to reserves[1] as part of the overall resource of a country; otherwise, they are resources sensu stricto. In recent years, the terminology concerning reserves and resources has been tightened and standardised internationally (UNFC 1997; SPE/WPC 1997; SPE/WPC/AAPG 2000; UNFC 2004; JORC 2004; CIM 2005; CRIRSCO 2006; SPE/WPC/AAPG/SPEE 2007; UNFC 2009). Whereas formerly expressions like "potential reserves" were used frequently, these are strictly "resources" today. Governments should have the interest to maximise the amounts of commercially recoverable quantities on the one hand and to optimise the return to their country on the other hand.

The boundary between reserves and resources is a dynamic one. Resources of today may be converted into commercially recoverable quantities by a change of prices, change of technology and change of commitment to develop, for example by a brilliant idea of a company taking over a resource which formerly was considered uneconomic or otherwise. Also, the legal framework has an influence on the resource/reserve boundary. What in one jurisdiction and one fiscal regime might be a reserve can be a resource in another. Taxation is often a cost element. Gross taxes and fees always are. "Brown taxes" where the government collects revenue by entering into a direct financial participation (by taking the financial role of a private party) are not. Most government revenue collection schemes are found between these extremes. High taxation costs make lower grade parts of a deposit uneconomic whereas a "Brown tax" will not. Therefore, governments must have a good database and a good understanding of the natural resources in their country to formulate a legal framework and a fiscal system for the natural resources industry to benefit their country. For example, in Germany, where the jurisdiction for natural resources lies with the states, the state of Lower Saxony abolished royalty payments for natural gas out of so-called tight gas reservoirs, i.e. reservoirs with very low porosities and permeabilities, to encourage the development of such fields. Another example from metal mining: Some countries such as formerly Ireland, for example, offered tax holidays for the first years of production to encourage mining investments, meaning offering an incentive for a conversion of resources into commercially recoverable quantities by reducing the costs through a tax rebate.

[1] While reserves are often used to designate the quantities to be recovered by commercial projects, the use of the term is diverse. This has led to a loss of precision in the use of the term and a replacement of it by other terms in the United Nations Framework Classification for Fossil Energy and Mineral Reserves and Resources 2009. The terms "reserves" and "resources" are used here in their generic and commonly understood way.

Of influence is also the speed with which a deposit is exploited. The longer an operation exploiting non-renewable reserves lasts, the longer infrastructure supplied by government can be used. Since all raw material prices show a cyclical behaviour, the longer an operation lasts, the higher the chances that parts of a deposit with marginal grades can be exploited economically in times of high prices. There are rules of experience for the speed of exploitation, for example the Taylor rule for metal deposits (Taylor 1977) or other rules of experience (Wellmer et al. 2008). For oil and gas deposits, similar rules have been used, first to conserve the value of the resources and subsequently to stretch the resource base in volume terms. The first conservation rules were introduced to balance supply and demand in the market and introduced by the Texas Railroad Commission in the early twentieth century. Today, OPEC practises this. Later, the conservation rules were motivated by technical concerns, where they have taken the form of setting maximum efficient rates from a reservoir to avoid inefficient recovery. In a broader resource management perspective, the rate of production is regulated through licensing or by staging development decisions to ensure inter alia, the efficient use and re-use of infrastructures. As an example of government's interest to prolong the lifetime of a mine, the Nanisivik lead-zinc-silver mine in the Arctic in today's Nunavut Territory of Canada could be mentioned. The Canadian Federal Government required a minimum lifetime of at least 12 years for its willingness to supply infrastructure like an airport. Actually, the mine was in operation for 26 years due to the extension of ore reserves (Bowes-Lyon et al. 2009).

There are other circumstances which might make it advisable to proceed on a staged development path. Frequently, it can be observed, especially in the development of large-scale so-called porphyry copper deposits, that at a later stage, enlargement of the mining and milling capacity takes place. At this stage, the key parameters for an economic operation are well known and under control, such as mining dilution and grade of the run-of-mine ore, mining losses, recovery in the mill, etc. The technical risk for a larger investment is reduced, therefore. Such capacity extension offers the chance to benefit from "economies of scale" (Wellmer et al. 2008) and thereby the possibility to exploit marginal parts of the deposit. This means a maximum of resources can be converted into commercially recoverable quantities not only to the benefit of the mining companies but also to the benefit of a country. This underlines governments' need to know and to understand.

The idea of extension of the ore reserves by successful exploration to optimise the lifetime of a natural resources operation and also the return on the investment was taken up by Wagner (1999, also Wagner and Wellmer 1997). He investigated mainly Canadian case histories of base metal mines. In mining, there exists a conflict of interest. On the one hand, one can optimise the return on investment by shortening the lifetime of a mine (Wellmer et al. 2008). On the other hand, to achieve exploration successes, one needs time for learning. The longer the operation, the longer the time to climb up a learning curve for understanding the ore deposit environment, resulting in increased chances of finding additional reserves. Should one, therefore, prolong operational life of a mine in the expectation of additional discoveries, or speed up extraction from the start in the interest of maximum economic return? Wagner's conclusion, however, was that the Taylor rule (Taylor 1977) could still apply as a rule for the optimal extraction rate of non-fossil fuel deposits.

The issue of reserves growth in oil and gas fields is similar and well documented (Klett and Tennyson 2007). Often, reserves are seen to grow more in large fields produced over a long time than in small fields that are depleted more rapidly.

Information is the goal of all exploration and represents an important part of the "infrastructure." Governments should take into account the learning and experience curve in exploration in their mining legislation and safeguard the information, regardless of whether it is collected by them or the industry. It should require companies to file all exploration results with government offices. Once the commercial role of the information has receded, the government should seek ways of making the data public. Then subsequently, other companies can use the data to

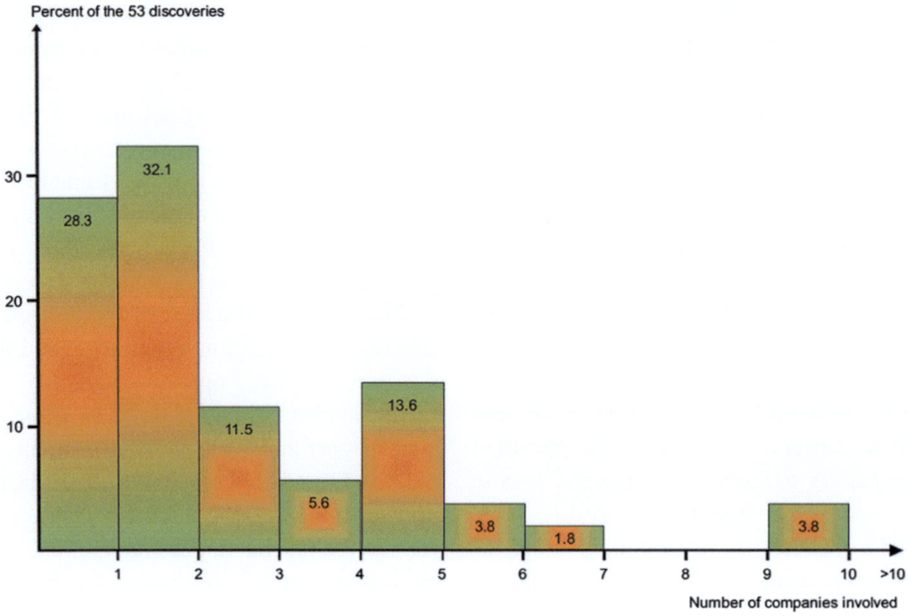

Fig. 1 The number of companies involved from the start of exploration to the successful end of exploration for 53 base metal and precious metal discoveries in the circum-Pacific region (After Sillitoe 1995)

generate new exploration ideas and not having to repeat exploration surveys executed before, i.e. rather fast and inexpensive "data mining" can be optimised. Analysis of exploration case histories shows that exploration follows learning curves with paradigm changes (Wellmer et al. 2003). Sillitoe (1995) published an interesting case history about exploration successes for precious and base metal deposits in the circum-Pacific region. He lists the number of companies involved in each prospect from the start of exploration to the successful end of exploration with an economically successful mining operation (Fig. 1). Each company represents a learning curve; a new company brings in new ideas and creates a change of paradigm. The analysis of Sillitoe (1995) shows that less than 50% of the prospects went through this learning curve only once; the extreme was ten times, i.e. several changes of paradigm were necessary to be finally successful.

It is critical to strike the right balance between early efforts and late rewards. Our analytical tools are reasonable when it comes to valuing the early efforts. They often do not have sufficient resolution with respect to the factors that impact value at distant times. They invariably discount the values here to near zero. Alternative methods to simple discounted cash flow analyses are now becoming available, taking better account of the opportunities and risks over time than what is possible with one standard exponential formula and a single discount rate (Laughton et al. 2008).

Concerning government policies to make optimal use of the natural resources within their boundaries, thereby stretching their availability, we have to distinguish between fossil fuel energy resources and non-energy resources especially metal resources. Fossil energy resources are consumed; the exergy, the usable part of the energy, is used up and cannot be recycled. On the contrary, metal resources are used but not consumed. They can be recycled. So for optimising the natural resources efficiency of metals, not only the primary sources but also the secondary ones, the metals in the technosphere, not only scrap but also waste, have to be considered in government policies. Waste is "raw material at the wrong place." Most industrialised nations today have waste avoidance acts to maximise re-use of secondary metals and other resources (Wagner and Wellmer 2009).

This introduction illustrates the obligation of governments to ensure that natural resources are exploited to the largest extent possible and government's need for a good information base for formulating policy rules. It paves the way for industry to perform the necessary tasks required profitably, securing supplies and long-term economic growth. In this chapter, we will highlight a system of structuring resource information for the mutual benefit of governments and industry. We will also give an example from the petroleum sector where government and industry have acted in concert on a countrywide basis on the principles mentioned, securing energy supplies, particularly for Europe.

The United Nations Framework Classification (UNFC-2009)

The UNFC-2009, as serviced by the United Nations Economic Commission for Europe (UNECE), was originally designed to make different national reserve/resource classification systems comparable, especially for coal. After the collapse of the centrally planned economies, the question arose in the international community how the vast resources in these countries could be compared to those in market economy countries (Kelter 1991). With the emergence and consolidation of global financial, energy and mineral commodities markets grew the need also for global standards.

The initial formulation of the UNFC, first published in 1997, proved to be a very strong one, well suited to become a global standard. This system has been developed further to incorporate all energy and non-energy non-renewable resources (ECE 2009b).

Today, it aims to serve the following four principal needs:

1. International energy and mineral studies to formulate robust and long-sighted policies
2. The needs of governments in managing their resources accordingly
3. The industries' needs for information while deploying technology, management and finance to secure fossil energy and mineral supplies and capture value efficiently within the established frameworks. This will serve its host countries, shareholders and stakeholders
4. The financial community's need for information to allocate capital appropriately, providing reduced costs and improved long-sightedness through the application of lower risk-compensated discount factors

The UNFC-2009 categorises recovery projects rather than the accumulations or deposits in which they are found. The projects are categorised with respect to economic and social viability, project feasibility and maturity as well as uncertainty with respect to the quantities involved. The categorisation of projects rather than of accumulations is a recent development in classifications that has taken place over the last 15 years or so. It provides coherence with other important management information such as production, cash flows, value, demand for various input factors, etc.

This key aspect of the UNFC reflects the critical relationship between the quantities that can be economically recovered and the recovery processes (projects) that must be implemented to achieve those recoveries. In so doing, it focuses on the impact of possible recovery projects, highlighting potential wastage of resources through flaring or inefficient recovery processes.

Many resource inventories are still based on a characterisation of the geologic endowment only. The UNFC is designed to also be a harmonisation tool, allowing these early inventories to be mapped to a UNFC inventory without loss of information. With use, these pre-existing inventories can be expanded to contain the project detail required for an efficient resource management.

To illustrate, Fig. 2 shows a normal value chain starting with exploration, proceeding to the evaluation of discoveries, design of one or more consecutive development projects, building of the facilities and, finally, extraction. At the building and extraction phase, there will normally not be any hindrances to extraction in the economic and social domain. A distinction is made between sales production and non-sales production. For petroleum projects, the non-sales production will normally be on-site fuel usage and flared gas. Non-sales production is included also to account

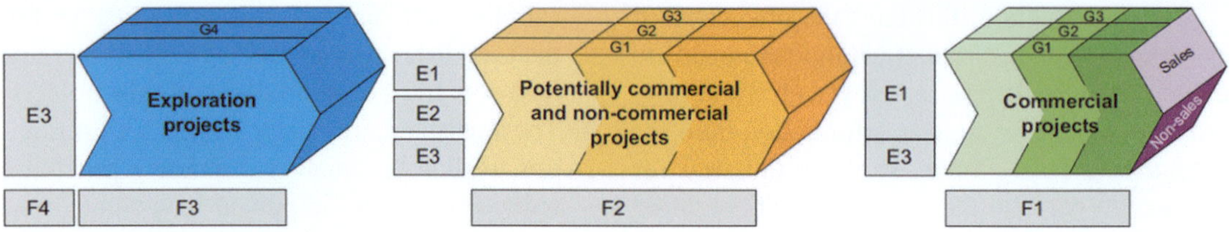

Fig. 2 The UNFC and the project value chain (for legend, see Fig. 3)

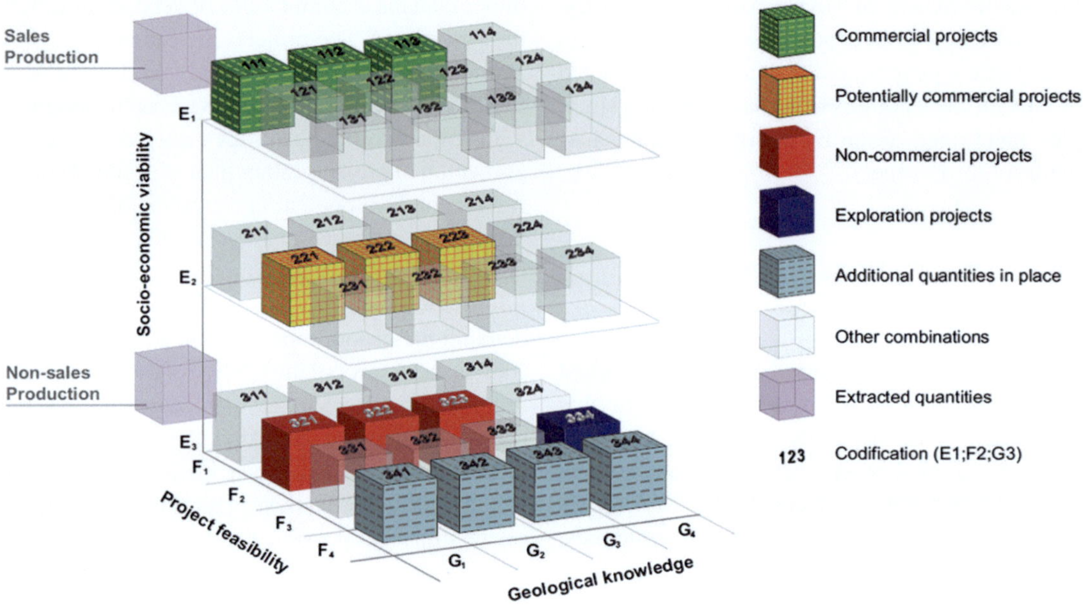

Fig. 3 The United Nations framework classification for fossil energy and mineral reserves and resources 2009

for other production to household economies that is not involved in the commercial buying and selling.

In the evaluation phase, there may be restrictions both in the technical and industrial domain and in the social and economic domain. The UNFC allows projects to be categorised independently with respect to maturity in both of these domains. The effects on recovery of improved social and economic framework conditions or of improved technical and industrial processes are then seen.

Figure 3 shows the condensed and formal representation of the UNFC-2009 with the unique and language-independent numbering system for the categories. Categories are the E categories in the economic and social domain, the F categories in the project feasibility and industrial domain and the G categories in the geological domain reflecting uncertainties in quantities.

The United Nations Economic and Social Council[2] considered the UNFC in 1997 and again in 2004. In its decision 2004/233 it states:

> At its 42nd plenary meeting, on 16 July 2004, the Economic and Social Council, recalling its decision 1997/226 of 18 July 1997, welcomed the endorsement by the Economic Commission for Europe of the United Nations Framework Classification for Fossil Energy and Mineral Resources and decided to invite the Member States of the United Nations, international organizations and the regional commissions to consider taking appropriate measures for ensuring worldwide application of the

[2] One of the six principal bodies of the UN (http://www.un.org/en/ecosoc/).

> Framework Classification. The Council noted that new classification for fossil energy and mineral resources, which now included energy commodities (for example, natural gas, oil and uranium), was an extension of the earlier framework developed for solid fuels and mineral commodities, on which the Council had taken similar action in 1997 upon endorsement and recommendation by the Economic Commission for Europe.

After 2004, a number of measures have been taken to ensure world wide application, taking advantage of the significant harmonisation that took place through adjustments of other commonly used classification (ECE ENERGY SERIES No. 33). This allowed a simplification of the UNFC in 2009 that was endorsed both by the UNECE Committee on Sustainable Energy and the Economic Commission for Europe after consultation with the other four UN Regional Commissions.

Introduction of the UNFC-2009 is taking place while preparations for detailed applications are made by the UNECE Expert Group on Resource Classification (EGRC). The process that is designed for its detailed application consists of a first phase where the stakeholders who will apply the classification to meet each of the four principal needs identified are invited to identify issues that require further specification (secondary rules) under the UNFC. The competent professional organisations and appointed experts will be invited to develop specifications to meet these requirements and present their proposal back to the stakeholders for approval and implementation. Financial reporting standard setters (the International Accounting Standards Board) and security regulators play an important coordinating role in this process. Their decisions with respect to global financial reporting requirements will affect all entities seeking capital in the international financial market. The introduction of an international financial reporting standard for extractive activities will represent the appropriate opportunity to introduce the UNFC-2009 on a broad basis down to the asset level, affording economising benefits with respect to communication generally.

In parallel with these activities, the EGRC is currently considering the potential for applying the UNFC to projects that are placing material in the subsoil for permanent or temporary storage. The injections of CO_2 or natural gas into recipient reservoirs are examples.

The Case of Norway Managing Its Oil and Gas Resources

The case of Norway demonstrates how principles mentioned in the introduction and supported by the UNFC-2009 have secured oil and gas supplies. This has been achieved by a concerted effort by governments and industries to achieve high recoveries.

Already in 1975, the importance of recovery efficiency was recognised in the Norwegian Royal Decree regulating petroleum activities. The text developed then has remained unchanged until today. The petroleum law in force states:

> Production of petroleum shall take place in such a manner that as much as possible of the petroleum in place in each individual petroleum deposit, or in several deposits in combination, will be produced. The production shall take place in accordance with prudent technical and sound economic principles and in such a manner that waste of petroleum or reservoir energy is avoided. The licensee shall carry out continuous evaluation of production strategy and technical solutions and shall take the necessary measures in order to achieve this.

These are not empty words.

The history is shown in Fig. 4. Developments started in the south and progressed towards the north, allowing early investments in infrastructure to be re-used. This helped to reduce costs and consequently to increase the well head values. Increased efforts to improve recovery efficiencies became economically viable. Bold technological advances with respect to recovery processes and offshore facilities were conceived, developed and implemented to recover the resources. With time, the framework conditions were also made more efficient. Elements of gross tax that act like costs in reducing well head values for the developer were removed. Neutral taxes that do not affect incentives were introduced. This took the form of abolishing royalties and of introducing a "Brown tax," the State Direct Financial Interest (SDFI).

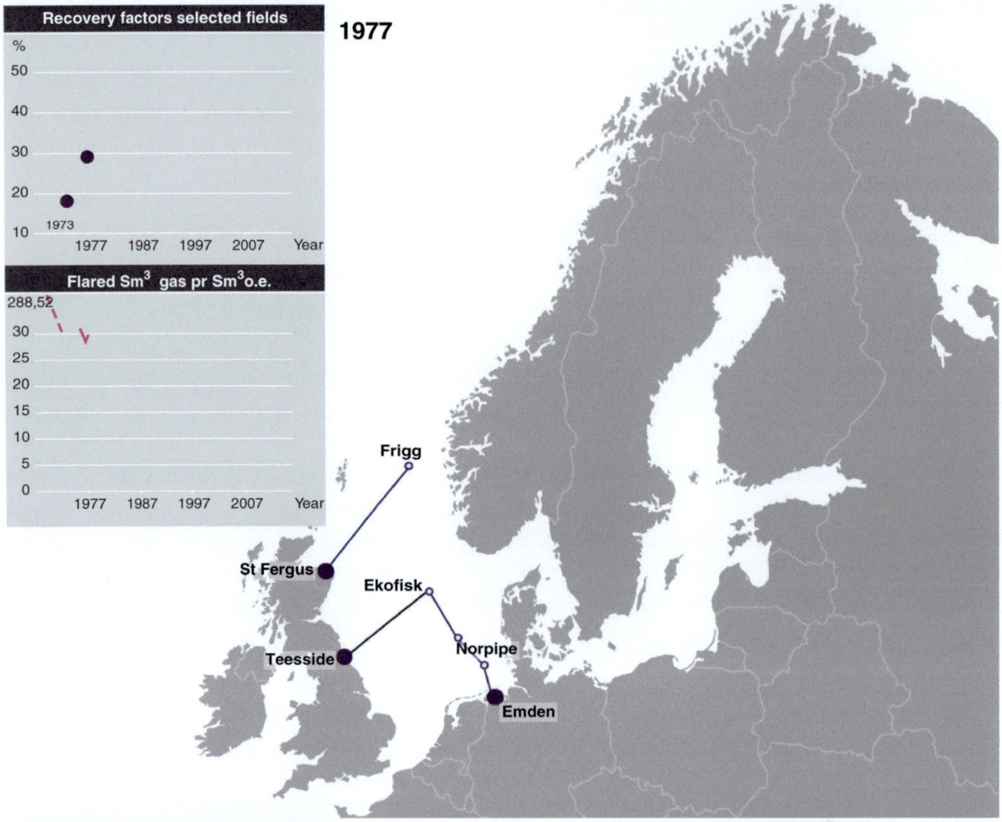

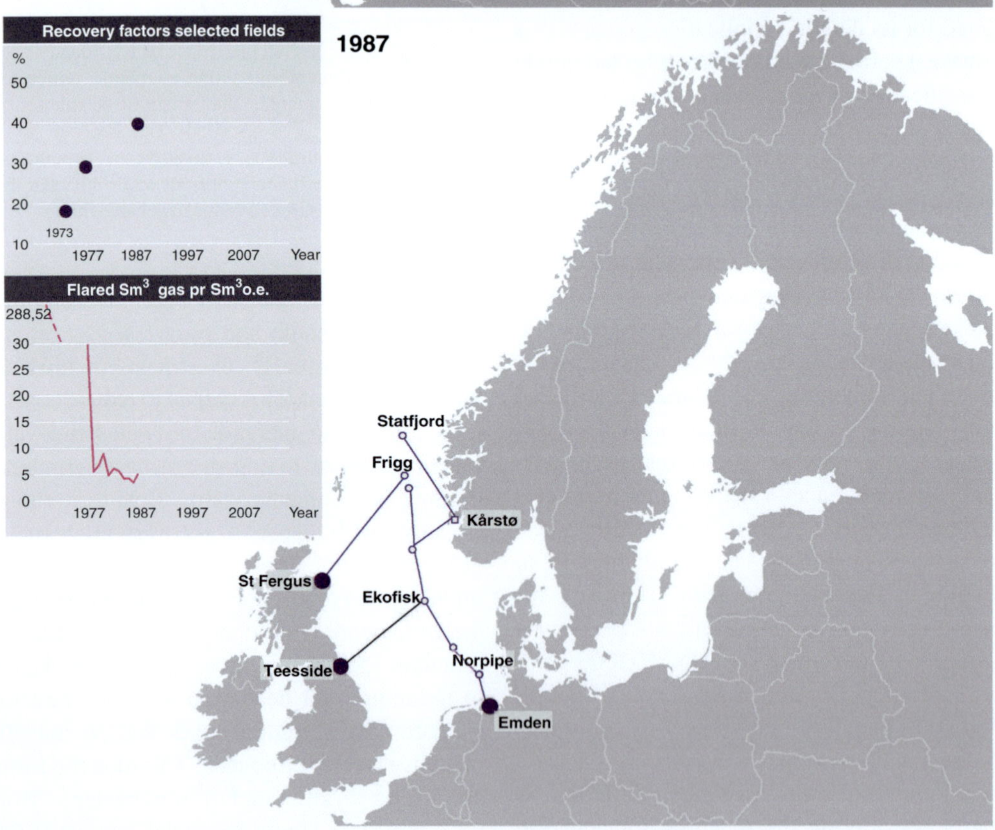

Fig. 4 Gas transport and terminals, recovery factors and flared gas (Source: Norwegian Petroleum Directorate) (One standard cubic meter (Sm^3) is defined as the volume of the substance at 15°C and 1 atm)

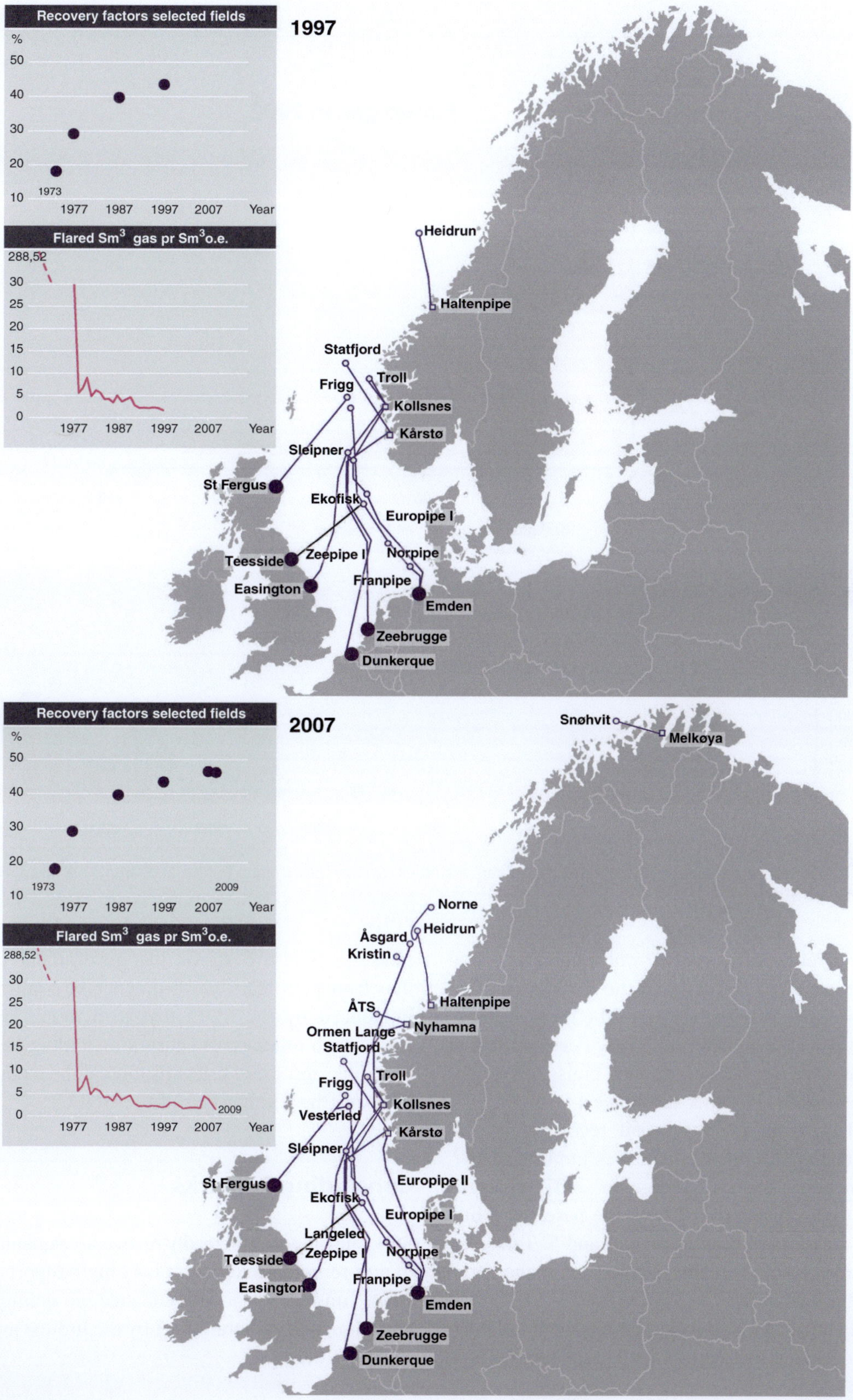

Fig. 4 (continued)

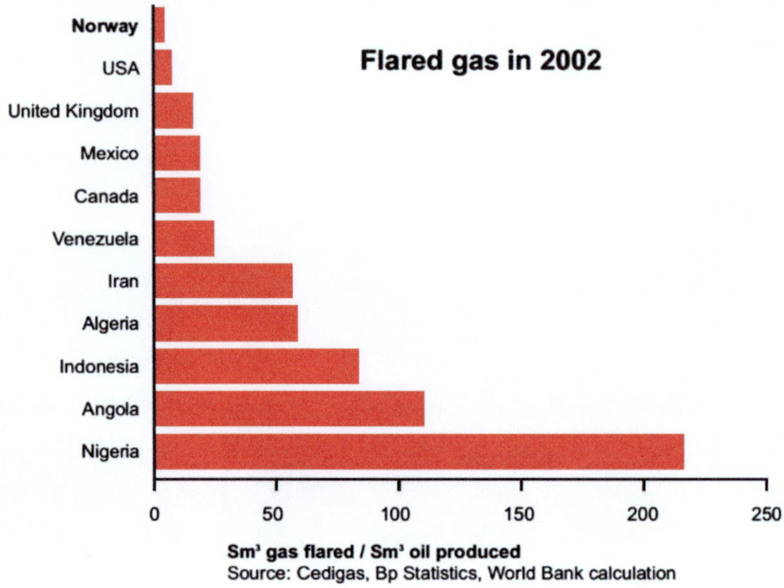

Fig. 5 Flared gas in 2002

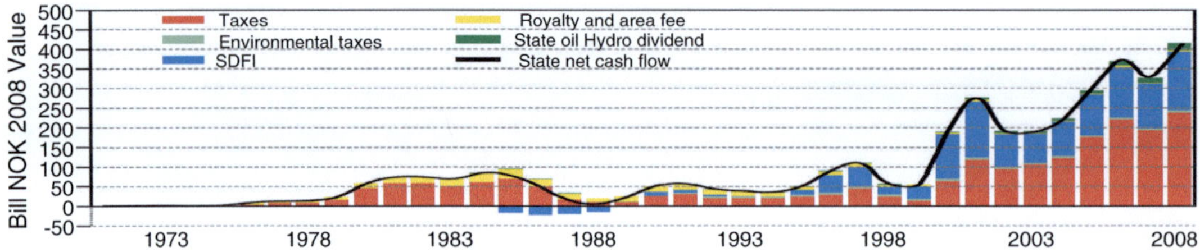

Fig. 6 The net government cash flow from petroleum activities (Source: Norwegian Public Accounts 2008 and account figures for SDFI)

The pipeline infrastructure was regulated to facilitate access to markets and the transfer of prices to well heads without loss in the form of excess profits and fees paid for its use. Fees and regulations were introduced to curb emissions of carbon dioxide and other substances.

The improvements in oil recovery from an estimated 18% of the quantities in place in 1973 to nearly 50% on average in 2007 represents nearly a tripling of the resource base. Very little gas was lost as seen in Figs. 4 and 5. Today, the avoided loss is valuable gas, located in the region where it is used.

Finally, the Norwegian government has managed these achievements with great benefits to its treasury. Figure 6 shows the government cash flow from petroleum. Government take is strongly influenced by the SDFI that to a much lesser extent than other taxes cause a degrading of the economic and social categorisation of the recovery projects (the E categories of the UNFC).

Concluding Remarks

Stretching the commercially recoverable quantities of non-renewable resources is of high importance when considering how to enhance the quality of life within the constraints set by the limited availability of them.

We do not know with great precision what the recoverable quantities are or how demand will grow. Both depend on human actions. We do know that they are in great demand and that they are finite. We also know that we cannot recover 100% of what is in the ground and that the percentage will depend on the recovery efforts made.

The recovery processes are for the most part physically irreversible processes. The implication is that the amount that we can recover and use depends on the entire history of past efforts in addition to future efforts. Said in plain words, if we fail to invest early for high recoveries in the long term, commercially recoverable quantities will be lost.

Decisions to invest for high recoveries in the long term are decisions to secure supplies. The real negotiations are between governments and industry together on one side confronted with the geological constraints on the other.

Immediate investments to gain production in the longer term are based on the current opinion of future values of the resources at the source (well head or mine gate). The higher and the more predictable they are, the easier it is to commit to making the required efforts to recover the substantial quantities of resources that may initially be economically marginal. It is important that decisions with respect to these enduring projects are informed by valuation models that reflect the values at distant times correctly.

A comprehensive approach is required addressing the economic and social conditions affecting values at the source, the efficiency and cost of recovery operations and the geological conditions. With the globalisation of the financial, energy and commodity markets, the international community of governments, industry and the financial community all influence the recoverable quantities. They can increase it substantially if they act in concert.

The UNFC-2009 holds the potential to facilitate these joint actions through its incorporation and enhancement of many of the elements of commonly used classifications. It is well suited to communicate clearly the efficiencies to be gained by changes in the social and economic domain, in the technical and industrial domain and by making visible the losses and wastes that different projects produce. It allows the communication of uncertainties that each party needs to know in order to manage its opportunities and risks.

Most importantly, the UNFC-2009 provides a vehicle for accurate communication in global markets. Today, this is hampered at government level by the use of different classifications by the different governments. This is further aggravated in international companies engaged in multiple partnerships with other international companies and in a multitude of countries. It is clearly intolerable to not have a common terminology when reporting to the international financial market.

A common terminology is a clear advantage. The value that each player can derive from it depends on the information that is communicated. It is the privilege of the information owner to decide what to communicate. With a common terminology, the efforts in communicating accurately are reduced, and the chance that the information communicated is understood and properly used by the receiver is significantly enhanced. This facilitates higher quality decisions.

Acknowledgements This chapter summarises intense efforts over many years and by many good teams of professionals and friends. The teams of the Norwegian Petroleum Directorate, of The Ministry of Water, Energy and Minerals of Tanzania, of Statoil, of the Oil and Gas Reserves Committee of the Society of Petroleum Engineers and most importantly of the UNECE's Sustainable Energy Division and the UNECE Ad Hoc Group of Experts on the Harmonization of Fossil Energy and Mineral Resources Terminology[3] have all made their important contributions to the understanding expressed here. Great thanks are due to Professor Richard Sinding-Larsen for taking the initiative to produce this chapter and for his constructive encouragement and comments. The views are, however, those of the authors and not necessarily those of the institutions named. Special acknowledgement is due to the UNECE for placing many of the issues presented in this chapter on the agenda for their 63rd Session in 2009.

---

[3] Now the UNECE Expert Group on Resource Classification.

References

Bowes-Lyon LM, Richards JP, McGee TM (2009) Socio-economic impacts of the Nanisivik and Polaris mines, Nunavut, Canada. In: Richards JP (ed) Mining, society and a sustainable world. Springer, Berlin/Heidelberg/New York, etc, pp 371–396

CIM (2005) Canadian Institute of Mining, Metallurgy and Petroleum: CIM definition standards on mineral resources and mineral reserves, adopted by CIM Council on 1 Dec 2005. www.cim.org/committee/guidelinesSTANDARDS_main.cfm

CRIRSCO (2006) Committee for Mineral Reserves International Reporting Standards International Reporting Template July 2006. www.crirsco.com/crirsco_template_first_ed_0806.pdf

ECE (1997) United Nations International Classification for Reserves/Resources - Solid Fuels and Mineral Commodities - Final Version (Finalized and submitted by the United Nations Task Force) - ENERGY/WP.1/R.77 - Printed at United Nations, Geneva GE.97-31650-September 1997

ECE (2004) The United Nations Framework Classification (UNFC) for Energy and Mineral Resources (January 2004)

ECE (2009a) Mapping of the United Nations Framework Classification for Fossil Energy and Mineral Resources. ECE ENERGY SERIES No. 33 and ECE/ENERGY/71. (http://www.unece.org/energywelcome/areas-of-work/unfc-and-resource-classification/publications.html). Accessed 4 Jan 2012

ECE (2009b) Economic Commission for Europe, Committee on Sustainable Energy: United Nations Framework Classification for Fossil Energy and Mineral Reserves and Resources 2009, ECE ENERGY SERIES No. 39, United Nations, New York and Geneva. See also: ECE/ENERGY/GE.3/2009/L.5. (http://www.unece.org/fileadmin/DAM/energy/se/pdfs/UNFC/unfc2009/UNFC2009_ES39_e.pdf). Accessed 4 Jan 2012

JORC (2004) Joint Ore Reserve Committee: Australasian code for reporting of exploration results, mineral resources and ore reserves—The JORC code 2004 edn. (http://www.unece.org/energywelcome/areas-of-work/unfc-and-resource-classification/publications.html). Accessed 4 Jan 2012

Kelter D (1991) Classification systems for coal resources-a review of the existing systems and suggestions for improvements. Geol Jb A 127:347–359

Klett T, Tennyson ME (2007) An approach to the classification of potential reserve additions of giant oil fields of the world. USGS Open-File Report 2007–1438. (http://www.unece.org/fileadmin/DAM/energy/se/pdfs/UNFC/oct07/TimKlett_USGS.pdf). Accessed 4 Jan 2012

Laughton D, Guerrero R, Lessard D (2008) Real asset valuation: a back-to-basics approach. J Appl Corpor Finance 20(2):46–65, A Morgan Stanley Publication

Sillitoe RH (1995) Exploration and discovery of base- and precious-metal deposits in the Circum Pacific region during the last 25 years. Resour Geol Spec Issue Tokyo 19:1–119

SPE/WPC (1997) Petroleum Reserves Definitions. Society of Petroleum Engineers, World Petroleum Council, 1997. (http://www.spe.org/industry/docs/Petroleum_Reserves_Definitions_1997.pdf). Accessed 4 Jan 2012

SPE/WPC/AAPG (2000) Petroleum Resources Classification System and Definitions. Society of Petroleum Engineers, World Petroleum Council, American Association of Petroleum Geologists. (http://www.spe.org/industry/docs/Petroleum_Resources_Classification_2000.pdf). Accessed 4 Jan 2012

SPE/WPC/AAPG/SPEE (2007) Petroleum Resources Management System. Society of Petroleum Engineers, World Petroleum Council, American Association of Petroleum Geologists, Society of Petroleum Evaluation Engineers (http://www.spe.org/industry/docs/Petroleum_Resources_Management_System_2007.pdf). Accessed 4 Jan 2012

Taylor HK (1977) Mine valuation and feasibility studies. In: Mineral industry costs. Northwest Mining Assoc, Spokane, pp 1–17

Wagner M (1999) Ökonomische Bewertung von Explorationserfolgen über Erfahrungskurven. Geol Jb SH12:225

Wagner M, Wellmer FW (1997) The optimum lifetime of a mine taking into account the learning curve of exploration. In: Pei Y (eds) Proceedings of the 30th international geological congress, Beijing (China) 1996, vol 9. VSP Int Sci Publ, Utrecht, pp 257–262

Wagner M, Wellmer FW (2009) A hierarchy of natural resources with respect to sustainable development –a basis for a natural resources efficiency indicator. In: Richards JP (ed) Mining, society and a sustainable world. Springer, Berlin/Heidelberg/New York, etc, pp 91–121

Wellmer FW, Leifeld D, Röhling S, Wagner H (2003) Mineral and energy resources: economic factor and motor for research and development. Z Deut Geol Ges 154(1):1–27

Wellmer FW, Wagner M, Dalheimer M (2008) Economic evaluations in exploration. Springer, Berlin/Heidelberg/New York, etc, 250 p

Secondary Raw Material Sources for Precious and Special Metals

Christian Hagelüken

Abstract

Special and precious metals play a key role in modern societies as they are of specific importance for clean technologies and other high-tech equipment. The use of these "technology metals" has accelerated significantly over the past 30 years, and their sufficient future availability is crucial for building a more sustainable society with the help of technology. Recycling can contribute significantly to secure access to these metals, conserve metal resources, and mitigate potential temporary scarcities. If conducted in state-of-the-art processes, recycling of technology metals – which mostly occur in low ore concentrations only – offers as well considerable benefits compared to mining, with respect to energy, land, and water requirements.

While today, very efficient metallurgical processes exist to recover base and precious metals from both primary concentrates and various secondary materials, special metals recovery is usually coupled to the former processes and in many cases still shows potential for improvement. An eco-efficient recycling of technology metals from complex products cannot be achieved without the use of high-tech processes that make use of economies of scale and sophisticated metallurgical flow sheets. The actual achievable recycling rates thereby depend on the setup of the entire recycling chain, from collection and sorting over dismantling/preprocessing down to the final metallurgical metal recovery steps. Decisive factors for the success of such a recycling chain are – in addition to the applied technologies – stakeholder cooperation and the management of interfaces.

The biggest challenge however is to secure that end-of-life products are entering into the most appropriate recycling pathways. Today, a large share of old consumer goods like electronics or cars is – partly illegal – traded

C. Hagelüken (✉)
Umicore, Rodenbacher Chaussee 4,
Hanau D-63457, Germany
e-mail: christian.hagelueken@eu.umicore.com

R. Sinding-Larsen and F.-W. Wellmer (eds.), *Non-Renewable Resource Issues: Geoscientific and Societal Challenges*, International Year of Planet Earth,
DOI 10.1007/978-90-481-8679-2_10, © Springer Science+Business Media B.V. 2012

across the globe and escapes recycling or ends up in backyard recycling operations with low recovery rates and dramatic impacts on health and environment. This chapter provides an overview on the recycling of technology metals; it elaborates the factors impacting success and shows that legislation can be supportive but that consequent enforcement and new business models are essential to close the loop for consumer products.

Keywords
Recycling • Technology metals • Metals scarcity • Life cycles • Urban mining • Waste electronics

Significance of Precious and Special Metals

Special and precious metals are of specific importance for clean technologies and other high-tech equipment. Important applications are information technology (IT), consumer electronics, as well as materials used in sustainable energy production such as solar cells, wind turbines, fuel cells, and batteries for hybrid cars. They are crucial for more efficient energy production (in steam turbines), for lower environmental impact of transport (jet engines, car catalysts, particulate filters, sensors, control electronics), for improved process efficiency (catalysts, heat exchangers), and in medical and pharmaceutical applications. Figure 1 provides an overview of these main applications areas for each metal and illustrates their significance for modern life. For example, electronic products can contain up to 60 different elements, and in their entity are major demand drivers for precious and special metals: just the annual sales of mobile phones and computers account, for example, for about 3% of the world mine production of gold and silver, 15% of palladium, and over 20% of cobalt (Hagelüken and Meskers 2008). Driving forces for the booming use of these "technology metals[1]" are their extraordinary and sometimes exclusive properties, which make many of these metals essential components in a broad range of applications. Building a more sustainable society with the help of technology hence depends to a large extent on sufficient access to technology metals.

Their use has experienced a tremendous growth over the last years. Eighty percent or more of the cumulative mine production of platinum group metals (PGM), gallium, indium, rare earth elements, and silicon, for example, has occurred over the last 30 years. For most other special metals, more than 50% of their use took place in this period, and even for the "ancient metals" (gold, silver) use from 1978 onward accounts for over 30% (Fig. 2). In many cases, the booming demand especially from consumer mass applications drove up metal prices significantly. For example, the significant increase in demand and prices of platinum and palladium was mainly caused by automotive catalysts (50% of today's platinum/palladium demand) and electronics (Fig. 3). New technological developments are expected to push their demand further.

The Need for Recycling

Efficient recycling of technology metals will gain an increasing importance and will be a future necessity to secure metal supply. Even if for most metals, an absolute depletion of geological resources is not likely to be expected in the foreseeable future – as has been discussed in other chapters of this book – a significant secondary metals supply contribution is crucial to overcome

[1] The term technology metals is used in this chapter as a synonym for the precious metals (Au, Ag, and the PGMs Pt, Pd, Rh, Ru, Ir) plus the special/specialty metals (among others In, Ga, Ge, rare earth elements, Sb, Se, Si, Te). The latter group is also sometimes called "minor metals," in distinction from "major" or base metals such as Al, Cu, Ni, Pb, and Zn.

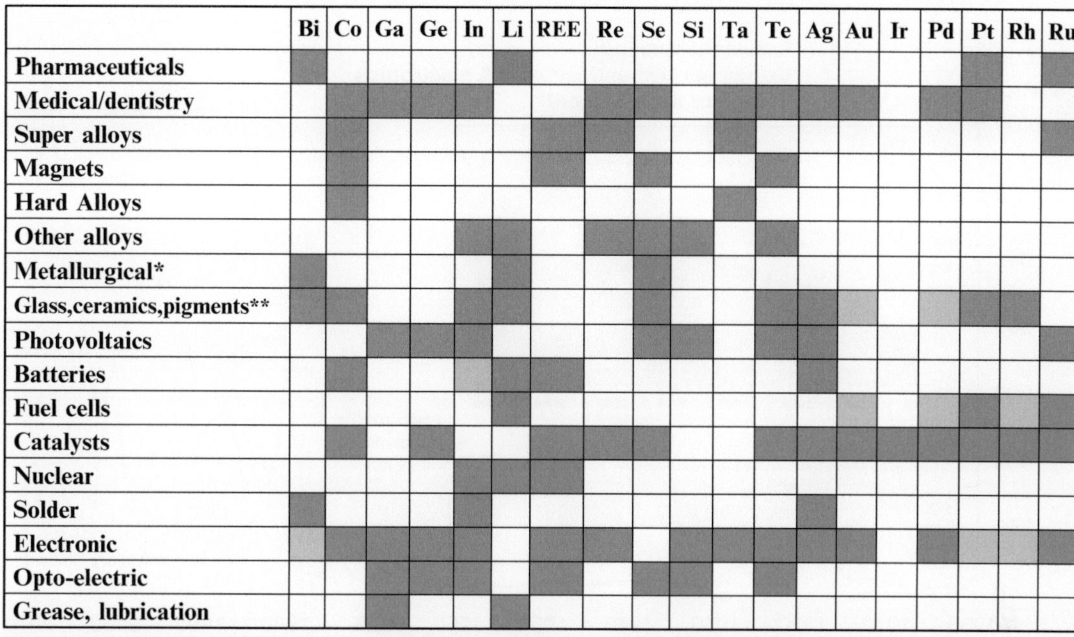

* additives in smelting, ..., plating. ** includes Indium Tin Oxide (ITO) layers on glass

Fig. 1 Important applications for technology metals (Hagelüken and Meskers 2010)

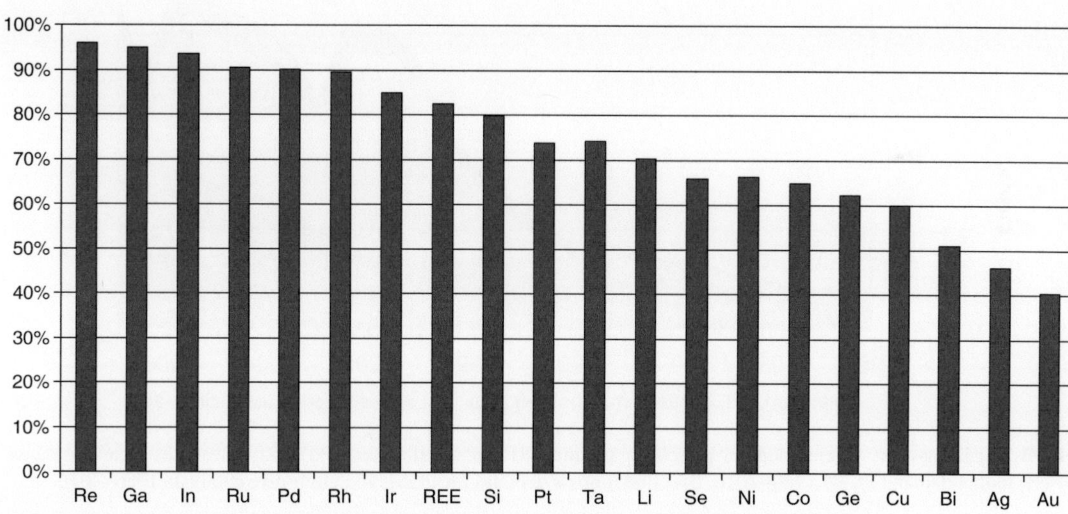

Fig. 2 Share of technology metals mined since 1978 compared to cumulated production between 1900 and 2007; copper and nickel included for comparison (Modified after Hagelüken and Meskers 2010)

temporary and structural resource scarcities. *Temporary or relative scarcity* is a phenomenon which has been already widely experienced. In this case, metal supply is for a certain period in time not able to meet the demand. Reasons can be manifold. New technological developments, strong market growth in existing applications, or speculative buying can drive up the demand significantly within a short time so that mine supply lags behind. Also the supply can be disrupted by political developments, armed conflicts, natural disasters, or other constraints in the mining

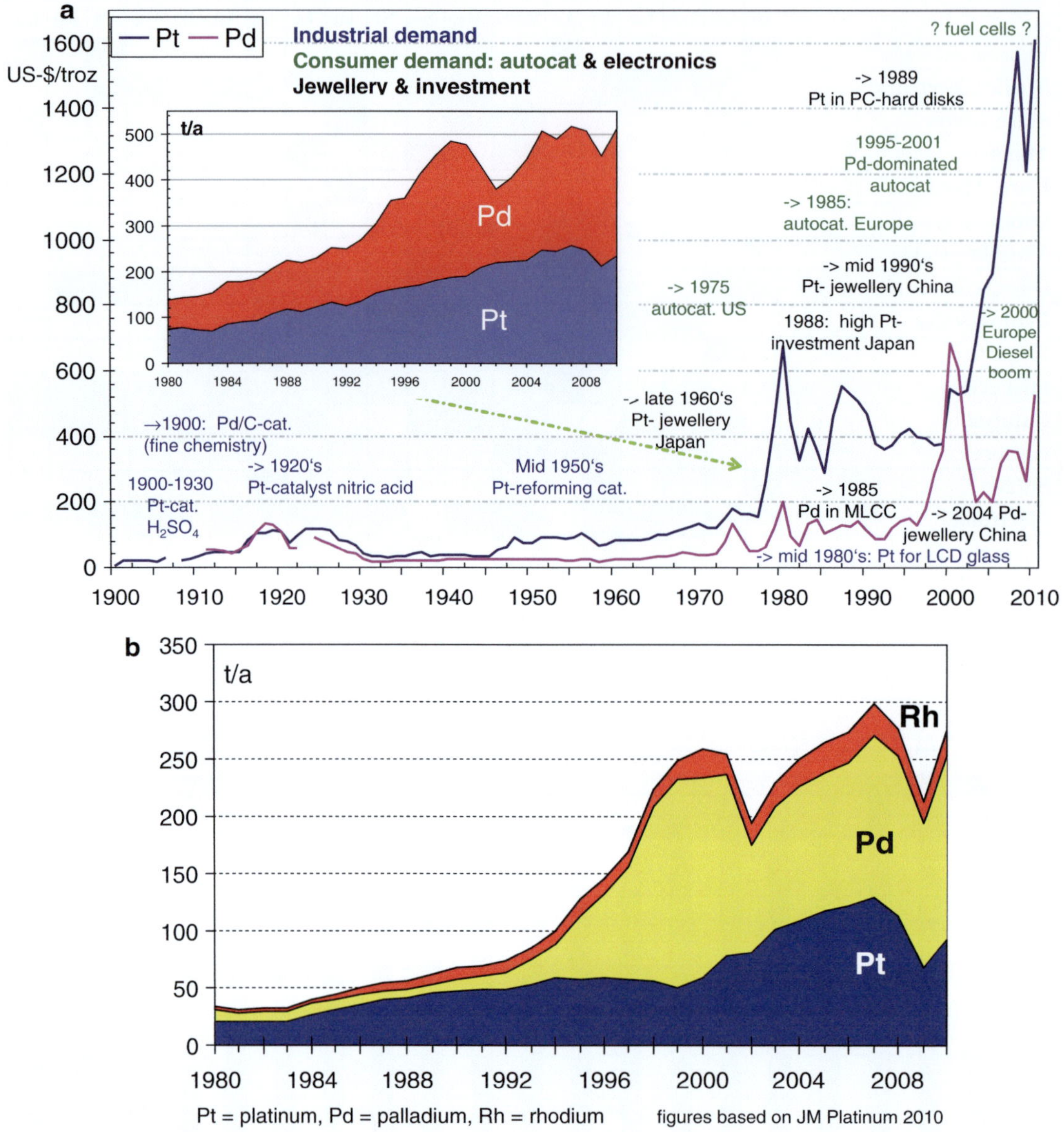

Fig. 3 (**a**) Long-term development of prices for platinum (Pt) and palladium (Pd) and milestones in applications. Insert: global demand Pt, Pd 1980–2010. (**b**) Global gross demand for PGMs in automotive catalysts 1980–2010

countries themselves, within the transport of ore concentrates, or also at major smelters/refineries. Temporary scarcities are a main reason for the sometimes extreme price volatility in (precious) metal markets. The risk of temporary scarcities increases with increasing concentration of the major mines or smelters in few and/or unstable regions, or in few companies. Often, different factors come together and then accelerate the development. For instance, in the first quarter of 2008, a soaring demand for PGMs from automotive catalysts and (speculative) investment coincided with a reduced supply from South African mines due to shortages in electric power. The prices of platinum and rhodium went to record heights within a short time as South Africa contributes to over 75% of the primary platinum and rhodium supply. Speculation about potential depletion of indium resources started when the sales boom of LCD devices (monitors, TVs, mobile phones, etc.), which use indium-tin oxide (ITO) as transparent conductive layer, drove up indium prices significantly. In future, a take off in thin film photovoltaics would boost the demand for tellurium, indium, selenium, and gallium; mass applications of electric vehicles will require large amounts of lithium, cobalt, and some rare earth elements; and fuel cell cars would need significantly more platinum than is used today in a catalytic converter. Developing and expanding mining and smelting capacities is highly capital intensive, risky, and it takes many years to materialize. Hence, temporary scarcities are likely to happen more often in the future, and comprehensive recycling is needed to mitigate such effects and counteract primary supply shortages.

The *structural scarcity* is most severe for many technology metals, which are often not mined on their own but occur only as by-products from major or carrier metals. Indium and germanium, for example, are mainly by-products from zinc mining, gallium from aluminum and selenium and tellurium from copper (and lead). The PGMs occur as by-products from nickel and copper mines and as coupled products in own mines (for the distinction by-products/coupled products, see Wellmer 2008). Within the PGMs ruthenium and iridium are by-products from platinum and palladium. Since the by-product ("minor metal") is only a very small fraction of the carrier ("major") metal, here the usual market mechanisms do not work. An increasing demand will certainly lead to an increasing price of the by-product metal, but as long as the demand of the major metal does not rise correspondingly, mining companies will not produce more because this would erode the major metal's price. In this respect, the supply of by-product metals is to a large extent price inelastic.

Moreover, many technology metals are an important ingredient for multiple emerging technologies (Fig. 1), so a competition between applications becomes likely, and increasing demand from various segments will intensify the pressure on supply. Substitution is not likely to become the solution for many of these metals either since the required functional properties can often be met only by metals from the same metal family. For example, substituting platinum by palladium in catalytic applications will just shift the problem from one temporary/structural scarce metal to the other, which was experienced in the second half of the 1990s, pushing the before cheaper palladium to record heights in 2000/2001. In emerging opto-electronics, the crucial metals are silicon, tellurium, gallium, selenium, germanium, and indium. They can partially substitute each other, though this will not really mitigate the problem. It can only be overcome by increasing the efficiencies in the primary supply chain (possibly leading to considerable gains) and, above all, by comprehensive recycling efforts as pointed out hereafter. These consequences of the by-product character of many technology metals are often not considered sufficiently in the discussion about resource availability.

Independent of whether or not supply constraints are likely, the impact of mining of lower-grade ores and from more challenging locations must not be overlooked. It will inevitably lead to increasing costs, energy demand, and raising emissions; it will impact the biosphere (if mining in rain forests, arctic regions, oceans); and it can increase the dependence on certain regions ("battle for resources"). This can imply significant constraints on emerging technologies, unless effective life cycle management enables the increasing use of recycled (secondary) metals in the forthcoming years (Hagelüken and Meskers 2010).

Recycling: Basic Considerations

Metals are not consumed; they are only transferred from one manifestation into another, moving in and between the lithosphere and the technosphere. Thus, the latter becomes our future "renewable" resource in society. Thoroughly extracting "urban mines" is the only sustainable solution to overcome long-term supply disruptions. Metal combinations in products often differ from those in primary deposits, which results in new technological challenges for their efficient recovery. In products such as electronics or catalysts, the precious metals (Au, Pt, Pd, etc.) have become

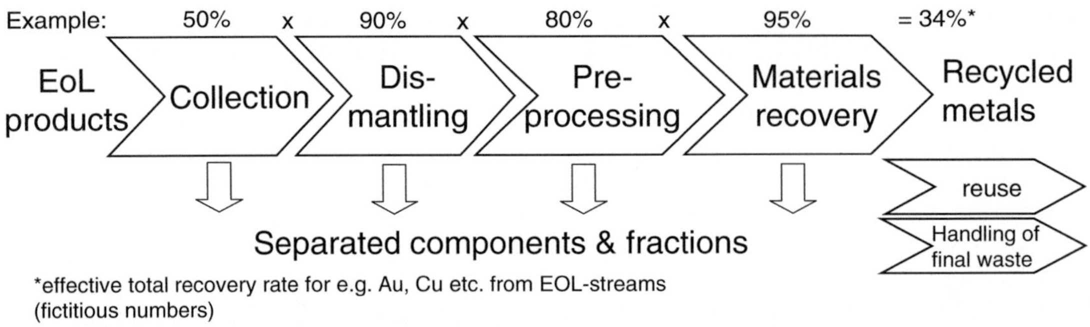

Fig. 4 Principle main steps in a recycling chain for consumer products; metallurgical processes are used for final recovery of metals (end processing); *EOL* end-of-life

the economic drivers for recycling ("paying metals"), while many special metals (Se, Te, In, etc.) can be recovered as by-products when state-of-the-art treatment and refining operations are used. Very low concentration of technology metals in certain products or dissipation during product use sets economic and technical limits in many cases, and technical challenges exist especially for complex products such as vehicles and computers.

Recycling technologies for precious metals and base metals like copper or lead have been developed over centuries and in most cases today are capable in achieving high recovery yields when it comes to the final metallurgical recovery step. Recycling of special metals is usually connected to base and precious metals metallurgy but still offers potential for optimization in many cases. However, recycling of consumer products and some industrial products is much more than metallurgy and requires a complete chain, starting with collection, sorting, and dismantling/preprocessing to separate components containing valuable metals or to upgrade relevant fractions for subsequent metallurgical end processing. In this final step, technology metals are extracted by pyro- and/or hydrometallurgical processes and purified to pure metals which are delivered back to the market for a new product life. Thus, the real physical metals recycling takes place at the very end of the chain, but the preceding steps are decisive to direct effectively and as comprehensively as possible the technology metals containing fractions into final recovery.

Metals that during their path through the recycling chain are dissipated – for example, into dust fractions – or that are diverted by unintended coseparation into the wrong end processes, such as precious metals directed into steel plants or aluminum smelters, are lost (Meskers et al. 2009b). The total efficiency of a recycling chain is the product of the individual efficiencies of each subprocess; hence, the weakest step in the recycling chain has the biggest impact on the overall recovery rate. In practice, today the biggest precious metal losses do indeed happen due to insufficient collection or other downstream problems prior to end processing, and although state-of-the-art metallurgical processes usually achieve over 95% precious metals recovery rates for many consumer products, the total chain efficiency is mostly well below 50% only (Fig. 4). For special metals, the situation usually is even much worse. A report of the UNEP resource panel provides a comprehensive overview on the current knowledge about metal recycling rates (UNEP 2011).

There are three main influences on the recyclability of metals:

- The intrinsic metal value of the material, in dependence of its absolute metal content and the relevant metal prices: This determines the economical attractiveness of recycling and sets a benchmark for recycling technology and overall recycling costs. Materials containing precious metals often make recycling rather attractive.

- The material composition of the product: This influences the technical recovery processes, the recovery costs, as well as the technical recovery yields that can be achieved.
- The application segment of a product and the way how it is used: This influences the probability for products/metals to enter into an appropriate recycling channel at their end-of-life (EOL).

Effective recycling could, in theory, lead to an infinite metal cycle without a decrease in quality since for metals in principle no "downcycling" occurs (other than for paper or plastics). In practice, however, metals are lost from the life cycle because, for various reasons, they cease to be accessible for recovery. The role of recycling is to minimize these metal losses, which can take place on all levels of a metal/product life cycle.

Primary metal production is usually not included when recycling is discussed. Nevertheless, the improved treatment of tailings, slag, or other side streams from mining, smelting, and refining can contribute significantly to resource efficiency and supply of technology metals. Improved efficiency in the primary production combined with the reprocessing of historic primary stocks will provide a large and rather easy accessible additional metal source for most by-product metals like indium, germanium, molybdenum, rhenium, or gallium. For many coupled metals like the PGMs, these inefficiencies have been largely overcome already due to their higher economic value.

In *manufacturing*, the challenges to recycle production scrap usually increase when moving further downstream in the production process.[2] For some "high-tech" manufacturing technologies – like sputtering – scrap arisings are significant (>50%) and offer a huge recycling potential for many special metals. In the case of indium, germanium, and ruthenium, this has been increasingly applied over the last years. Metal losses occurring during the *product use phase* are hardly recyclable due to their mostly dissipative nature.[3]

The recycling of *EOL products* will be the key in achieving a sustainable use of metals. Expressions like "urban mining" or "mine above ground" refer to the resource potential in our "wastes." This has also been recognized by governmental bodies such as the European Commission, which strives to make Europe a "recycling society" and seeks to prevent the creation of waste and to use waste as resources. Supportive legislative measures, like the Directive on End-of-life Vehicles (ELV) from September 2000 and the Directive on Waste Electrical and Electronic Equipment (WEEE) from January 2003, underline this approach. End 2008, the EU Commission launched the Raw Materials Initiative, focusing on an enhanced access to important metals for the European industry (EU COM 2008).

Does this mean that everything now is on the right track? Could the closed loop for most metals be expected to become a reality soon? How well does this all fit to the recycling of technology metals?

Recycling Processes and Frame Conditions

The success by which metals will finally be recovered from EOL products depends on a set of main impact parameters as well as on the setup of the recycling chain as a whole. No single universal recycling process exists. Depending on the products and materials involved, various logistical and technology combinations are required, and many different stakeholders are involved. Success factors are interface optimization between the single recycling steps, specialization

[2] For example, indium recycling gets increasingly more difficult for: target manufacturing → spent ITO target → scrapings from the sputtering chamber → broken or out-of-spec LCD glass → entire out-of-spec or obsolete LCD monitor.

[3] One example is PGM loss from car catalysts: In contrast to earlier conditions, today's autocatalysts under European or American driving conditions emit hardly any PGMs during the use phase. However, under typical "developing country" driving conditions (e.g., bad roads, low car maintenance, misfires, bad petrol quality), a catalyst is likely to be mechanically destroyed, and PGMs with broken catalyst ceramic are blown out from the exhaust and dissipated along the roadside.

on specific materials, and utilization of economies of scale. The key impact parameters comprise technology and economics, societal/legislative factors, and the life cycle structure of a product (Hagelüken 2007b).

Technical Impact Factors on Metal Recycling Rates

Impact of Product Properties

Technical capability to recover metals effectively from products needs to be evaluated under a systems perspective that considers the entire recycling chain. Technical conflicts of interest cannot be fully avoided, and from complex streams, some metals will not be recovered. Setting the right priorities is important. Considerations in this context are:

- Complexity, that is, the variety of substances in a product: Cars and electronic devices are examples of highly complex products, each consisting of a large number of complex components. Many substances are used in numerous combinations, often closely interlinked, and comprise both valuable and hazardous substances. Precious and special metals are frequently key elements in such products or in one of their key components.
- Concentration and distribution of metals: The recovery of technology metals that occur on a ppm level (e.g., in circuit boards, catalysts, or LCD screens) is technically more challenging than the recovery of copper from a cable, aluminum from a wheel rim, or lead from a car battery since the latter is highly concentrated in these components.
- Coupled recovery: Similar to coupled production in primary production, a limited number of valuable "paying" metals provide the economic incentive for recycling, enabling the additional recovery of "by-product" metals with subeconomic value or concentration. For printed circuit boards, for example, the drivers for recycling are gold, silver, palladium, and copper, however, various special metals can also be corecovered with appropriate technologies.
- Product design and accessibility of components: A good "design for recycling" eliminates the use of (hazardous) substances that hamper recycling processes (e.g., mercury in backlights of LCD monitors) and ensures the accessibility of critical components. An example of an easily accessible component is the car catalyst, which can be cut from the exhaust system prior to shredding and fed into the appropriate recycling chain. The opposite is the case for most car electronics, which are widely distributed over the vehicle and thus are seldom removed prior to shredding. Consequently, most technology metals contained in car electronics are lost during the shredder process.

Metal Recovery: Smelting and Refining

Complex products require a well-organized and dedicated process chain (Fig. 4), involving different stakeholders. Especially for the efficient recovery of technology metals in low concentrations from complex components, high-tech metallurgical processes are required. For example, a large "integrated smelter refinery" in Belgium recovers 14 different precious and special metals together with the major metals copper, lead, and nickel, which are used as metallurgical collector (Fig. 5). For precious metals from circuit boards or catalysts – in spite of their low concentration – yields of over 95% are realized, while simultaneously, tin, lead, copper, bismuth, antimony, indium, selenium, and others are reclaimed (Hagelüken 2006). In other dedicated processes, the company recovers cobalt, nickel, and copper from batteries (nickel metal hydride, lithium ion, lithium polymer type) (Meskers et al. 2009a); germanium from waver production scrap; and indium from indium-tin oxide (ITO) sputtering targets. Research is ongoing to extend further the range of feed materials (e.g., into photovoltaic applications) and additionally recovered special metals (e.g., Ga, Li). For metals that already follow other metal streams or can be separated from the off gas or effluents, recovery might be achieved through affordable adjustments of the flow sheet and/or the development of dedicated after-treatment steps. On the contrary, for metals

Fig. 5 Simplified flow sheet of Umicore's integrated smelter-refinery process to recover precious and special metals (Hagelüken 2006)

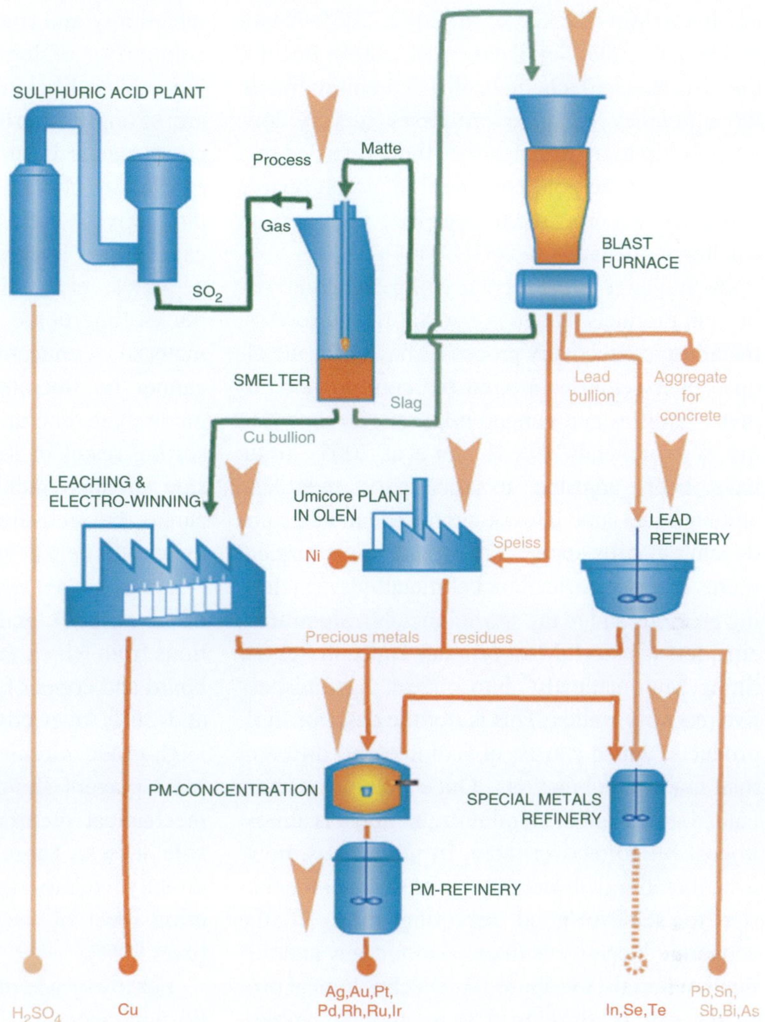

that oxidize easily and are dispersed as a low-grade slag constituent, economic recovery can become extremely difficult or even thermodynamically impossible.[4]

The combination of metals, toxic, and organic substances with halogens in many EOL products requires special installations and considerable investments for off-gas and effluent management to secure environmentally sound operations (prevention of heavy metal and dioxin emissions, etc.). In practice, many recovery plants, in particular in Asian transition countries, are not equipped with such installations. Electronic scrap is "industrially" treated in noncompliant smelters or leached with strong acids in hydrometallurgical plants with questionable effluent management; here the focus is primarily on recovery of (only) gold and copper. The largest part of electronic scrap is handled in the informal sector in thousands of "backyard recycling" operations. This involves open-sky incineration to remove plastics, "cooking" of circuit boards over a torch for desoldering, cyanide leaching, and mercury

[4] Examples are tantalum or rare earth elements used in electronic applications. Present only in very low concentrations (e.g., in circuit boards), they dilute even more into the slag. Due to their dispersion/dilution, the additional energy needed to recover and recycle the metal can exceed the energy requirement for virgin extraction.

amalgamation (Puckett et al. 2002, 2005; Kuper and Hojsik 2008; Sepúlveda et al. 2009). Besides the disastrous effects on health and environment, the efficiency of these processes is very low. An investigation in Bangalore, India revealed that only 25% of the gold contained in circuit boards is recovered, compared to over 95% at integrated smelters (Rochat et al. 2007).

A challenge is the recovery of metal combinations in products that do not occur in nature. Most metallurgical recovery processes have been developed over centuries around the combinations of metal families and gangue minerals, as shown in the "web of metals" (see Reuter et al. 2005). Some have been adjusted to secondary materials, although the same laws of chemistry and thermodynamics still apply. The main metallurgical routes are: copper/lead/nickel metallurgy (including precious and many special metals), aluminum, zinc, ferrous/steel. Most primary concentrates are fitting "automatically" into one of these respective recovery routes. This is not the case for EOL products, which consist of (completely) different man-made combinations. Once precious metals enter a steel plant or aluminum smelter, it is almost impossible to recover them. In most cases, however, metallurgical technology itself is not the barrier to achieve good recycling rates. If the economic incentive is there, also for new and difficult materials, the appropriate technological processes can be developed (e.g., mobile phones, lithium-ion batteries, gas to liquid catalysts, diesel particulate filters, or fuel cells).

Preprocessing

The goal of preprocessing is to generate appropriate output streams for the main smelting and refining processes (see above) by disintegration ("shredding") and sorting. It must be able to deal with a feed which changes over time and consists of many different models and types of equipment (Reuter et al. 2005). While this largely works for major metals, it is much more difficult to achieve for technology metals. Quantification of the losses during preprocessing for the different routes and the impact of material combinations (product design) is necessary to evaluate the efficiency of processes and improvement possibilities, mainly in the technical interface between metallurgy and (mechanical) preprocessing. The complexity of high-tech EOL devices leads to incomplete liberation of the materials, as these are strongly interlinked. For example, the precious metals in circuit boards occur with *other metals* in contacts, connectors, solders, hard disk drives, etc.; with *ceramics* in multilayer capacitors, integrated circuits (ICs), hybrid ceramics, etc.; with *plastics* in circuit board tracks, interboard layers, ICs, etc. Small-size material connections, coatings, and alloys cannot be liberated during shredding. Hence, incomplete liberation and subsequent incorrect sorting result in losses of technology metals to side streams (including dust) from which they cannot be recovered during metallurgical treatment (Meskers et al. 2009b; Reuter et al. 2006).

An industrial test indicated that the percentage of silver, gold, and palladium reporting to fractions from which they could be recovered (circuit board and copper fractions) was only 12%, 26%, and 26%, respectively (Chancerel et al. 2009). High-grade circuit boards and cell phones or MP3 players should thus be removed prior to mechanical preprocessing to prevent irrecoverable losses. These components/devices can be directly fed into a smelter-refinery process, recovering most of the metals with high efficiency (over 90%).

For low-grade materials, such as small domestic appliances or most audio/video equipment, the direct smelter route is usually not applicable, and some degree of mechanical preprocessing is required. Instead of intensely shredding the material, a coarse size reduction followed by manual or automated removal of circuit board fractions can be a valid alternative. Trained workers in this context are often able to remove more selectively certain complex target components than automated sorting technologies. Whenever there is local access to affordable and reliable manual labor, like in many developing and industrializing countries, manual dismantling, sorting, and removal of critical fractions, such as circuit boards or batteries, combined with state-of-the-art industrial metals recovery processes (which then are mostly located abroad), can be a valid alternative. Such "best of two worlds" approaches are currently being investigated in the framework of the StEP initiative[5] by the United Nations University. Also in industrialized countries, the outsourcing of well-defined dismantling steps to local social initiatives can contribute to improved overall recycling efficiencies.

Altogether, a holistic optimization of product design, sorting method, and depth and destination of the various fractions produced can lead to a substantial increase in overall yields, especially for technology metals. Often, new developments in products lead to new challenges. An example is the middle- and upper-class car of today. This "computer on wheels" contains numerous electronic components containing in total a lot of technology metals. However, the electronic parts are scattered throughout the vehicle; current shredder technology will not be able to recover these and also cost-effective manual dismantling seems difficult.

Collection

Effective collection systems are a prerequisite for metals recovery, and the respective infrastructure must be adjusted to the local circumstances. Collected EOL products are sorted into several categories, which are (at least in the EU) determined by legislation at a country level. For logistic purposes, a reduction of the number of categories is often attempted; however, too much reduction results in a very heterogeneous mixture of high-grade and low-grade materials, which reduces the effectiveness of the preprocessing and recovery processes following collection. A balance must be struck between too many and too little categories, to maximize the overall recovery of technology metals along the recycling chain.

Economic Impact Factors on Metal Recycling Rates

High prices for metals and resources offer, in principle, a solid basis for stimulating recycling efforts. The scrap value is determined by the intrinsic monetary value of its contained substances and the total costs that are needed to realize that value. Thus, value is determined by metal market prices as well as the variety and yields of recoverable substances. The costs comprise logistics, treatment in the subsequent steps of the recycling chain, and costs for environmentally sound disposal of unrecoverable fractions/substances. Complexity of a product and hazardous substances contained therein drive the costs, but "traces" of expensive precious and special metals or base metals in higher concentrations push up the value.

To illustrate the incentive to recycle complex products/components, consider the following. In mid-2009, the net intrinsic metal value for an "average" mobile phone without battery was about 8,000–10,000 USD/t; for a computer board, 4,000–6,000 USD/t; and for a catalytic converter, 40–80 USD per piece.[6] These values already include costs at the smelter, in compliance with today's strict emission legislation. At lead prices of over 2,000 USD/t, commercial lots of car batteries are attractive to a lead smelter; the same applies at high cobalt prices for lithium-ion batteries treated in dedicated smelters to recover cobalt, nickel, and copper. Taking into account steel and copper prices, EOL vehicles can offer a good scrap value. The more efficient the entire recycling chain can be setup, the more an EOL product value can ultimately be realized.

Although only contained in small amounts per device, precious metals often constitute the biggest share of the intrinsic value. For example, precious metals in a mobile phone, a computer circuit board, or other high-grade devices, which account for less than 0.5% of the weight, contribute to over 80% of the value, followed by copper (10–20% of the weight; 5–15% of the value). Iron, aluminum, and plastics, which dominate the weight for these devices, make only a small

[5] *S*olving *t*he *E*-Waste *P*roblem (www.step-initiative.org).

[6] At metal price levels, as of Oct. 2009: net value = recovered metals value minus smelting and refining charges, but without consideration of collection, preprocessing, and shipment costs in the preceding recycling chain. Value can vary significantly depending on specific quality/type (especially for autocatalysts).

contribution to the overall value. With few exceptions (e.g., Co in rechargeable batteries), the value share of most special metals on a unit level is still negligible, due to their very low concentrations and (compared to precious metals) relatively low prices. Many recovery plants focus, therefore, only on copper and precious metals. This means that a large amount of special metals is lost that could technically be recovered. However, in state-of-the-art large-scale recovery plants, special metals recovery can on a yearly base generate attractive additional revenue. This "by-product recovery" is comparable to investments in by-product recovery for primary materials.

Societal and Legislative Factors

It is evident that the awareness to recycle consumer goods is of the utmost importance. Legislation, public campaigns (e.g., from authorities, nongovernmental organizations (NGOs), manufacturers), and an appropriate infrastructure for handing in old products are important prerequisites. Europe (in particular, the Scandinavian, Benelux, and German-speaking countries) has progressed quite far in developing a general "recycling mentality." Although many people are used to trading or returning old goods to collection points for reuse, some items (e.g., mobile phones or "high-price" electronics as computers) require incentives to mobilize them out of "hibernation" or obsolescence. A consumer survey indicated that only 3% of the people return old mobile phones for reuse or recycling, whereas 44% have them stored at home (Nokia 2008). The amount of EOL products continues to increase and is influenced, for example, by both the consumer behavior related to product lifetime and the general consumption of materials. The product lifetime is determined by durability as well as functional, technical, and aesthetical obsolescence; these are, in turn, determined by the product design and social factors such as fashion of the day and lifestyle. It appears that for some first-time owners, the lifetime of a product becomes increasingly shorter, in particular, for fashion- and technology-sensitive items like mobile phones, computers, and MP3 players.

Most people in Europe look for a proper solution when they are ready to discard their products. Nevertheless, a lot of goods handed in for recycling or reuse do not enter the appropriate channels. This is not due to missing awareness or legislation, but rather to weaknesses in control and enforcement, as well as in structural deficits. In many developing countries, in India and China, collection by the informal sector often works well and sometimes builds on a long tradition, due to the attraction of an even low economic value. The problems arise here when the collected and dismantled items are then locally treated for metals recovery by the informal sector or substandard "industrial" processes.

Impact of Legislation on Technology

Legislation has limited impact on recycling *technologies*. Counterexamples are:

- Mandatory removal of certain parts from EOL devices (catalysts, circuit boards, batteries). Legislation should, however, not prescribe *how* (e.g., manually) certain parts have to be removed since this restricts possible (future) technological solution. Important is the result, that is, that the desired substances are directed in an identifiable way into the most appropriate treatment process or output stream.
- Classification of electronic scrap into certain qualities for collection (sorting) is necessary in order to obtain optimized streams for further downstream processing.
- Defining technical and environmental treatment standards is vital for the recycling industry because they help create a level playing field and promote innovation. The control and enforcement, that is, through certification, of these standards is crucial, especially with respect to recycling plants in developing and transition countries. There are no objections against the recycling of, for example, European scrap in a non-European country as long as the environmental standards of the European legislation are met (which, in practice, is often not the case).

However, legislation can also negatively impact recycling results by setting wrong priorities which might lead to the use of less appropriate technologies.

One example is the obligation to meet certain weight-based recycling rates, as it promotes the recycling of the main product constituents, which are not necessarily the most important ones from an economic and environmental perspective. Technology metals with low concentrations are not taken into account here at all. There are examples of technology developments, for example, for mechanical processing of mobile phones, aiming at increasing separation of plastics (high weight percentage), but hazarding the consequences of high losses in precious metals (see above). Although this might meet the legal recycling goal, it is counterproductive from an eco-efficiency point of view.

Impact of Legislation on Recycling Economics

At mid-2011 prices, products such as mobile phones, computers, and cars have a positive net value if handled in professional recycling chains. Other products (e.g., a TV or monitor, most audio/video equipment, and small household appliances) still have a negative net value. However, if the true costs of landfill and environmental damage caused by nonrecycling would be accounted for, then on a macroeconomic level, proper recycling most probably is viable for such products as well. Legislation can, and does, provide ways to finance the recycling costs of these "negative goods." Thus far, legislation has focused on hazardous substances but not on technology metals contained in such products. Waiting for the market to regulate itself by further increasing special metal prices, which 1 day would generate enough recycling incentives, cannot be an acceptable approach. Due to a delayed reaction time of metal prices, too many secondary technology metal resources will, in the meantime, inevitably be lost. From a national economy's point of view, consideration should be given to provide more legislative support, especially for special metals recycling (e.g., in goods with a negative net value).

Structural Factors: The Product Life Cycle

In view of the discussion on economic, legislative, and technical factors, one could expect that car catalysts, mobile phones, computers, and cars are products that (at least in Europe) achieve a very high recycling rate because (1) efficient technologies and sufficient capacities to recycle these goods in an environmentally compliant way exist; (2) legislation, consumer awareness, and a collection/recycling infrastructure are widely in place; and (3) economic incentives for recycling are attractive.

However, for all these products, actual recycling rates are well below 50%. The most eye-catching example in this context is the valuable car catalyst (Hagelüken 2007a). On a global level, only about 50% of the PGMs are finally recovered. In Europe, this level is even below 40%, due to its large exports of end-of-life vehicles. This happens although (1) it is easy to identify and remove the catalyst from a scrap car at the dismantler (required by the ELV directive); (2) a more than sufficient number of catalyst collectors is aggressively chasing autocatalysts at dismantlers, scrap yards, and workshops, paying high prices per piece; and (3) appropriate smelting and refining technologies are able to recover more than 95% of the platinum group metals (PGM) contained in a catalyst. Something must go essentially wrong with additional factors playing a negative role during the life cycle.

The Significance of Life Cycle Structures

A comprehensive research project investigated the structural factors that play a role in the life cycle of PGMs in detail[7] (Hagelüken et al. 2005).

[7] The focus and system limit were the FR Germany. Global conditions for the materials flow of PGM were, however, adequately considered in the study. Areas of investigation include all relevant application segments for PGMs: automotive catalysts; chemical and oil refining catalysts; glass manufacturing; dental applications; electronics; jewelry; electroplating; fuel cells, etc.

Structural factors investigated were product lifetime, sequence of product ownership, sequence of locality of use, system boundaries/global flows, and structure of the recycling chain. Two distinctly different life cycle structures were identified – "closed cycles" and "open cycles," commonly referred to as direct (closed loop) and indirect (open loop) systems. The structural factors identified for PGMs can be extended to other metals in industrial and consumer products in general.

Closed Cycles: Recycling from Industrial Processes

Closed loops prevail in industrial processes where metals are used to enable the manufacture of other goods or intermediates. Examples are PGM process catalysts (e.g., oil refining catalysts) or PGM equipment used in the glass industry. For PGMs, the manufactured goods do not typically contain PGMs themselves (e.g., petrol produced in an oil refinery with a Pt catalyzed reforming process). Instead, the metals are part of an industrial product which is owned by and located at the industrial facility and thus has a high economic value that facilitates recycling. Changes in ownership or location are well documented, and material flows are kept transparent. All stakeholders in the life cycle work closely together in a professional, industrial manner. As a result, closed loop systems are inherently efficient, and more than 90% of the PGMs used in industrial processes are typically recovered. A "closed loop" however does not mean that no metal losses occur; this is just thermodynamically not possible. Key is that the industrial structure of such life cycles "drives" old products into the best fit recycling processes, which is not the case for most consumer goods in "open cycles" (see below).

A long product lifetime does not negatively affect the achieved recovery rate. Oil refining PGM catalysts with a lifetime over 10 years are still recycled. Thus, the attractive intrinsic value (of PGMs) combined with the frame conditions of an industrial cycle is the driver for success. Recycling of industrial products without precious metals, such as sputtering targets or production scrap in general, can be less economically attractive, but the other fundamental frame conditions remain similar. Old industrial infrastructure and machinery offer a significant future recycling potential for steel, copper, and many other metals. Whereas massive infrastructure is difficult to relocate, and thus is a good target for "urban mining," it has been reported that second-hand machinery is also increasingly leaving Europe (Janischewski et al. 2003).

Open Cycles: Recycling from Consumer Durables

Open loop systems (Fig. 6) are prevalent in the recycling of EOL consumer products, such as cars and electronics. Their complex structure and lack of supportive frame conditions evokes inefficient/failing metal recovery. As recycling rates for valuable PGM-containing catalysts are below 50%, it can be assumed that for most technology metals, this situation is even worse. Many participants in the life cycle are not aware of the (economic) metal value in EOL consumer goods. Although the concentration (and value) per product is low, the huge product sales represent a significant metal resource and economic value in total.

Consumer products often change ownership during their life cycle, and with each change, the connection between the manufacturer and owner becomes weaker. This is compounded by the fact that change of owner often means a change of location and that highly mobile consumer goods are spread all over the globe. Trading of old equipment and donations to charities have led to steady but nontransparent flows of material to Eastern Europe, Africa, and Asia (Buchert et al. 2007).[8] A clear distinction between an EOL

[8] It is estimated that about 50% of used IT electronics leave Europe one way or another. For mobile phones, less than 5% of the theoretical recycling potential is currently being realized globally in a compliant way. For 2008, monitoring results for ELV in Germany showed that out of 3.2 million deregistered passenger cars, only 420,421 were recycled in Germany, while 1.75 million were exported as "used cars." A gap of cars addresses mainly

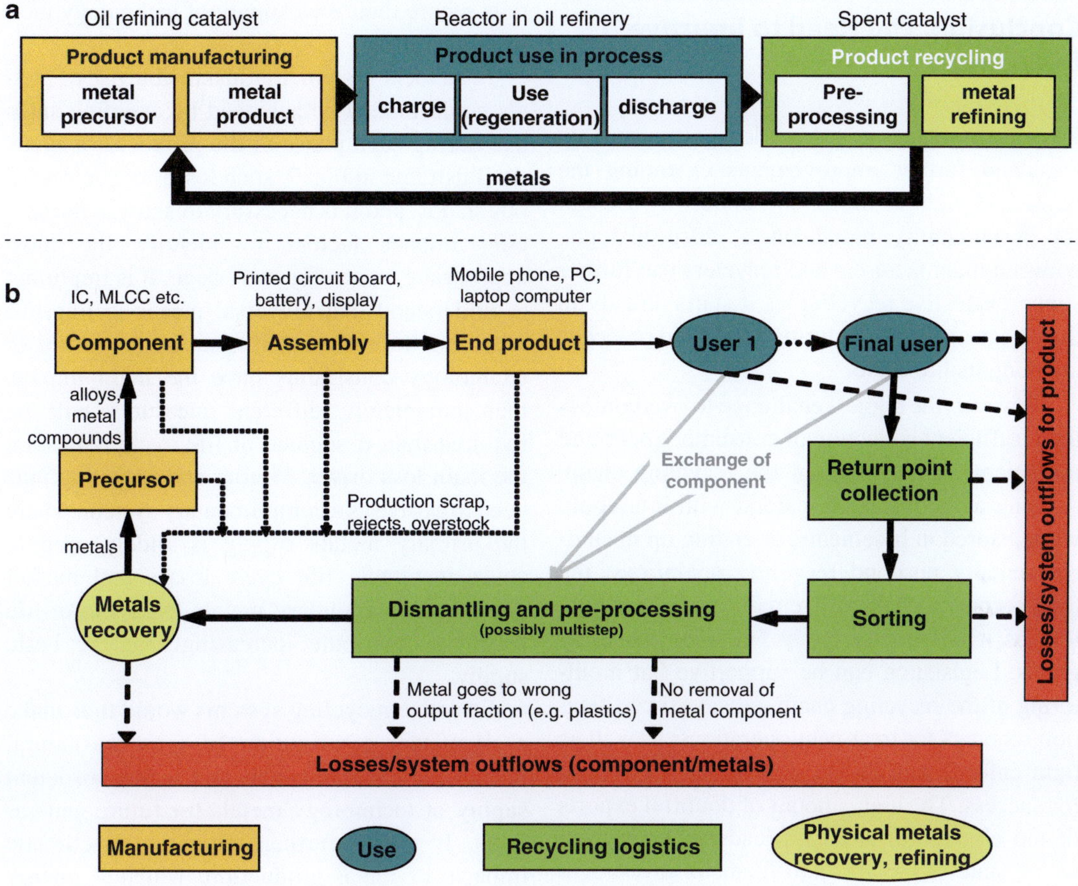

Fig. 6 (**a**) Closed loop systems for industrial applications (e.g., process catalyst) versus (**b**) Open loop systems for consumer goods (e.g., electronics)

product for recycling and for reuse is dependent on location (i.e., waste in Europe=reuse in Africa). Traders take advantage of this by exporting for reuse, although a fair amount of these exports evade the Basel Convention waste export procedures. Thus, old products collected in good faith for recycling or reuse can escape, in a dubious way, only to resurface in primitive landfills or disastrous backyard "recycling" operations in developing countries. In practice, the probability of effective recycling at final EOL in developing and transition countries is rather low, as appropriate recycling infrastructure is not in place, or only some valuable (precious) metals are recovered at very low efficiency (Rochat et al. 2007). The insufficient cooperation along the life cycle and recycling chain (although "extended producer responsibility" has been implemented), combined with insufficient tracking of product and material streams along the entire chain, explains why inefficient open cycles continue to exist.

unregistered exports. A recycling rate of 89.2% was reported (Umweltbundesamt 2010), but this refers only to the 420,221 cars scrapped in Germany. Calculated on the 3.2 million deregistrations, Germany's recycling rate would fall to 13.1%. Although 1.5 million of the exported cars go primarily into other (mainly Eastern) EU states, it can be assumed that a big portion will ultimately leave Europe. The export of about 2.6 million cars represents a secondary materials potential of 1.3 million tons of steel, 180,000 t aluminum, about 110,000 t of other nonferrous metals, and about 6.25 t of PGMs. Significant quantities of ELVs are also exported from other European countries (Buchert et al. 2007).

Conclusion: The Need to Improve Recycling

Recycling technology has made significant progress, and further improvements extending the range and yield of metals are underway. Design for sustainability based on a close dialogue between manufacturers and recyclers can further support effective recycling as it starts already in the design and manufacturing phase and proceeds all along its life cycle.

However, the biggest challenge to overcome is the insufficient collection of consumer goods and inefficient handling within the recycling chain. As long as goods are discarded with household waste, stored in basements, or ending up in environmentally unsound recycling operations, the total recovery rates will remain disappointingly low, as it is the case today for most consumer goods. Legislation can be supportive but monitoring of the recycling chain, advanced certification schemes for treatment operations, as well as tight enforcement of the regulations are crucial for success. The high amount of doubtful exports of old electronics and cars leads to a situation where state-of-the-art, high financial investment recycling facilities in industrialized countries are underutilized because "recycling" and the associated environmental burden of environmentally unsound treatment is "outsourced" to the developing world. The usually primitive "backyard recycling processes" applied in these countries focus on (highly inefficient) gold and copper recovery, while most other technology metals are completely lost. The "urban mine" is thus wasted irreversibly.

To effectively close the loop for consumer products, new business models need to be introduced that provide strong incentives to hand in products at their end-of-life. This can include deposit fees on new products or product service systems (PSS) like leasing or other approaches. Especially for emerging technologies (electric vehicles, photovoltaics, etc.), setting up "closed loop structures" will be essential, and manufacturers who put successful models in place can thus secure their own supply of technology metals in the future.

In an ideal system, the sustainable use of metals could indeed be achieved by avoiding spillage during each phase of the product life cycle. As illustrated in Fig. 7, such losses occur at various stages, and it is necessary to analyze the specific impact factors to identify the most appropriate means for each stage. It is important to understand that universal means to improve recycling do not exist; if material properties or technology constraints have the main impact, then completely different measures will be required than if societal or life cycle issues are the main loss driver. Mining and recycling thus need to evolve as a complimentary system, where the primary metals supply is widely used to cover inevitable life cycle losses and market growth, and secondary metals from end-of-life products contribute increasingly to the basic supply.

Effective recycling systems would thus make a significant contribution to conserve natural resources of scarce metals and secure sufficient supply of technology metals for future generations. It would further mitigate the climatic impacts of metal production, which is energy intensive, especially in the case of precious metals mined from low concentrated ores (e.g., Au mined at 5 g/t from 3,000 m underground). The mining of annually 2,500 t of gold worldwide generates some 17,000 t of CO_2 per ton of gold produced (based on ecoinvent 2.0 database of EMPA/ETH Zurich) or 42 million tons CO_2 in total. For PGMs, the ore grade and specific CO_2 impact is of the same magnitude, while copper mining "only" causes 3.5 t CO_2/t Cu, but adding up to 56 million tons for the production of 16 million tons copper annually. Some mass products are relatively rich "bonanzas" in comparison to mines, for example, a computer motherboard with 200–250 g/t of gold, a mobile phone handset with 350 g Au/t, or an automotive catalytic converter with some 700 g/t of PGM. If effective collection systems and state-of-the-art recovery processes are used, then the secondary metal production from such products

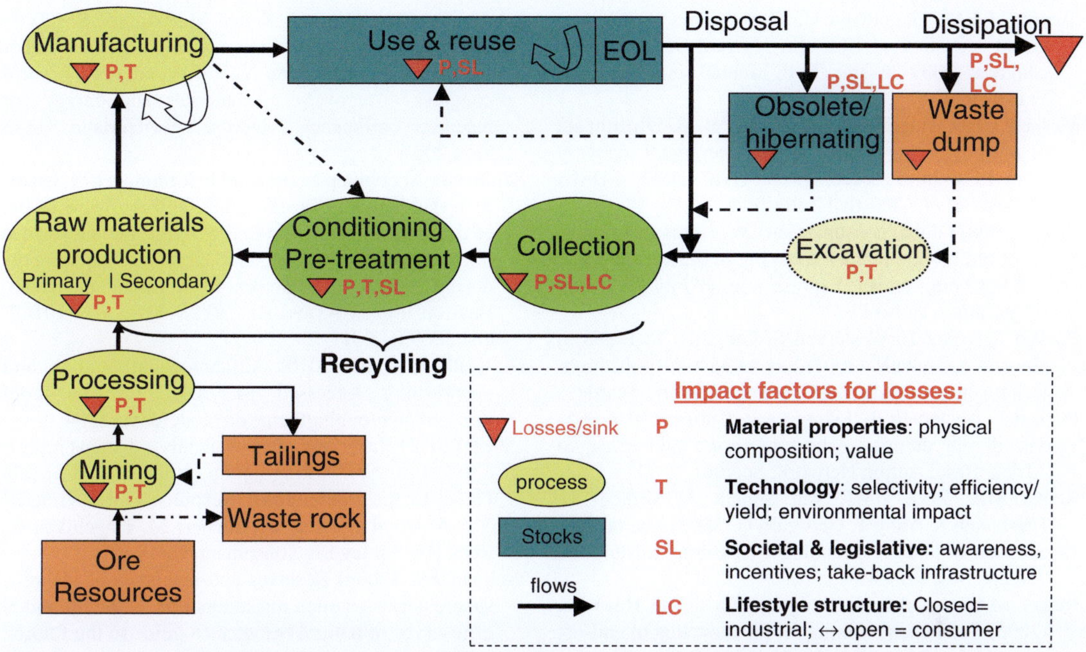

Fig. 7 Life cycles of metals/products and impact factors for losses at various stages (McLean et al. 2010)

will only require a small fraction of energy/CO_2 compared to mining (Hagelüken and Meskers 2008).

To summarize, efficiently recycling our end-of-life products today is insurance for the future. It will mitigate metal price surges and secure a sustainable and affordable supply of metals needed for our products of tomorrow.

References

Buchert M, Hermann A, Jenseit W, Stahl H, Osyguß B, Hagelüken C (2007) Optimization of precious metals recycling: analysis of exports of used vehicles and electrical and electronic devices at Hamburg port. Federal Environmental Agency of Germany, Dessau

Chancerel P, Meskers CEM, Hagelüken C, Rotter VS (2009) Assessment of precious metal flows during preprocessing of waste electrical and electronic equipment. J Ind Ecol 13(5):791–810

EU COM (2008) 699 The raw materials initiative – meeting our critical needs for growths and jobs in Europe, Brussels, SEC (2008) 2741

Hagelüken C (2006) Recycling of electronic scrap at Umicore's integrated metals smelter and refinery. World of Metallurgy - Erzmetall 59(3):152–161

Hagelüken C (2007a) Closing the loop – recycling of automotive catalysts. Metallurgy 61(1–2):24–39

Hagelüken C (2007b) The challenge of open cycles. In: Hilty LM, Edelmann X, Ruf A (eds) R'07 world congress – recovery of materials and energy for resource efficiency, Davos, Switzerland, 3–5 Sept 2007

Hagelüken C, Meskers CEM (2008) Mining our computers – opportunities and challenges to recover scarce and valuable metals. In: Reichl H, Nissen N, Müller J, Deubzer O (eds) Proceedings of electronics goes green conference 2008. Fraunhhofer IRB, Stuttgart, pp 585–590

Hagelüken C, Meskers CEM (2010) Complex lifecycles of precious and special metals. In: Graedel, TE, van der Voet E (eds) Linkages of sustainability. Strüngmann Forum Report 4. MIT Press, Cambridge, pp 165–197

Hagelüken C, Buchert M, Ryan P (2005) Materials flow of platinum group metals. GFMS, London

Janischewski J, Henzler M, Kahlenborn W (2003) The export of second-hand goods and the transfer of technology: a study commissioned by the German Council for Sustainable Development. Adelpi Research GmbH

Kuper J, Hojsik M (2008) Poisoning the poor: electronic waste in Ghana. Greenpeace International, Amsterdam

McLean HL, Duchin F, Hagelüken C, Halada K, Kesler SE, Moriguchi Y, Mueller D, Norgate TE, Reuter MA, van der Voet E (2010) Mineral resources: stocks, flows, and prospects. In: Graedel, TE, van der Voet E (eds) Linkages of sustainability. Strüngmann Forum Report 4. MIT Press, Cambridge, pp 199–218

Meskers CEM, Hagelüken C, Van Damme G (2009a) Green recycling of EEE. In: Howard SM (ed) EPD congress 2009 at the TMS annual meeting, San Francisco, CA

Meskers CEM, Hagelüken C, Salhofer S (2009b) Impact of pre-processing routes on precious metal recovery from PCs. In: Proceedings of EMC 2009, GDMB Medienverlag, Clausthal Zellerfeld

Nokia (2008) Global consumer survey reveals that majority of old mobile phones are lying in drawers at home and not being recycled (press release July 8). Nokia Corporation, Helsinki

Puckett J, Byster L, Westervelt S, Gutierrez R, Davis S, Hussain A, Dutta M (2002) Exporting harm – the high-tech trashing of Asia. Basel Action Network, Seattle

Puckett J, Westervelt S, Takamiya Y, Gutierrez R (2005) The digital dump – exporting re-use and abuse to Africa. Basel Action Network, Seattle

Reuter MA, Boin UMJ, Van Schaik A, Verhoef E, Heiskanen K, Yang Y, Georgalli G (2005) The metrics of material and metal ecology. Elsevier, Amsterdam, 720 pp

Reuter MA, Van Schaik A, Ignatenko O, de Haan GJ (2006) Fundamental limits for the recycling of end-of-life vehicles. Miner Eng 19(5):433–449

Rochat D, Hagelüken C, Keller M, Widmer R (2007) Optimal recycling for printed wiring boards in India. In: Hilty LM, Edelmann X, Ruf A (eds) R'07 world congress – recovery of materials and energy for resource efficiency, Davos, Switzerland, 3–5 Sept 2007

Sepúlveda A, Schluep M, Renaud FG, Streicher M, Kuehr R, Hagelüken C, Gerecke AC (2009) A review of the environmental fate and effects of hazardous substances released from electrical and electronic equipments during recycling: examples from China and India. Environ Impact Assess Rev 30:28–41. doi:10.1016/j.eiar.2009.04.001

Umweltbundesamt (2010) Altfahrzeugaufkommen und -verwertung Accessed 14.1.2012. www.env-it.de/umweltdaten/public/theme.do?nodeIdent=2304

UNEP (2011) Recycling rates of metals – a status report: a report of the Working Group on the Global Metals Flows to the International Resource Panel. Graedel TE, Allwood J, Birat JP, Buchert M, Hagelüken C, Reck BK, Sibley SF, Sonnemann G, 48 pp

Wellmer FW (2008) Reserves and resources of the geosphere, terms so often misunderstood. Is the life index of reserves of natural resources a guide to the future? Z dt Ges Geowiss 159(4):575–590

Certified Trading Chains in Mineral Production: A Way to Improve Responsibility in Mining

Gudrun Franken, Jürgen Vasters, Ulrike Dorner, Frank Melcher, Maria Sitnikova, and Simon Goldmann

Abstract

Mining in Central Africa has been associated with violent conflict, mistreatment of artisanal miners, illegal trading and the diversion of state funds. In 2002, the UN Group of Experts on the Democratic Republic of the Congo stated that the plunder of natural resources and other forms of wealth of the Democratic Republic of Congo was fuelling conflict in the region. Especially the sector of artisanal and small-scale mining, which provides livelihood for millions of people in Central Africa, is part of the informal and illegal trading chain of minerals. In November 2006, the eleven member states of the International Conference on the Great Lakes Region (ICGLR) signed the Protocol against the Illegal Exploitation of Natural Resources, which includes the aim of implementing a mechanism for the certification of natural resources in its Article 11. The Federal Institute for Geosciences and Natural Resources (BGR) developed two initiatives aiming at transparency and responsibility in the trading chains of metal ores linked to financing conflicts: first the development of a geochemical fingerprint for coltan ores and second a concept for a chain of custody assurance system (Certified Trading Chains – CTC), based on the establishment of transparent, traceable and ethical trading chains. Since the year 2000, on the one hand Central African coltan (tantalum ore concentrate) has developed into a significant supplier for the world market, on the other hand profit from ore production and trade has been financing military groups. The fingerprinting method aims at identifying the origin of a concentrate by comparing its mineralogical and chemical characteristics with samples of known provenance that are stored in the BGR coltan database. Adaption of the method to tin and tungsten ores, and application either as a forensic instrument, or in conjunction with certified trading

G. Franken (✉) • J. Vasters • U. Dorner • F. Melcher • M. Sitnikova • S. Goldmann
Bundesanstalt für Geowissenschaften und Rohstoffe, Stilleweg 2, Hannover D-30655, Germany
e-mail: Gudrun.Franken@bgr.de

R. Sinding-Larsen and F.-W. Wellmer (eds.), *Non-Renewable Resource Issues: Geoscientific and Societal Challenges*, International Year of Planet Earth,
DOI 10.1007/978-90-481-8679-2_11, © Springer Science+Business Media B.V. 2012

chains, would allow for the control of a significant portion of mineral exports from Central Africa's conflict region. The geochemical fingerprint is imbedded in CTC as an additional checking instrument for traceability (beyond the documentary system). The concept of Certified Trading Chains found entry to the protocol of the G8 summit in Heiligendamm/ Germany in 2007 as part of efforts in the mining sector to assure greater responsibility for reducing poverty and preventing conflicts. As a result, a CTC pilot project in Rwanda commissioned by the German Government was initiated concerning the feasibility of a certification system for selected raw materials (tin, tantalum and tungsten). Consultations with stakeholders from Rwanda paved the way for national implementation currently underway, however integrated in the regional policy of the International Conference of the Great Lakes Region. Major features of CTC are a set of five principles with 20 standards based on OECD guidelines to be applicable in the artisanal and small-scale mining sector and which were adapted to local conditions by a national consultation process in Rwanda. Key elements are the voluntary participation of the companies, the lead by national governmental institutions as well as an independent audit of an internationally accredited auditor. The proper implementation of certification will foster good governance, where mineral resources are produced and traded legally and transparently and in ways, which protect workers, communities, and the environment. Certification will also progressively transform and formalize informal mining. Formalization is a precondition for achieving transparent recording of production and trade, to improve governance and reduce conflicts associated with the mining sector.

Keywords
Mineral certification • CTC • Conflict minerals • Coltan • Analytical fingerprint • AFP • Due diligence • Artisanal mining • ASM • Democratic Republic of Congo • Rwanda

Resources and Conflicts

Introduction

Natural resources can play a key role in triggering, prolonging and financing conflicts. Since the mid of the 1990s, there has been growing research on this topic. Recent findings show that natural resources are never the only source of a conflict. Poverty, ethnic or religious grievances, unstable governments and rebel financing play major roles. Natural resources often linked to civil conflicts are largely oil and high-value minerals like coltan, diamonds, gold and gemstones. In most of the conflicts, multiple resources are involved.

Resource-related conflicts are especially relevant for African countries (Bannon and Collier 2003). Three of the world's worst wars of recent years were related to natural resources and took place in Sierra Leone, Liberia and the Democratic Republic of Congo (DRC). The eastern part of the DRC (North and South Kivu) has been suffering from armed conflicts for more than 12 years. In North and South Kivu, armed groups and the Congolese national army control the trade of cassiterite (tin ore), gold, columbite-tantalite (coltan), wolframite (tungsten ore) and other minerals. The warring parties have unrestricted access to these minerals and have been able to establish lucrative trading networks because of a lack of state control and the unregulated nature of the mining sector.

The Case of the Democratic Republic of Congo

Mineral production in the DRC depends to a large amount on the artisanal and small-scale mining (ASM) sector. Current estimates indicate that about two million people are working in this sector and about ten million people are depending on the earnings generated here – this is 20% of the total Congolese population. In the early years of this decade, the artisanal mining sector in the DRC produced between 80% and 100% of the total Congolese mineral production (Garrett and Mitchell 2009) due to difficult operating environments for industrial mining operations and lack of infrastructure.

The Congolese population suffers not only from reprisals of the warring groups but also from wanting working conditions in artisanal mining. Beyond that, in the initial phase of operation, artisanal miners commonly receive a trader's credit. During the pay down of the credit, they are forced to sell their production under unfavourable conditions to the lender. This results in a long-term dependency of the miner and resembles modern peonage.

The two main warring parties, the Forces armées de la République du Congo (FARDC, Congolese national army) and the Forces Démocratiques de Libération du Rwanda (FDLR), the predominantly Rwandan Hutu armed group, control highly mineralised areas (North and South Kivu) where they benefit from mineral production and trade. They impose illegal 'taxes' on miners and traders. The FDLR obtains a significant proportion (up to 75%) of its income from gold mining and trading. Informed estimates suggest that the Congrès National pour la Défence du Peuple (CNDP), a political armed militia engaged in the Kivu conflict, earns up to 15% of its revenue from the mineral trade, and the FARDC up to 95% (Garrett and Mitchell 2009).

The analysis of export data from the Kivu provinces suggests that under-declaration and smuggling are the major problems. The trade starts at remote sites, from where the minerals are transported by foot, car, truck and plane to the main export centres. From there, the exported material passes through Burundi, Kenya, Rwanda, Uganda and/or Tanzania. Differences between Congolese and Rwandan statistics suggest that traded minerals are either declared as transit goods or untaxed at the DRC border and then declared in Rwanda. Independent research (Garrett and Mitchell 2009) indicates that 28,700 ton of cassiterite has been exported from the DRC and Rwanda in 2008, with 87% originating from the DRC.

The conflict dynamics in Eastern DRC are more complex than a simple cause and effect connection between military groups, mineral production and the trade in minerals. Weak institutions and a dysfunctional army enable military groups to pose security problems to mining and trade for their benefit. Stopping or interrupting mining and trade is less promising than figuring out how to tackle the insecurity problems and formalise the sector. On the other hand, it is essential to improve the situation of the workers depending on ASM in order to strengthen their position.

The Role of Artisanal and Small-Scale Mining

In our modern post-industrial society, the importance of the artisanal and small-scale mining sector is frequently underestimated. However, it is a worldwide reality. In addition to large-scale mining, ASM production contributes significantly to the raw materials supply of industrialised countries. There is no general accepted definition of artisanal and small-scale mining yet. In the following context, the term artisanal and small-scale mining is used to describe extraction that is manual and very labour-intensive, using only shovels and basins or somewhat mechanised, using heavy machinery on a small scale (Fig. 1).

Artisanal and small-scale mining provides livelihood to an unprivileged part of society in

Fig. 1 Artisanal cassiterite mining, Nemba (Rwanda)

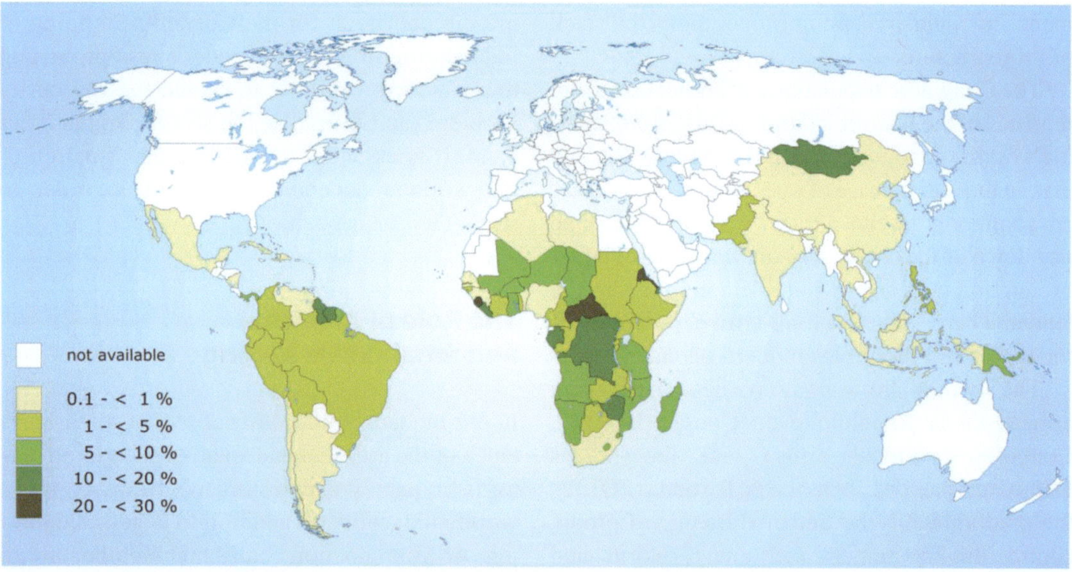

Fig. 2 Percentage of the population depending on ASM (CASM 2009)

many developing countries. Worldwide, about 15 million people make their living in the ASM sector. About 100 million people – workers and their families – depend existentially on ASM (Fig. 2) compared to about seven million people worldwide in industrial mining. Earnings in the ASM sector – at least in gold and diamond extraction – are considerably higher than the average income of comparable agricultural regions (about 3 USD/d in ASM, against about 0.6 USD/d for a farmer in central Africa).

The share of mineral production by ASM differs considerably between different raw materials. Gold as well as the high-tech and electronic metals tin, tantalum, tungsten and cobalt are produced to a high degree in ASM operations.

Nevertheless, the redistribution of income from ASM shows that only the smallest part of the extracted wealth eventually reaches the miner. This is also due to the fact that ASM is part of the informal and illegal trading chain of minerals. Although artisanal and small-scale mining is dealt with in most mining regulations, the actual surveillance of this sector by the mining authorities is low. Working conditions are generally poor, and child and forced labour are particularly abundant in weak governance regions. Missing or neglected mine safety procedures are the cause of frequent work accidents; environmental impacts can be devastating.

In addition, the lack of free market access for artisanal mineral production, caused by insufficient infrastructure or by legal restraints, is another obstacle for the thriving of this sector. It is commonplace that mineral buyers and traders can determine their prices. Because the local and intermediary trade of minerals needs a secure business environment especially in conflict areas, the traders have to buy security support from local political and military rulers or private companies. Part of the revenue from mineral trade is lost by means of corruption or is used to finance conflicts.

Up to now, there exists no generally accepted mechanism to ensure transparency in the mineral production of high-value minerals in ASM (apart from the so-called Kimberley Process for diamonds) and also to ensure adherence to internationally accepted minimum standards. Both are the prerequisite for responsible mining and production and due diligence also for the consumer industries.

A differentiation of the production regarding compliance to internationally accepted standards would add value to the ASM production and provide direct access to the mineral markets in industrial countries. For industrial countries, the access to ASM production without fearing of reputational loss would help to improve the supply security since the flexibility of the sector with regard to price changes would help to reduce the consumer's need for stock-keeping of minerals.

A new certification mechanism developed for the mineral trading chain, which is described in detail in section "Certified Trading Chains (CTC) – Background and Objectives", addresses the problems mentioned above. A pilot project has been conducted in Rwanda (see section "CTC in Practice") and focuses on artisanal mining of tantalum, tin and tungsten. Besides gold, these are the minerals of interest in central-east Africa.

Resources of Interest: The 3 T´s – Tin, Tungsten and Tantalum

High-commodity value metals such as gold and the 3 T´s, tin, tungsten and tantalum, are especially linked to financing conflicts in Eastern DRC. Artisanal mining of tin and tantalum significantly adds to their world supply (Table 1). Tantalum, tin and tungsten deposits are abundant in the Kibaran orogenic belt of central-east Africa (a major orogenic belt of Mesoproterozoic age extending from Katanga in the south to Uganda in the north), where they are extracted exclusively by artisanal mining.

Within the past 10 years, tantalum production (ca. 1,900 tons in 2008) was mainly focused on Western Australia (30–40%), Brazil (12%), Asia (13%) and Africa (35%) (Fig. 3). The economic

Table 1 Estimation of the share of ASM in the global production of selected minerals and metals (Wagner et al. 2007)

Metals	
Antimony	4%[a]
Beryl	5%
Lead	2.5%
Chromite	5%
Iron	4%
Gold	10%
Cobalt	30%
Copper	0.5%
Manganese	11%
Platinum	1%[b]
Mercury	18%
Silver	7%
Tantalum	40%[c]
Tungsten	6%
Zinc	1%
Tin	30%

[a]Exclusive PR China
[b]Supposed production of ASM in Russia and Colombia
[c]Mining in African countries (DRC, Mozambique, Rwanda, Ethiopia and Burundi)

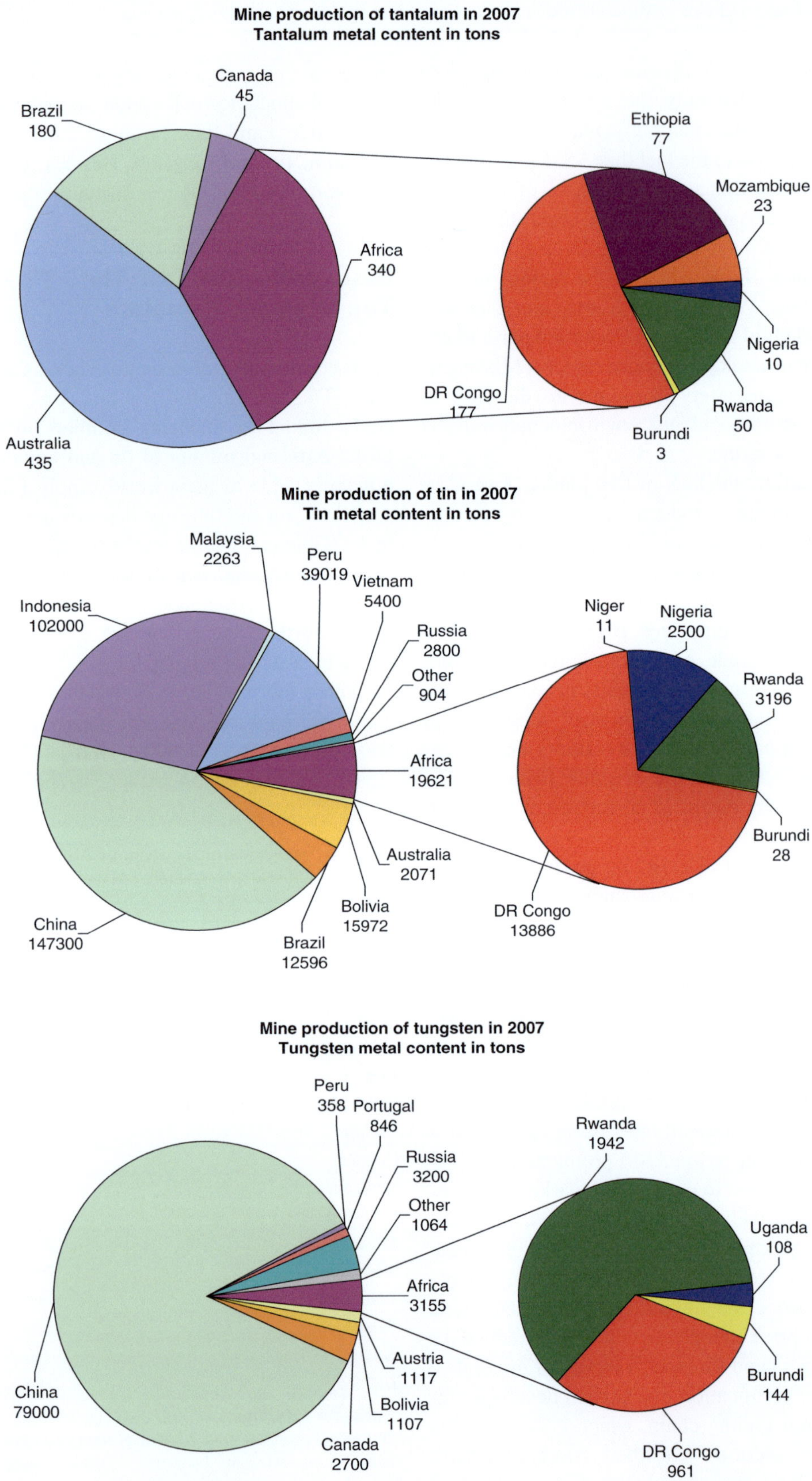

Fig. 3 World mine production (metal contents) and African production of tantalum, tin and tungsten according to the USGS (2009) and the BGR (2009b)

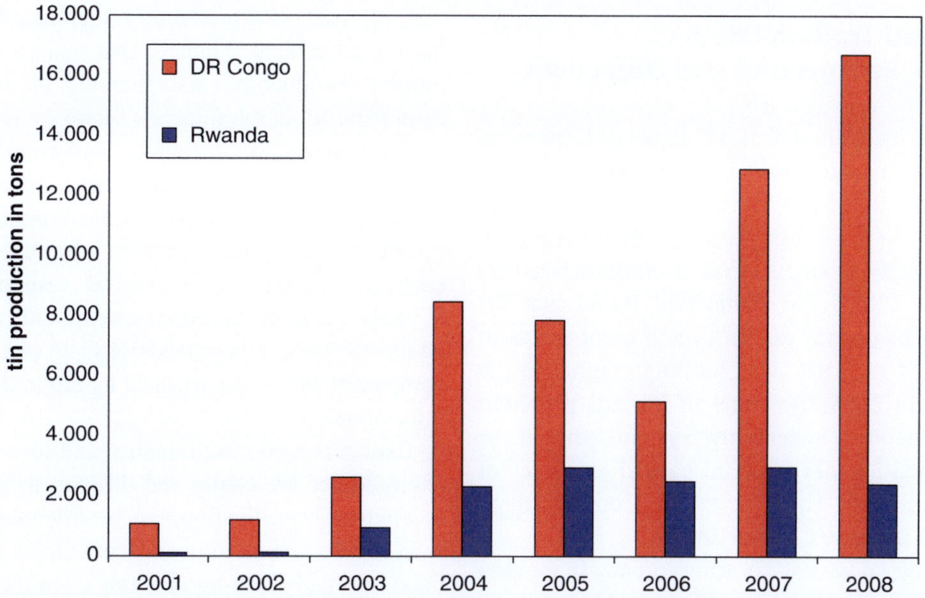

Fig. 4 Production of tin from Rwanda (*blue columns*) and the Democratic Republic of Congo (*red columns*) from 2001 to 2008. Tin production was calculated from exports of cassiterite concentrates, using a grade of 65% tin metal in concentrate

crisis in 2009 has lead to temporal mine closures in Australia and Canada, with Africa, Brazil and China remaining the principal producers. Major African producers are Ethiopia, Mozambique, Rwanda and the DRC, followed by Nigeria, Zimbabwe and others (Fig. 3). After a tremendous rise of tantalum prices in 2000, African production increased from <20% to >30% of the world market share.

The global tin production strongly increased from 1990 until today, with 350,000 tons of metal content mined in 2007 (Fig. 3). Major producers are China and Indonesia, accounting for more than 70% of the production, followed by Peru. Africa's mine production in 2007 was estimated to contain approximately 20,000 tons tin, representing a 6% share of the world production (Fig. 3). Tin exports as cassiterite concentrates from central Africa have strongly increased from 2001 to 2008 (Fig. 4); especially in 2008, about 17,000 tons of tin content were produced in the DRC and 2,400 tons in Rwanda (Fig. 4).

With 52,000 tons of tungsten and 81% of the market share, China is by far the largest producer of this commodity, followed by Canada, Russia and Rwanda. Austria and Portugal have a 3.2% market share (BGR 2009a, b) (Fig. 3). Only 3% (3,155 tons of tungsten metal content) were originating from Africa in 2007. The tungsten production in Africa is restricted to Kibaran deposits in the Great Lakes Region with Rwanda (61%), the DRC (31%), Burundi (4%) and Uganda (4%) as the major producers.

Tin, tungsten and tantalum play a crucial role for our modern society. Tin is mainly used as a solder material in the electronics industry (53%), as corrosion-resistant tinplates (16%), in chemicals and in the manufacturing of bronze. The major use for tungsten is in the production of heat- and corrosion-resistant hard metals that consumes about 60% of the global tungsten consumption. Tantalum is an important electronic metal, widely used for the production of small capacitors that make up about 60% of the current tantalum demand. Such capacitors are used in consumer electronics, e.g. in cars, laptops, cell phones, digital cameras and pagers. In many products of daily use, tantalum may be substituted by cheaper metals such as niobium and aluminium. However, increasing demand is expected mainly due to its use in aerospace and aviation industries.

Certified Trading Chains (CTC) – Background and Objectives

In 2002, the United Nations' Panel of Experts on the Illegal Exploitation of Natural Resources and Other Forms of Wealth of the Democratic Republic of the Congo presented their findings to the Security Council (United Nations Security Council 2002). The Panel had found that the plunder of natural resources and other forms of wealth of the DRC was fuelling conflict in the region. In 2005, the Panel of Experts proposed that enhanced traceability systems should be developed for all important natural resources of the Democratic Republic of the Congo (United Nations Security Council 2005).

Taking up the call for transparency in mineral production, the Federal Institute for Geosciences and Natural Resources, the Geological Survey of Germany (BGR), started two research projects in 2006. The first research project aimed to test the feasibility of 'fingerprinting' coltan samples based on the mineralogical and chemical characteristics of specific ore concentrates (see section "The CTC Principles and Standards"). In the second research project, a chain of custody assurance system was developed, based on the establishment of transparent, traceable and ethical trading chains. This concept of Certified Trading Chains (CTC) found entry into the preparatory discussions for the G8 summit in Heiligendamm in 2007. The summit protocol stressed the need for action in the artisanal and small-scale mining sector and acknowledged the potential of certification systems to increase 'transparency and good governance in the extraction and processing of mineral raw materials … to reduce environmental impacts, support compliance with minimum social standards, and resolutely counter illegal resource extraction' (Article 85). It also expressed support for 'a pilot study… concerning the feasibility of a designed certification system for selected raw materials' (Article 86) (G8 Summit 2007). To this end, the German government has taken the initiative to support such a pilot project for an implementation in Rwanda.

CTC in mineral production is an instrument to implement ethical standards and transparency in mineral production and thereby to improve responsibility in the minerals sector by introducing a concept of voluntary self-commitment among the partners within the value chain. Since highly mechanised and large-scale mining (LSM) operations commonly operate within acceptable corporate social and ecological responsibility (CSER) standards, the approach explicitly focuses on artisanal mining organisations and small-scale companies which use artisanal labour in developing countries.

Export bans on high-quality and low-volume commodities are costly and difficult to enforce. In contrast, certification can be oriented at the same goals, whilst allowing companies to continue mining or buying minerals from the region on the basis that they have demonstrated to achieve their social and environmental responsibilities, as required by the certification system. In this way, responsible buyers can use their buying power to effect positive change by remaining engaged in the mineral supply chains rather than disengaging. By ensuring traceability along the trading chain, CTC serves as an instrument

- to ensure that the trade of certain mineral resources is conducted legally and does not support belligerent groups in the region,
- to assure that process and production methods at the mine site adhere to minimum social and ecological standards.

CTC on the one hand aims to increase the contribution of the mineral sector to poverty reduction and the political stabilisation of developing nations. On the other hand, it aims to improve supply security for the processing industry and fosters responsibility in industrialised economies thus strongly implicating the dual objectives of producer and consumer benefits. The proper implementation of certification will support areas of good governance, where mineral resources are produced and traded legally and transparently and in ways which protect workers, communities and the environment. Certification will also progressively transform and formalise informal mining. Formalisation is a precondition for achieving transparent records of production and trade to improve governance and to reduce conflicts associated with the mining sector.

Certification can only be regarded as the second best option and would be superfluous in case of total conformity with national and international laws, regulations and standards. However, law enforcement and institutional capacity are often weak in the mineral sector of many developing countries; CTC tries transitionally to fill the gaps between the ideal world and the reality of sectoral governance.

The CTC Principles and Standards

On the basis of company audits at different mining sites in Rwanda, a set of standards was drafted based on a number of international 'integrity instruments' as well as national law. Each standard was derived from specific provisions in the Organisation for Economic Cooperation and Development's (OECD) Guidelines for Multinational Enterprises (2000) and Risk Awareness Tool for Multinational Enterprises in Weak Governance Zones (2006), as well as from some of the International Finance Corporation's Performance Standards and the Voluntary Principles on Security and Human Rights.

The set of standards was drafted with the focus on transparency along the trading chain including certification of origin as well as assuring corporate social and environmental responsibility. Certification of origin on the one hand is based on documentation and plausibility checks related to production, trade and export as well as – in case of doubt – the additional checking instrument of the geochemical fingerprint for certain minerals (see section "Geochemical Fingerprint").

After consultation on the content of this standard at the 8th annual conference of the World Bank's initiative Communities and Small-Scale Mining in Brasilia, Brazil, October 2008, five basic principles (Table 2) were established each referring to a thematic cluster and two additional standards added, one referring to gender issues and the other on handling influx migration.

A further step in the CTC consultation process was a workshop on the revised standard held at the conference of Fatal Transaction and the Bonn International Centre for Conversion (BICC) 'Digging for Peace – Private Companies and Emerging Economies in Zones of Conflict' in Bonn, November 2008. The workshop enhanced the dialogue with the civil society on the content of the CTC. In March 2009, an implementation workshop with national stakeholders in Kigali, Rwanda, started the national CTC implementation. The standards were related to Rwandan legislation as well as revised and adopted specifically for Rwandan conditions.

An important element of the CTC procedure is the traceability of the traded minerals. Therefore, an innovative procedure using geochemical fingerprinting for the identification of the locality of origin of the traded minerals with a high degree of accuracy was developed. In the following section, the method of the geochemical fingerprint is described in more detail.

Geochemical Fingerprint

'Coltan' (a trade name for ores of tantalum associated with columbium from Africa) has been identified as an appropriate object of study to develop a geochemical fingerprint because an estimated 40% of the world's tantalum production originates from African countries, especially from the DRC. Furthermore, rising tantalum prices coincided with civil wars in the DRC, resulting in huge exploitation efforts for raw materials. Sales of coltan, tin, tungsten and gold were used to finance the various armed groups in the region to a significant extent.

In the DRC, coltan is produced by artisanal methods from numerous small-scale mining operations in the Kivu provinces, Ituri, Maniema and Katanga. In neighbouring Rwanda, Burundi and Uganda, artisanal coltan production is smaller but also significant. Heavy mineral concentrates are usually produced by panning

Table 2 Principles and standards of CTC

Principle	Standard
1. Origin and volumes of produced and traded goods as well as company payments to host government are transparent	1.1 Origin and production volume of minerals from the pilot mine site throughout the trading chain are traceable
	1.2 Meet fiscal obligations required by host government law
	1.3 Publish all payments made to government according to internationally accepted standards
	1.4 Actively oppose bribery and fraudulent payments
2. The company does not use child labour and ensures fair remuneration and work conditions as well as continual improvement of health and safety measures for all employees	2.1 Maintain salary or payment levels equal to or greater than those in comparable enterprises within Rwanda
	2.2 Ensure that no child labourers (age under 16) work on company sites
	2.3 Support workers' organisations and collective bargaining
	2.4 Provide essential protective and production services to support the work of artisanal miners
	2.5 Ensure occupational health and insurance in all company operations
	2.6 Provide training for employees and contractors on safety, health and effective use of on-site facilities
3. The company ensures security on company sites whilst respecting human rights	3.1 Provide sufficient and adequately trained security forces
	3.2 Undertake security risk assessments
4. The company consults communities in which it operates and contributes to their social, economic and institutional development taking into account gender-sensitive aspects	4.1 Interact regularly with communities and local governments to address grievances and other common concerns
	4.2 Support local enterprises to supply company operations
	4.3 Implement integrated development programmes in nearby communities for livelihood security, social and physical infrastructure and capacity building
	4.4 Obtain free, prior and informed consent before acquiring land or property
	4.5 Understand the situation and perspectives of the women in the company's area of influence and design and implement company's operations in a gender-sensitive way
	4.6 Carry out an assessment on human migratory streams created by company operations and develop an influx migration action plan
5. The company seeks continual improvement of its environmental performance	5.1 Carry out an environment impact assessment as the basis for developing an environmental management and protection plan and strategy
	5.2 Properly treat or dispose of hazardous material and waste from its site(s)
	5.3 Make provisions for the full cost of rehabilitation upon closure

and sieving and may undergo limited additional processing steps (magnetic separation) before they are exported. The mines operate on eluvial and alluvial deposits associated with small lenses of pegmatite rocks. The host rocks of pegmatites in the area of interest are metasediments and metaigneous rocks of the Kibara Province.

Comprehensive investigations have been carried out by the BGR on behalf of the German Federal Ministry for Economic Cooperation and Development (BMZ) from 2006 to 2009. The goal was to develop analytical methods to verify the origin of coltan. Based on an evaluation of geological and mineralogical factors, the following measurable properties were identified as critical parameters (Fig. 5):

- Mineralogical composition and variability of ore concentrates
- Geochemical composition of tantalum minerals in the concentrates
- Radiometric age of tantalum minerals

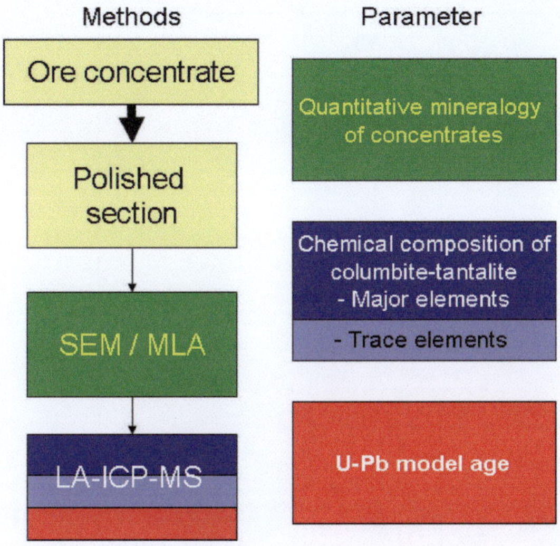

Fig. 5 Parameters and methods used to fingerprint the origin of coltan

Several methodological approaches were tested. A three-stage analytical protocol is proposed, taking into account the time available in routine work (Fig. 5):

- Preparation of polished sections
- Quantification of the mineralogy of ore concentrates using automated scanning electron microscope techniques (SEM/MLA)
- Determination of the chemical composition of about 50 single tantalum mineral grains, including their uranium and lead isotopic composition for radiometric dating, using laser ablation ICP-MS

Using these methods, a certification laboratory could analyse about 1,000 concentrate samples per year. The results are entered into a relational database, and the samples are classified using statistical methods, e.g. by machine learning techniques. At present, the reference database consists of more than 25,000 analytical data sets representing 343 samples from 33 countries.

The combined mineralogical and geochemical signatures allow for the definition of nine large tantalum provinces in Africa. These range in age from the Archean (>2,500 Ma) to the Mesozoic (younger than 225 Ma) (Melcher et al. 2008a, b). The accuracy of allocating a sample to a province or sub-province is calculated by applying error-validated classification methods. Blind experiments revealed excellent hit rates.

Within the Kibara province, the fingerprint of coltan concentrates relies on two parameters only because uranium-lead model ages are similar for most pegmatite-derived ores (between 930 and 980 Ma). However, distinct differences in the mineralogy of concentrates and in the geochemical composition of single tantalum mineral grains enable a discrimination down to the scale of an ore district such as Gatumba in Rwanda (Fig. 6) (Lehmann et al. 2008). Automated scanning electron microscope–based methods such as mineral liberation analysis (MLA) allow for a rapid quantitative measurement of a polished section prepared from an ore concentrate. Even rare phases are detected with a high probability. Once identified by MLA, the compositions of single tantalum mineral grains are then determined by laser ablation ICP-MS, resulting in a comprehensive data set for each concentrate. In columbite-tantalite minerals, the concentrations of the major elements (Fe, Mn, Nb, Ta) and of the minor to trace elements (Li, Sc, Ti, Zr, Y, Sn, Hf, W, U, rare earth elements) vary systematically within a given ore body (Melcher et al. 2008a, b). In combination with textural attributes such as intergrowths, inclusions and zoning, chemical data are used to identify the host pegmatite and to discriminate it from neighbouring occurrences.

Thus, for the first time, an analytical method is available to fingerprint the origin of tantalum ore concentrates. The method may be applied as a forensic instrument, or in conjunction with certified trading chains in mineral production. Extension of the analytical fingerprint to tin and tungsten ores is currently underway. If necessary, this would allow for the control of a significant portion of mineral exports from Central Africa's conflict region.

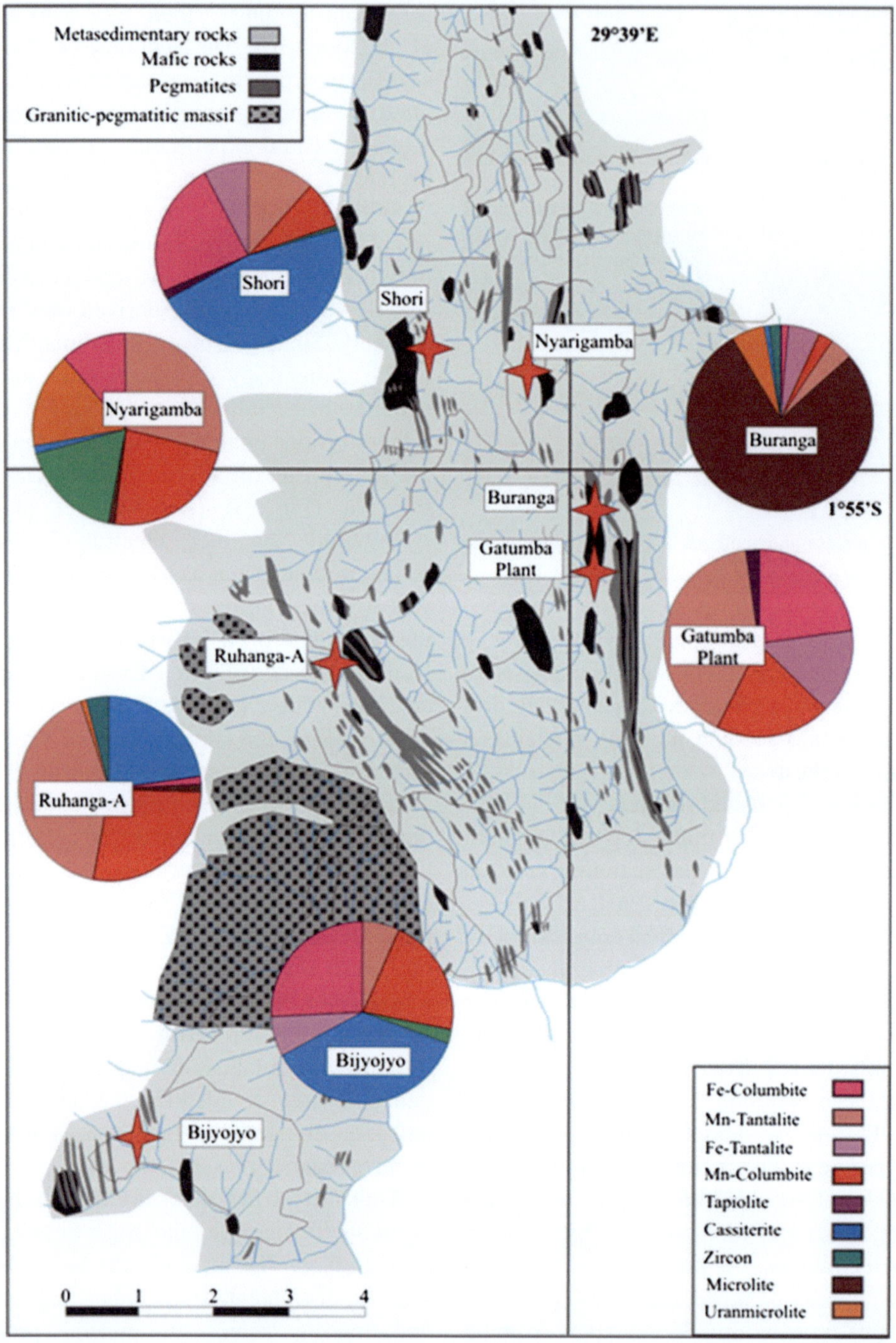

Fig. 6 Location of samples of mineral concentrates from the larger Gatumba district. The geological map is redrawn from Dewaele et al. (2008). Modal mineralogy from mineral liberation analysis (*MLA*). Note that the Gatumba sample is from the Gatumba treatment plant and represents a mixed concentrate

CTC in Practice

The pilot project on CTC in Rwanda was started within the framework of a technical cooperation programme to strengthen the competitiveness of the Rwandan mineral sector by developing best practice and enhancing transparency. The pilot project was implemented in cooperation with the Office de la Géologie et des Mines du Rwanda (OGMR) and private mining and processing companies.

Rwanda supports the desire to establish Certified Trading Chains (CTC) through the strengthening and supervising of the mining sector in the country. In order to implement socio-economic and environmental best practice in its mining industry, Rwanda is setting and will later enforce good standards in its mining sector to guarantee a sustainable market for its minerals. Preliminary findings suggest that high-value metals produced in and exported from Rwanda including tantalum (coltan), tin and tungsten offer a leverage to handle poverty alleviation, prevent conflict as well as supply security.

First steps were bilateral consultations with the Rwandan government and other national stakeholders, especially from the industry. A workshop with stakeholders from government institutions and the mining industry started implementation in March 2009. After a period for expression of interest, three mining companies, local producers of cassiterite, wolframite and tantalite concentrates who cooperate with or engage ASM, volunteered to join the initiative: Natural Resource Development (NRD), Gatumba Mining Corporation (GMC) and Eurotrade Ltd. In a later stage, another two companies also expressed their interest to participate. An independent auditor conducted a baseline assessment of the companies, their concessions and trading chains to develop indicators for the certification scheme, to assess the actual status and to give recommendations for improvement. For each standard, a set of indicators for verification adopted to the situation of mining in Rwanda has been developed. An official audit of the participating companies was conducted in 2010. Two out of four companies received CTC certificates of compliance.

As a further step, technical cooperation between BGR and the Congolese Ministry of Mines with the aim of introducing a certification system for coltan, cassiterite, wolframite and gold has started in October 2009. The cooperation combines the pilot implementation of certification (with a focus on transparency of origin) at selected mining sites in South Kivu and neighbouring provinces with the capacity building of sector institutions so that they can fulfil their mining oversight function (Fig. 7).

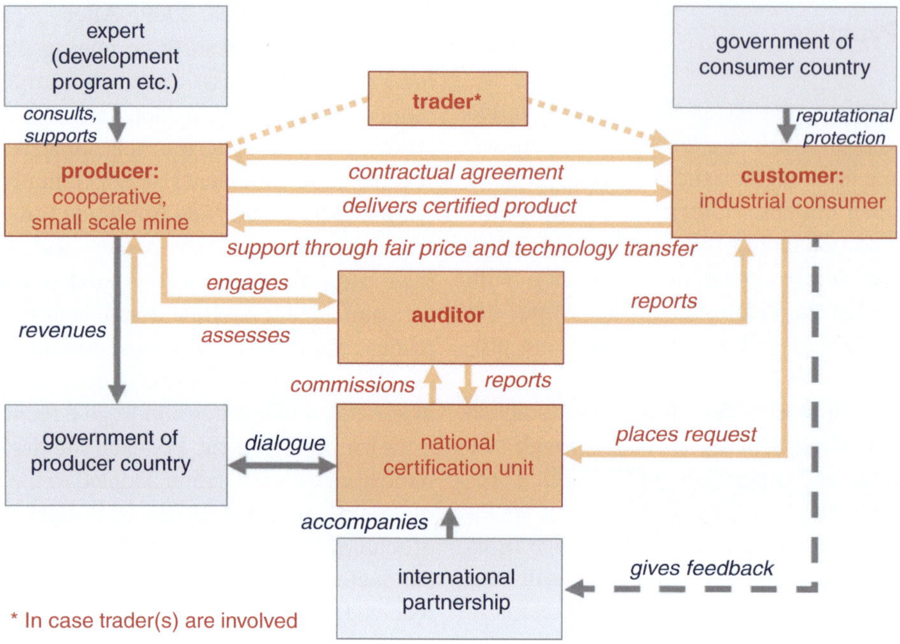

Fig. 7 Conceptual flowchart of a CTC

Outlook: Regional and International Integration

In November 2006, the 11 member states[1] of the International Conference on the Great Lakes Region (ICGLR) signed the Protocol against the Illegal Exploitation of Natural Resources, which includes the aim of implementing a mechanism for the certification of natural resources in its Article 11. By implementing the exemplary pilot project in Rwanda and supporting launching the certification in DRC, CTC aims for regional stability and peace building, as designed by the ICGLR, and for the support of the intergovernmental body in implementing certification. The results of the workshop and of the first steps of implementation were brought to the ICGLR meetings in April and in September 2009 as a Rwandan contribution to the development of a regional certification mechanism. ICGLR is currently developing a scheme for a regional certification mechanism.

The issue of certification of mineral production and trade not only is a national or regional effort. Support is also coming from the EU Special Representative for the Great Lakes who initiated a Joint Action Plan to curb illegal exploitation of and trade in natural resources in Eastern DRC. As part of this initiative, the OECD has initiated a working group to promote responsible investment through enhanced due diligence in the mining sector on the basis of the OECD Risk Awareness Tool for Multinational Enterprises in Weak Governance Zones and the OECD Guidelines for Multinational Enterprises. A consultation and implementation process with the private sector has been started, also including companies that participate in the CTC pilot project.

Certification as a tool for development can be relevant especially in artisanal and small-scale mining communities. Therefore, the CTC initiative is in close dialogue with the CASM network (Communities and Small-Scale Mining). Relevant other initiatives in the sector provide exchange and lessons learned such as the Kimberley

[1] Member states: Angola, Burundi, Central Africa Republic, Republic of Congo, Democratic Republic of Congo, Kenya, Uganda, Rwanda, Sudan, Tanzania, Zambia.

Process for diamonds, the Alliance for Responsible Mining (ARM) for the development of a fair trade gold standard or the diamond development initiative. Ongoing other emerging initiatives in this sector show that the issue of responsible mining and certification might develop beyond the current relevance for limited commodities (diamond, jewellery, gold) as well as beyond the large-scale mining sector and increase its relevance along the trading chain.

References

Bannon I, Collier P (eds) (2003) Natural resources and violent conflicts. Options and actions. The World Bank, Washington, DC

BGR (2009a) Bundesrepublik Deutschland. Rohstoffsituation 2008. Rohstoffwirtschaftliche Länderstudien Heft XXXVIII. Bundesanstalt für Geowissenschaften und Rohstoffe, Hannover

BGR (2009b) BGR-databank. Bundesanstalt für Geowissenschaften und Rohstoffe, Hannover

CASM (2009) Community on small-scale mining. Mapping ASM. URL: http://www.artisanalmining.org/casm/mappopulation, 10.01.2012

Dewaele S, Tack L, Fernandez-Alonso M, Boyce A, Muchez P, Schneider J, Cooper G, Wheeler K (2008) Geology and mineralization of the Gatumba area, Rwanda (Central Africa): present state of knowledge. Etud Rwandaises 16:6–24

Garrett N, Mitchell H (2009) Trading conflict for development. Utilising the trade in minerals from eastern DR Congo for development. Resource Consulting Service. URL: http://www.dfid.gov.uk/r4d/PDF/Outputs/Crisis States/trade-conflict-development.pdf, 10.01.2012

G8 Summit (2007) Growth and responsibility in the world economy. Summit declaration (7 June 2007). URL: http://www.g-8.de/Content/EN/Artikel/__g8-summit/anlagen/2007-06-07-gipfeldokument-wirtschaft-eng,templateId=raw,property=publicationFile.pdf/2007-06-07-gipfeldokument-wirtschaft-eng.pdf, 10.01.2012

International Conference on the Great Lakes Region (2006) Protocol against the illegal exploitation of natural resources, 30 Nov 2006. www.icglr.org

Lehmann B, Melcher F, Sitnikova MA, Ruzindana Munana J (2008) The Gatumba rare-metal pegmatites: chemical signature and environmental impact. Etud Rwandaises 16:25–40

Melcher F, Sitnikova MA, Graupner T, Martin N, Oberthür T, Henjes-Kunst F, Gäbler E, Gerdes A, Brätz H, Davis DW, Dewaele S (2008a) Fingerprinting of conflict minerals: columbite-tantalite ("coltan") ores. SGA News 23:1–14

Melcher M, Graupner T, Henjes-Kunst F, Oberthür T, Sitnikova M, Gäbler E, Gerdes A, Brätz H, Davis D, Dewaele S (2008b) Analytical fingerprint of columbite-tantalite (coltan) mineralization in pegmatites: focus on Africa. In: Proceedings, ninth international congress for applied mineralogy (ICAM) 2008, Brisbane, Queensland. Australasian Institute of Mining and Metallurgy, pp 615–624

United Nations Security Council (2002) Final report of the Panel of Experts on the Illegal Exploitation of Natural Resources and Other Forms of Wealth in the Democratic Republic of the Congo, UN Security Council, S/2002/1146 (No. UN document S/2002/1146). United Nations, New York

United Nations Security Council (2005) Letter dated 26 July 2005 from the Chairman of the Security Council Committee established pursuant to resolution 1533 (2004) concerning the Democratic Republic of the Congo addressed to the President of the Security Council, UN Security Council, S/2005/436. United Nations, New York

USGS (2009) Mineral commodity summaries. U.S. Geological Survey, Denver, CO URL: http://minerals.usgs.gov/minerals/pubs/mcs/

Wagner M, Franken G, Martin N, Melcher F, Vasters J (2007) Zertifizierte Handelsketten im Bereich mineralischer Rohstoffe. Projektstudie, Bundesanstalt für Geowissenschaften und Rohstoffe, Hannover

Issues and Challenges in Life Cycle Assessment in the Minerals and Metals Sector: A Chance to Improve Raw Materials Efficiency

Mohan Yellishetty, Nawshad Haque, and Alain Dubreuil

Abstract

A critical review was conducted on existing literature concerning life cycle assessment (LCA) and its application to the minerals and metals sector. This extensive literature search uncovers many of the issues that require immediate attention from the scientific community involved with LCA. The methodological drawbacks, mainly problems with inconsistencies in LCA results for the same situation under different assumptions and issues related to data quality, are considered to be the current shortcomings of LCA. In the minerals and metals sector, it is important to increase the objectivity of LCA by way of estimating and reporting those uncertainties; for example, whether land use has to be considered in detail or at a rough level. In regard to abiotic resource characterisation, the weight and time scales to be considered become a very critical issue of judgement. How the temporal and spatial dimensions should be incorporated into LCA is one of the biggest challenges ahead for those who are concerned. Addressing these issues will enable LCA to be used as a policy tool in environmental decision making. There has been enormous unresolved debate with respect to land use impacts, abiotic resource depletion, allocation procedure open-loop recycling and spatial and temporal dimensions. An example case has been presented for Australian iron ore using SimaPro software based on

M. Yellishetty (✉)
Civil & Environmental Engineering, School of Applied Science & Engineering, Monash University, Gippsland Campus, Churchill 3842, VIC, Australia
e-mail: Mohan.Yellishetty@gmail.com

N. Haque
CSIRO Process Science & Engineering, Minerals Down Under Flagship, Clayton, VIC 3169, Australia

A. Dubreuil
Natural Resources Canada, CANMET Mining and Mineral Sciences Laboratories, Ottawa, ON, Canada

R. Sinding-Larsen and F.-W. Wellmer (eds.), *Non-Renewable Resource Issues: Geoscientific and Societal Challenges*, International Year of Planet Earth, DOI 10.1007/978-90-481-8679-2_12, © All Rights Reserved 2012

published inventory data to demonstrate that uniformity is required. Discussions aimed at bringing consensus amongst all the stakeholders involved in LCA (i.e. industry, academia, consulting organisations and government) have been presented. In addition, a commentary of different points of view on these issues has been provided. This review brings into perspective some of those contentious issues that are widely debated by many researchers. Finally, the authors conclude with their views on the prospects of LCA for future research endeavours.

Keywords
LCA • Minerals and metals sector • Acidification • Toxicity • Land use • Abiotic resource depletion • Allocation • Spatial variability • Temporal differentiation

Introduction

Life cycle assessment (LCA) is a major method and tool for assessing sustainability in industrial sectors including minerals and metals. LCA has proven to be a valuable tool for evaluating the potential environmental impacts of products and materials. This is why many companies are selecting LCA as one of the important tools in evaluating their environmental performance. This increased global interest in using LCA, and the consequent benefits associated with it as a tool for achieving environmental stewardship, has attracted the attention of the mining, minerals and metals sectors.

The mining sector provides materials that are essential elements in a wide range of goods and services, which create value by meeting human needs. Mining and processing activities are an integral part of most complex material cycles, and therefore the application of LCA to minerals and metals has gained prominence. In the past decade, increased use of LCA in the minerals and metals sector has advanced scientific knowledge through the development of scientifically valid life cycle inventory databases. Though scientifically valid, LCA still needs to depend on the goal and scope, technical assumptions and value judgments of the matter being investigated. Figure 1 illustrates the difficulty for LCA to make comparative assertions.

In this chapter, a critical review of selected issues and challenges concerning LCA, in general, and issues related to the mining and metals sector in particular, is presented. This work builds further on the similar review conducted by Yellishetty et al. (2009). For example, measuring the environmental burden related to abiotic resource depletion and land use impacts within the LCA is widely debated issues. Also, incorporating spatial and temporal sensitivities in LCA,

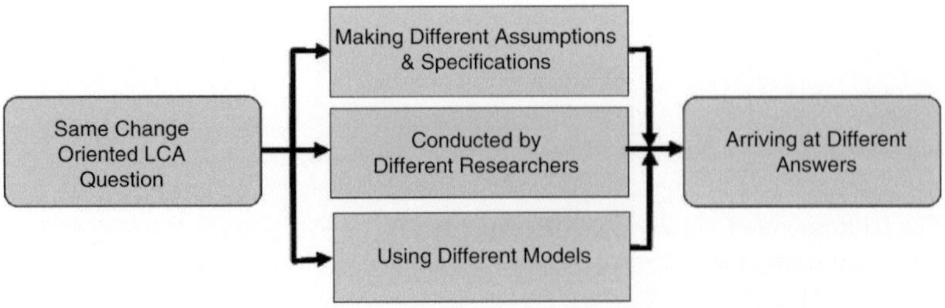

Fig. 1 Schematic representation of missing objectivity in current LCA (Adopted from Yellishetty et al. 2009)

to make it a more consistent scientific tool, is yet to be resolved. This review article discusses existing LCA methods and proposed models on selected issues in relation to the minerals and metals sector.

Overview of Life Cycle Assessment

LCA addresses the environmental aspects and potential environmental impacts, human health and resource concerns for a product system. The product's life includes raw material acquisition through production, use, end-of-life treatment, recycling and final disposal (International Organization for Standardization, ISO 14044: 2006). Life cycle impact assessment is different from risk assessment. In life cycle assessment, inventories are defined by the functional unit, and therefore unlike risk assessment, both the strength and exact location of process emissions, for example, are missing. However, LCA is still a fairly new tool, and the methodology is still under development. Since it was first used, there have been several modifications and alterations in its structure, methodology and protocols. A schematic of the classification of LCA is presented in Fig. 2. Accordingly, before an LCA is conducted, it is very important to understand the purpose of LCA and its limitations.

LCA was developed directly from a desire to limit the energy used in manufacturing processes in the 1970s, especially in the manufacture of plastics, steel, pulp and paper and petroleum refining (Huppes 1996). However, until the early 1990s, no standardisation of methods was undertaken. One of the major limitations was the lack of objectivity, in the sense that when LCA was conducted under different approaches, there was wide variation in results (Fig. 1). This could be due to varying demarcations of system boundaries, process assumptions and input data. One important problem that needs to be addressed is 'bringing harmony between the impacts predicted by LCA and the expected occurrence of actual impacts'. This calls for standardisation of LCA procedures (Huppes 1996). A summary of the historic developments is presented in Table 1. The result of this historical sequence of events has led to the publication of a series of international standards, particularly ISO 14040:2006 and ISO 14044:2006, which describe the essential elements of LCA studies (The international standards ISO 14040:2006 and ISO 14044:2006 are replacing ISO 14040:1997, ISO 14041:1998, ISO 14042:2000 and ISO 14043:2000). A number of technical reports (TR) and technical standards (TS) (ISO/TR 14047:2003, ISO/TR 14049:2000, ISO/TS 14048:2002) have been published. A number of national standards and guidelines have been developed, adopting the ISO standards. The European Commission launched the International Reference Life Cycle Data System (ILCD) Handbook that is based on ISO 14040:2006 series of standards. The aim of the ILCD is 'to facilitate quality assurance and coherence of life cycle methods and data' (Pennington et al. 2010). During LCA studies, the material and energy resources consumed and wastes emitted are accounted for through the application of mass and energy balance checks, thus making it a scientific tool. This holistic scientific tool considers the inclusion of both the environmental effects directly due to the processing of the products under study and the effects of processing other products.

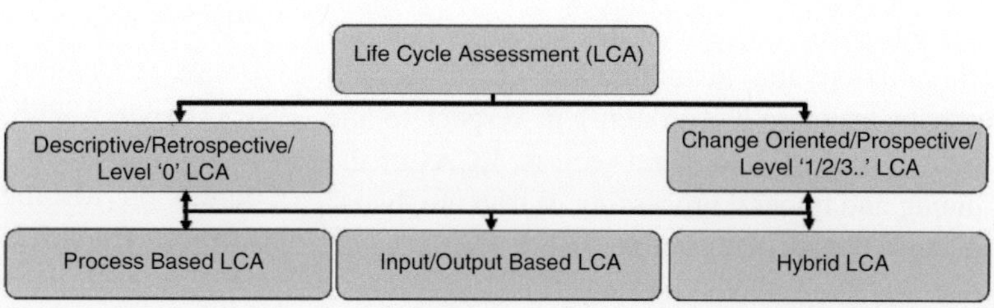

Fig. 2 Classification of LCA (Adopted from Yellishetty et al. 2009)

Table 1 Historical landmarks driving the LCA events (Modified from Yellishetty et al. 2009)

Year	Major events in LCA development as a technique
1960s	LCA was performed by the Midwest Research Institute (MRI) (Giudice et al. 2006)
1970s	The US Department of Energy commissions' studies on 'Energy Analysis' titled 'Resources and Environmental Profile Analysis' (REPA) (Guinée 2002)
1980s	'Green Movement' in Europe brings focus back on emissions and the need for recycling. European industries study their pollution releases and begin comparing alternatives (Guinée 2002)
1989–1990	SETAC is first involved in LCA. The first-ever SETAC workshop was held to define LCA (Guinée 2002)
1992	First Dutch guide on LCA was published (Guinée 2002)
1994	First involvement of International Organization for Standardization (ISO) (Giudice et al. 2006)
1995	UNEPs' first involvement through publication first document – 'LCA: what it is and how to do it'. Subsequently, in 1996, releasing 'Towards the Global Use of LCA' (Giudice et al. 2006)
1997	First-ever ISO 14040 standards series on LCA brought into force (Giudice et al. 2006)
2002	UNEP/SETAC Life Cycle Initiative began (Guinée 2002) International Council on Mining and Metals (ICMM) and Natural Resources Canada (NRCan) joined hands with UNEP-SETAC (Dubreuil 2005)
2010	The European Commission/Joint Research Centre launches the International Reference Life Cycle Data System Handbook (http://lct.jrc.ec.europa.eu/publications)

Current Status of LCA Research: A Minerals Sector Perspective

LCA is increasingly used in the mineral and metal industry and is seen as a tool for materials choice and environmental decision making, with the goal of achieving sustainable development. This is evident from the very fact that the International Council on Mining and Metals (ICMM) is continuously striving to facilitate the development of LCA protocols and guidance to increase their relevance to the minerals and metals sector. Accordingly, ICMM in collaboration with United Nations Environment Programme/Society of Environmental Toxicology and Chemistry (UNEP/SETAC) Life Cycle Initiative is involved in a project to improve the scientific validity of life cycle impact assessment methodologies.

This increased use of LCA has motivated organisations around the world to undertake extensive research and development work in the LCA area, particularly with regard to the generation and maintenance of life cycle inventories (compilation and quantification of inputs and outputs of a product system) in respect to the mining, mineral and metal sectors. LCAs of the mining and mineral processing of iron ore, bauxite and copper concentrate were carried out, focusing on embodied energy and greenhouse gas emissions in Australia. The results showed that loading and hauling make the largest contributions to the total greenhouse gas emissions for the mining and processing of iron ore and bauxite. In the case of copper ore, the crushing and grinding steps make the largest contribution to the total greenhouse gas emissions for the production of copper concentrate. These results indicated that efforts to reduce the increased greenhouse gas emissions from mining and mineral processing, anticipated in the future as a result of falling ore grades and more finer-grained deposits, should focus on loading and hauling for iron ore and bauxite, while for copper ore, the focus should be on grinding (Norgate and Haque 2010). In this chapter, inventory data from their study of iron ore has been used to compare a few impact methods using LCA software (SimaPro 7.1.8, PRé Consultants).

Minerals and metals have different characteristic features compared to other materials such as wood, paper and plastics, as is shown in Table 2. Understanding these important characteristics is opening a different perspective in eco-efficiency and thus helping in developing material dependant strategies. The International Council on Mining and Metals (ICMM) has become a founding member of UNEP/SETAC Life Cycle Initiative (Dubreuil 2005) and is

Table 2 Characteristics of main material families (Modified from Yellishetty et al. 2009; Young et al. 2001)

	Details		
Characteristics	Metals	Plastics	Wood and paper
Resource stock	Metal ore/minerals	Crude oil	Forests
Source	Lithosphere	Lithosphere	Biosphere
Extraction method	Mining: opencast/underground	Drilling and extraction	Harvest
Material structure	Elemental (metals/alloys)	Molecular	Cellular
Material renewal	Recyclable	Recyclable but quality degrades	Recyclable but quality degrades
Final fate	Elements are permanent and may remain for long time or released to atmosphere depending on bioavailability	Combustion, degradation and landfill	Biodegradation, combustion and landfill
Losses	Corrosion, wear, process loss	May return to carbon cycle	May return to carbon cycle
Time scale of material stock	Theoretically unlimited	Days to years	Days to decades
Applications	Beverage cans, consumer products, auto components, motors, structures, etc.	Furniture, coatings, packaging, appliances, auto parts, etc.	Building, books/newspapers, packaging, furniture, etc.

looking for the application of robust and appropriate concepts and tools for assessing potential environmental impacts of minerals and metals.

Generally, the main selected environmental impact categories for mining and minerals sectors are reported to be gross energy requirement, which is used to estimate global warming potential (GWP) or greenhouse gas emissions in CO_2 equivalent units, solid waste burden and water usage per functional unit (Norgate et al. 2007). Recently, toxicity impacts of minor elements in some waste streams in the metal production chain have been included in the LCAs (Norgate and Haque 2009), as well as a review of the methodology for including socio-economic impacts in LCAs such as using quality and well-being of human health indicators.

There are many other impact categories but some are quite challenging to quantify and include in LCA studies due to a number of limitations. Some of these impact categories and models are discussed here. In the following sub-sections, we will present and discuss the areas of concern in current LCAs to mineral and metal industries, such as impacts due to acidification, toxicity, biodiversity, land use, abiotic resource depletion, allocation issues related to end-of-life recycling and finally the spatial and temporal issues.

Acidification

Acidification is an impact category caused by the emission of acidifying substances to the air. Those substances increase the acidity of the soil or the water. Acidification is regarded as a regional effect. Coal combustion and the roasting/smelting of sulphidic ores are the main sources of sulphur dioxide (SO_2), an airborne acidifying substance. For this impact category, the acidic depositions on the oceans are not considered. Acidification has regional variability because the type of soil and its buffering capacity are considered in the impact assessment model. As a special case, the disposal of the residue from the alumina production, the 'red mud' is considered by ecoinvent. The model takes into account the high pH (Althaus and Classen 2005).

Toxicity

In life cycle assessment, toxicity is related to the relative potential impact of chemicals including metals on the human health and the environment. Human toxicity and eco-toxicity are impact categories that are of interest for the minerals and metals sector.

Eco-toxicity is associated with the terrestrial, freshwater and marine environment. In the mining and minerals sector, toxicity related with the minor elements can be an issue depending on the location and their bioavailability (i.e. capable of being taken up by an organism). Base metal ores contain low levels of hazardous elements such as arsenic, antimony, bismuth, cadmium, lead and mercury. Given the large tonnage of base metal ores mined and processed each year globally, the accumulated mass of minor elements introduced into the technosphere is large. Waste rock and tailings containing sulphide minerals not properly managed can generate acid and solubilise metals (Moncur et al. 2009; Price 2009). That process, acid mine drainage (AMD), can potentially have significant environmental impact on freshwater and marine environment. If proper technology is used, AMD can be mitigated and the quantity of waste rock and tailing can be removed from the life cycle inventory. In the current practice, due to the lack of reliable LCA model, the impact of AMD is not accounted for in ecoinvent (Althaus and Classen 2005).

The difficulty of impact assessment of complex flows such as land filled with end-of-life products or tailings is that life cycle impact assessment (LCIA) relates to single substances and energy flows. In order to ensure the increase of environmental relevance, it is necessary to completely model until emissions of single substances enter the natural environment. For example, inventory of 'tailings' is not environmentally relevant, while the leaching of specific metals from the tailings by AMD has to be modelled and inventoried as 'emissions to water' (European Commission 2010).

As mentioned earlier, LCA is a methodology that has been developed to assess the environmental impacts of material supply chains over their entire life cycle. However, assessing the toxicity impacts of metallic wastes and incorporating these impacts into LCA methodology is not a simple task, given the complexity of the issue, but research and methodology development is in progress internationally. Concerns about the bioavailability of metals, as opposed to their total concentrations, in waste streams have tended to limit the inclusion of toxicity impacts in LCAs. Numerous leaching tests have been developed by several different groups and specialists to help determine the 'hazardousness' of a waste material. Many of these tests have also been used to characterise the mobility (i.e. bioavailability) of metals in waste streams. Despite the limitations concerning the lack of a coherent framework for characterising the toxicity impacts of metals, including issues such as bioavailability, a study was carried out to give first estimates of the relative toxicity impacts of various waste streams from metal production processes. The methodology can be applied to give first approximations of the relative impacts of various metal production wastes, provided these limitations and associated assumptions are fully recognised (Norgate and Haque 2009).

A group of metals specialists, including LCA, eco-toxicity and chemistry met in order to improve the LCIA toxicity models. The clearwater consensus (Diamond et al. 2010) describes stages and considerations for calculating comparative toxicity potentials in the context of life cycle impact assessment for nonferrous metals. A strong focus was given on quantifying the bioavailable fraction of a substance. Recommendations were given to use a geochemical model. The mathematical description of the fate of the metal (fate factor) needs to incorporate speciation, size fractions and dissolution rates of metal complexes.

Biodiversity

Biological diversity is the variation in life forms of indigenous species and communities occurring in a given region. Biodiversity is the variety of all life forms – the plants, animals and micro-organisms – the genes they constitute and the ecosystems they inhabit. It includes genetic diversity, species diversity and ecosystem diversity. Genetic diversity refers to the variety of genetic information contained in all living things. There are genetic variations within and between populations of species as well as between (inter-species) and within species (intra-species). Species diversity refers to the variety of living species. Ecosystem diversity relates to the variety of habitats, biotic communities and ecological processes, as well as the diversity present within ecosystems in terms of habitat differences and the variety of ecological processes. In terms of physical location factors, it can be dry and sub-humid land, forest, inland water, marine and coastal biodiversity.

Mining activities often occur in remote environments where local communities engage in subsistence agricultural practices or sustainable livelihoods dependent on surrounding natural resources. In these circumstances, the human (social and economic) dimensions of biodiversity are of critical importance. This is particularly the case in the rural areas of developing countries, where entire communities are directly dependent on biodiversity and ecosystem services and therefore more vulnerable to their degradation (Commonwealth of Australia 2007). Including biodiversity impact within LCA study is problematic due to lack of robust metrics and quality data available for such assessment. Legislations in relation to biodiversity conservation can be used to develop objective metrics for using in the LCA database. A report recently provided a framework for dialogue between ecologists, LCA practitioners, government and industry by indentifying and discussing specific issues associated with the incorporation of land use impacts on biodiversity into LCA (Hamilton et al. 2008). In some current impact methods, such as Eco-indicator 99 and IMPACT 2002+, only one aspect of biodiversity is recognised (i.e. vascular plant species richness) and model parameters are derived from limited empirical data. It was clear from their conclusion that there are substantial impediments to the incorporation of these impacts into LCA in Australia as there is no universally agreed upon LCA method or suite of suitable biodiversity indicators. Additionally, there is a substantial variation in the scale at which relevant data are available.

Land Use Impacts

Mining is a site-specific activity and therefore has to take place where economic mineral occurrence exists. Mining is a land-intensive activity, that is, it generates high impact on a small area. Historically, most of the minerals exploitation occurs in locations remote from major population centres. Land use by the mining industry, though temporary in nature, can lead to substantial local impacts, particularly on biodiversity and soil quality crucial to ecosystem sustainability. Coincidentally, most of the land where mining takes place is also important habitats for highly diverse flora and fauna. Any mining and/or exploration activities could cause considerable, and at times irreparable, damage to the local land and the biodiversity of the region, thereby significantly affecting local natural resources. In recent times, there has been enormous debate on land use impact assessment in LCA, particularly in mining, which is a high-land-demanding sector. Two SETAC working groups started to frame the issue of land use impacts (Udo de Haes et al. 1996; Lindeijer et al. 2002). However, until now, no widely accepted assessment methodology has been developed to encompass the impacts due to mining land use into LCA.

Many references focus on suggesting indicators to include the effects of land use on biodiversity and biomass production (Köllner 2000; Schenck 2001; Weidema and Lindeijer 2001; Brentrup et al. 2002a, b; Bauer and Zapp 2005; Mila i Canals et al. 2007). Up until now, there has been debate going on regarding the level of detail at which LCA should assess land use impacts. A rough level assessment may allow the detection of hotspots from a life cycle perspective, whereas a more detailed assessment might allow the distinction between land management modes. Table 3 summarises the suggested impact assessment models based on both midpoint and damage approaches (Jolliet et al. 2004).

Although most of land uses by the mining industry are temporary in nature, they can lead to substantial local impacts, particularly on biodiversity and soil quality crucial to ecosystem sustainability. In recent years, land use impacts have received greater attention in LCA, but a consistent framework that brings consensus amongst all the stakeholders, and development of a robust method is still lacking. Additionally, the exclusion of land use impact issues may hamper the credibility of LCAs particularly in the context of mining as it can damage the land to a very great extent. Therefore,

Table 3 Examples of impact pathways from land use including requirements of LCI information (Adopted from Yellishetty et al. 2009)

Impact pathways		Category indicators	LCI modelling requirements
Soil quality	Natural environment	Should be explored according to the affected impact pathways. (*end-point/damage approach*)	To be investigated, according to the affected impact pathways (e.g. global warming, toxicity, etc.)
		Combinations of nine indicators, such as pore volume, SOM (organic matter??), microbial activity, etc. (*problem-oriented/midpoint approach*)	Effects due to agricultural activities on nine soil quality indicators (Oberholzer et al. 2006)
Biodiversity		Potentially disappeared fraction (PDF) of species. Potentially affected fraction (PAF) of species. (*end-point/damage approach*)	It may be worthwhile to work with these indicators as they are currently used by eco-toxicity categories (Jolliet et al. 2004)
		% of threatened vascular plant species in region (*problem-oriented/midpoint approach*)	Description of the land use interventions to render a possible link to empirical data (number of vascular plant species per square kilometre) (Müller-Wenk 1998)
Biotic production potential (Natural resources)		Surplus energy (*end-point/damage approach*)	Requirements to restore soil quality through addition of organic amendments and other soil fractions (clay, sand); other interventions may include gaseous emissions from organic amendments (Milà i Canals et al. 2006)
		Eroded soil (*problem-oriented/midpoint approach*)	Calculated with empirical models of the soil-erosion process (e.g. RUSLE, MUSLE, USLE, etc.), requiring slope gradient, rainfall intensity, vegetation cove, soil type, etc.

until such a consistent framework based on indicators such as biodiversity, biotic production potential and ecological soil quality, acceptable to all the stakeholders is established, the discussions relating to land use impacts in LCA will continue.

Abiotic Resource Depletion

In the context of LCA and the present chapter, abiotic resources are defined as 'raw materials or means for production or consumption activities' (Stewart and Weidema 2005). Depletion of abiotic resources implies that the resources are consumed through intended or unintended physical disintegration or dissipation. Depletion leads to the reduced availability of the corresponding type of resource for future generations. Though geologic studies indicate that the quantity of most of the abiotic resources accessible for humans is extremely high, some researchers consider the current reduction of the easily usable part of certain natural resources as not negligible.

By virtue of their inherent qualities (including strength, chemical properties, electrical conductivity), minerals and metals create value by meeting human needs. Therefore, it is logical to quantify the resource depletion in LCA. However, no definitive approaches for quantifying the effects in this impact category have yet been developed. Recently, there seems to be a broad consensus emerging on impact category indicators in LCIA and their significant environmental issues, but there is no agreement on how significant the problem of abiotic resource depletion is and to what extent it should be on the agenda of LCIA (Steen 2006). According to Steen, the existing methods for characterisation and weighting of abiotic resources appear to be based on the following four types of problem definitions: (1) mining cost will be a limiting factor, (2) collecting metals or other substances from low-grade sources is mainly an issue of energy, (3) scarcity is a major threat and (4) environmental impacts from mining and processing of mineral resources are the main problem.

Table 4 Impact assessment methods of abiotic resources (Adopted from Yellishetty et al. 2010)

Group	Assessment method	Empirical formula
Group 1	Aggregation of natural resource extraction on mass basis (Lindfors et al. 1995)	$ADP_i = \sum_i CF_i \times m_i$ where $CF_i = 1$
Group 2	Aggregation and assessment based on energy impacts, which are based on substitution of the current extraction process or improved future processes (Steen 1999)	$ADP_i = \sum_i WTP \times m_i$
Group 3	Aggregation and assessment based on the exergy or entropy content or change (Bösch et al. 2007)	$ADP_i = \sum_i CF_i \times m_i$ where $CF_i = Ex_{ch,i}$
Group 4	Aggregation and assessment based either on the quantity of resource that is ultimately available or the part of the reserve base that can be economically extracted and the extraction rate at the time of the assessment (Guinée and Heijungs 1995; Guinée 2002)	$CF_i = ADP_i = \frac{DR_i}{(R_i)^2} \times \frac{(R_{ref})^2}{DR_{ref}}$
Group 5	Aggregation and assessment based on the change in the anticipated environmental impact of the resource extraction process due to lower-grade deposits that have to be mined in the future (Goedkoop and Spriensma 2001; Müller-Wenk 1998)	$ADP_i = \sum_i CF_i \times m_i$ where $CF_i = SPE_i$

ADP_i abiotic depletion potential of resource i (kg Sb-equivents/kg resource i), CF_i characterisation factor for abiotic depletion of resource i, m_i mass of resource i consumed in the process, *WTP* is the willingness to pay to restore impacts to the 'safeguard subject', in this case abiotic stock resources, expressed in environmental load units (*ELU*, €), $Ex_{ch,i}$ chemical exergy of resource i (MJ/kg), DR_i the extraction rate of resource i (kg/year), R_i the ultimate reserve of resource i (kg), R_{ref} the ultimate reserve of the reference resource antimony (kg), DR_{ref} the extraction rate of the reference resource antimony (kg/year), SPE_i is the surplus energy (*MJ*) needed to extract 1 kg of a resource from a lower-grade ore

Impacts from resource use have been a prominent impact category in most environmental impact assessment methods for LCA developed since the early 1990s. Udo de Haes (1996) argues that resource management is viewed as crucial for LCA studies. In accordance with the general concerns of the broader public at the time, emphasis has been on abiotic resources, specifically on energy and metallic minerals and on the extraction stage of the minerals life cycle. The numerous methodologies proposed for the impact assessment of resource use were reviewed by the SETAC Working Group IA-2 (Lindeijer et al. 2002). Based on a suggestion by Finnveden (1996), the methodologies were categorised into four main approaches. Two out of four methodologies focus on current consumption, while the other two types focus on future consequences. The major deficiency in these methodologies has been their lacking of ability to adequately reflect the loss in functionality related to their use.

For LCIA, various proposals have been made for impact categories and category indicators for abiotic resources. Several reviews are available, such as Lindfors et al. (1995), Weidema (2000), Guinée (2002), SETAC-Europe, Finnveden (2005), Weidema et al. (2005), Brent and Hietkamp (2006), Ciroth and Becker (2006), and Steen (2006). In principle, there are four types of indicators based on the following: (1) energy or mass, (2) relation of use to deposits, (3) future consequences of resource extractions and (4) exergy consumption or entropy production. Based on the discussion presented by Guinée (2002) and Strauss et al. (2006), the abiotic resource depletion impact assessment methods may be characterised into six groups, as given in Table 4.

The assessment of the quantity of global metal stock available for a long-term perspective cannot be precise (Dubreuil 2008). The stock is depending on geological considerations, technology and economic aspects (Tilton and Lagos 2007; Tilton and Skinner 1987; Tilton 2003). An example of the current thinking on how to assess the mineral resources availability is given by Lambert et al. (2009).

In 'Limits to Growth', Meadows et al. (2005) identified that the increasing cost of resources is becoming a major problem for society during the first and second decade of the third millennium, and their ideas were accepted by some. While the Club of Rome's report made a lot of impact in the 1970, the analysis of Turner (2008) suggests in opposition to Meadows that non-fuel materials will not create resource constraints when substitution with other metals and materials is considered. Therefore, resource depletion may need to be considered in LCAs from the perspective of time, environmental and economic aspects of mineral extraction and future consequences of decreased availability of mineral resources for a region.

Given the consistent historical patterns of growing mineral production, it would be possible to complete similar models for other minerals and metals, and, excluding dramatic changes in technology or demands, it is possible to forecast the timeframes of their respective resource futures and potential depletion. At present, there are different schools of thought that have been proposed around the world within LCA regarding the time frame to consider when defining resource scarcity. One school of thought is that abiotic resources scarcity is perceived to be the decreasing concentrations of mineral resources at some time in the future, therefore it is an issue to be addressed in LCA (Steen 2006). Conversely, another school of thought argues that replacement technologies will address decreased minerals concentrations, and hence mineral depletion may not be an issue (Stewart and Weidema 2005).

Allocation Procedure for Recycling

Metals' recycling is an essential process that promotes material sustainability. Metal scrap that is collected for recycling is not a waste but constitutes valuable resource. In fact, metals are recycled to a large extent. Recycled metals can substitute or displace the use of primary metal and, as a consequence, the necessity to mine and produce new metal. The economic value of alloys is not determined by their recycled content but by their utility in a given application. In order to promote the benefits of recycling, two approaches are used – 'recycled content' and 'end of life'. In order to select the proper approach, it is important to understand the properties of the materials, be consistent with scientific knowledge and technical practices and reflect economic realities without creating market distortions that impede environmental objectives (Atherton 2007).

The recycled content approach for recycling is used in environmental labelling and is represented by a specific Mobius loop on a product as described in ISO 14021:1999. The recycled content approach is based on a waste diversion perspective. By informing the consumer, it creates demand for recycled materials (oil, plastic, paper, glass) that would otherwise be incinerated or landfilled. For metals, all feedstocks for recycling are used, and market stimulation would be ineffective (Atherton 2007).

The 'end-of-life' recycling approach is taking a product life cycle and material stewardship perspective in agreement with ISO 14044:2006. Accordingly, all material losses after the use phase have to be replaced by primary production. This approach supports the design improvements, the management of end-of-life products and material stewardship. This is the preferred approach by the metals industry (Atherton 2007).

Dubreuil et al. (2010a) presented a guidance document on the application of ISO 14044 (2006) in relation to the application of the allocation procedure. Since LCA takes a product perspective, the discussions and analysis place an emphasis on post-consumer products. Generic 'rules' for metal recycling maps are provided. This methodology has been applied to an automotive auto part made from three families of alloys (Dubreuil et al. 2010b).

Spatial and Temporal Issues in LCA

LCAs deal with complex, interwoven networks of industrial, commercial, household and waste management activities dispersed over many locations and spanning many decades (Guinée 2002). The mechanisms governing the dynamics are numerous, and the mathematics governing them is non-linear and dynamic in nature. Spatial variabilities emerge from the obvious fluctuations in the real-life scenarios in the world. Even though there are natural variations between geographical sites, the environmental interventions are summed up in the impact assessment irrespective of their space (Bjorklund 2002). On the other hand, variations over time are relevant in both the inventory and impact assessment as the process and factors in the receiving environment vary naturally over short and long time scales (Bjorklund 2002).

So far, little attention has been paid in the literature to temporal differentiation in LCA (Guinée 2002). Owens (1996, 1997) criticised this lack of spatial and temporal considerations in the LCAs. These studies mentioned rates of emissions, duration and frequency of exposure and seasonal influences of temperature and sunlight as missing characteristics of time. Pleijel et al. (1999) studied the influences of climate and the time of the day of NO_x and volatile organic carbon (VOC) releases on ozone formations in the troposphere in the context of LCA study. However, the biggest question yet to be addressed is how to incorporate this into a practical LCA methodology.

In recent years, the issue of spatial and temporal differentiation has been attracting attention (Potting and Hauschild 1997; Heijungs and Sleeswijk 1999; Bare et al. 1999; Schulze and Matthies 1999; Potting 2000; Bjorklund 2002). However, no comprehensive approach for integration of spatial differentiation into LCA has been agreed upon. In order for it to be incorporated into

Table 5 Various tools available to address spatial and temporal inconsistency issues in LCAs (Adopted from Yellishetty et al. 2009)

Sr no	Tools
1	Higher-resolution models
2	Sensitivity analysis
3	Uncertainty importance analysis
4	Classical statistical analysis
5	Bayesian statistical analysis
6	Scenario modelling

LCAs, space-specific equivalence factors must be established. Potting (2000) started developing such factors for acidification and human toxicity, although the study was restricted to Europe. Heijungs and Sleeswijk (1999) distinguish three dimensions in life cycle characterisation of toxic substances: fate, exposure and effect. The spatial differentiation of these variables may be performed independently. Potting and Hauschild (2006) argue that spatial differentiation is relevant for all non-global impact categories and further propose three levels of spatial differentiation such as (1) site generic (i.e. all sources are considered to contribute to the same generic receiving environment), (2) site dependent (i.e. some spatial differentiation is performed where classes of sources and their receiving environment are distinguished from each other) and (3) site specific (i.e. detailed spatial differentiation is performed by considering sources at specific location).

Huijbregts (1998) argued that spatial variability is introduced into LCA at two different phases that are during the inventory collection due to regional differences in emissions and also during the characterisation phase due to regional differences in environmental sensitivities. Similarly, the temporal variability is introduced in the inventory phase due to differences in yearly emission inventories and in the characterisation phase due to the different time horizons and due to changes in environmental characteristics over time. Table 5 provides an overview of the tools and approaches to address the spatial and temporal variabilities within LCA (Huijbregts 1998).

The disposal stage of residual and end-of-life products is often overlooked in LCA. In the ecoinvent methodology, the time frame for

Table 6 Integration over time (20 years and infinite time) of the relative toxic impact of direct emissions of 1 kg of substances to agricultural soil and to seawater on the terrestrial, marine and sediment

Direct emissions to	Agricultural soil		Seawater		Seawater	
Relative impact on ecosystems	Terrestrial eco-toxicity		Marine eco-toxicity		Sediment eco-toxicity	
	Time horizon (year)		Time horizon (year)		Time horizon (year)	
Substances	20	Infinite	20	Infinite	20	Infinite
Pesticides						
Aldicrab	4.2E+03	4.2E+03	1.5E+04	1.5E+04	2.2E+04	2.2E+04
Atrazine	6.6E+00	6.6E+00	6.0E+02	6.0E+02	6.6E+02	6.6E+02
Glyphosate	9.6E–02	9.6E–02	3.3E+01	3.3E+01	3.0E+01	3.0E+01
Malathion	7.6E–02	7.6E–02	5.1E+03	5.1E+03	2.8E+03	2.8E+03
Simazine	2.9E+01	2.9E+01	6.7E+02	6.7E+02	1.0E+03	1.0E+03
Natural organic molecules						
Benzo[a]pyrene	2.3E+01	2.3E+01	1.2E+05	1.2E+05	3.7E+05	3.7E+05
2,3,7,8-TCDD (dioxin)	2.7E+04	2.7E+04	4.4E+08	5.0E+08	1.6E+09	1.9E+09
Elements						
Arsenic	6.7E+00	3.3E+03	4.0E+02	3.4E+05	8.1E+02	3.4E+05
Cadmium	5.1E+00	1.7E+02	5.8E+03	1.8E+06	1.5E+04	1.9E+06
Copper	3.1E–01	1.4E+01	4.0E+03	1.5E+06	8.8E+03	1.5E+06
Lead	5.4E–02	3.3E+01	3.7E+01	1.1E+04	1.0E+02	1.2E+04
Nickel	3.2E+00	2.4E+02	5.7E+03	5.8E+06	1.1E+04	5.7E+06
Selenium	2.3E+00	1.1E+02	2.4E+03	2.9E+07	2.1E+03	1.2E+07
Zinc	3.6E–01	2.5E+01	3.5E+02	1.1E+05	8.6E+02	1.1E+05

'uranium mining and milling sites' is 80,000 years (Frischknecht et al. 2005). In regard to the mineral and metal sector, LCA inventory rarely models the emissions from repositories of tailing, slags and process residues. In the ecoinvent methodology, the time frame for 'uranium mining and milling sites' is 80,000 years (Frischknecht et al. 2005). The long-term emissions of Cd^{+2} and Cu^{+2} from slag dump were assessed by Hellweg et al. (2005).

For example, it is quantified as a flow of material directed to the landfill, and the modelling of releases to the environment is absent. It has been proposed that emissions from landfill to the environment are split between (1) short-term (1,000 years) and (2) long-term (60,000 years) time horizon.

In LCA, the toxicity impact score is a product of the relative toxicity by the integration of the concentration over an infinite time. Usually in LCA, impacts are integrated with an infinite time horizon. Organic substances, including pesticides and synthesis molecules, have a limited persistence in the environment: they degrade. But metals are elements; their concentration in a given media can be reduced only by a transfer to another media.

Table 6 has been constructed using data compiled by Guinée (2002) for the multi-compartment model USES-LCA. The cumulative eco-toxicity impacts for 20 years and for an infinite time are presented for several pesticides, natural substances and elements. According to the model, the substances are emitted as a pulse of 1 kg to the agricultural soil and to the seawater compartments. The environmental concentration is attenuated in function of the time due to the degradation of the substances and to their transport in other environmental compartments. The potential environmental impact is a function of the integration of the environmental concentration over the time.

Due to their degradation, it is reasonable to expect that for pesticides, the toxic impact occurs within 20 years. The same phenomenon is also observed with natural organic substances. Even dioxin is almost completely decomposed after 20 years. On the other hand, for metals, the integration time for the impact makes a huge difference. Two to three orders of magnitudes in the toxicity potential are calculated for emissions to agricultural soil. For the potential impact on the sediment eco-toxicity, the selection for the time of integration has even more importance. The fate modelling of metals can introduce huge uncertainty because the time of integration of the impacts is infinite.

The temporal question raises ethical challenge: can environmental impacts be discounted? That question has been discussed in the LCA context by Hellweg et al. (2003) and Hellweg and Frischknecht (2004). While discounting is a common feature in finance and economics, applying a discount rate on environmental impact will unfairly affect future generations. How should those generations be compensated for the degradation of their environment? In the end, in 2003, there was no consensus if short-term impacts and long-term impact should be considered with equal weighting. How should the impact of pollutants that will occur during the next thousands of years be considered? How can a cost-benefit analysis be applied if a reasonable discount rate cannot be used? The question of long-term impact should be studied with the participation of environmental economists. How should inter-generational impact transfer be assessed[1]?

Due to the absence of spatial and temporal differentiation in LCA, prediction of environmental concentrations and bio-availability of hazardous elements becomes difficult. Consequently, it becomes difficult to evaluate whether a no-effect level is exceeded. Usually, most LCIA methods make only limited use of spatial and temporal information because they predict concentration increases rather than full concentrations. This is one of the reasons why some LCA studies show poor accordance between predicted and actual environmental impacts at local level. Consequently, LCA cannot make a judgement whether or not the threshold is exceeded, particularly for human toxicity assessment; it only considers that less is better than more. This character is a feature of LCA. The reliability and validity of LCA results can greatly be improved by introducing the spatial and temporal differentiation. Although the location and time-specific data are rarely available for all the processes within a product life cycle, the spatially differentiated assessment may be preferable for those processes for which the required information is available. Therefore, if the necessary resources and funds were available, it is recommended that efforts be directed towards the development of time-specific and location-specific databases. Also, it is important to consider developing a time-dependent analysis LCA model, where the functional unit can be altered with time.

Discussion

To demonstrate the uncertainties that prevail in the current LCIA methodology, a comparison of impacts calculated using three different methods is presented. The inventory data for Australian iron ore from Norgate and Haque (2010) has been used and are shown in Tables 7, 8, and 9 (calculated using the SimaPro 7.1.8 software). It may be possible to use a particular impact method with particular impact categories for any LCA study depending on the purpose of the study and the geographical location. However, as mentioned before, the results from an LCA study would be different for various reasons. The different system boundaries, input inventory data and assumptions are the main contributing factors for such differences. If the same impact category is expressed in different units, the absolute numbers would logically be different, as evident from Tables 7, 8, and 9. For example, in Table 7, carcinogens are expressed in DALY (disability-adjusted life years) and kilogram of chloroethylene equivalent into air with different results (i.e. $2.7 \cdot 10^{-8}$ and $7.6 \cdot 10^{-3}$, respectively). In comparison, for mineral extraction and minerals impact category, the unit is MJ surplus (i.e. additional energy requirement to compensate lower future ore grade), and the value is $1.7 \cdot 10^{-3}$ MJ, same for any impact method used here. Even the contribution of each unit on various impact categories would be

[1] Comment: it is already complex for two generations in the current strict monetary condition while the models are quite robust. This is a case of soft sustainability, that is, a transfer of natural capital to social and economic. The social and economic developments require natural resources including minerals and metals. One set of capital (environment) is reduced, while others are increased. Overall, we need to ensure that we can satisfy 'the needs of the present without compromising the ability of future generations to meet their own needs'. World Commission on Environment and Development (WCED). *Our Common Future*. 1987.

Table 7 Comparison of impact methods for Australian iron ore (per tonne) (Source: Norgate and Haque 2010)

		Impact methods		
Impact category	Unit	Australian impact method	Eco-indicator 99 (E)	IMPACT 2002+
Acidification/eutrophication	PDF*m^{2}year		0.13	
Aquatic acidification	kg SO_2			0.041
Terrestrial acid/nutri	kg SO_2			0.13
Aquatic eutrophication	kg PO_4 P-lim			7.4E−5
Eutrophication	kg PO_4 eq	0.0025		
Aquatic eco-toxicity	kg TEG water			4931.9
Terrestrial eco-toxicity	kg TEG soil			10.82
Carcinogens	DALY	2.7E−8	2.7E−8	
Carcinogens	kg C_2H_3Cl			0.0076
Climate change	DALY		2.66E−6	
Global warming	kg CO_2 eq	12.65		
Land occupation	m^{2}org.arable			0.0015
Land use	Ha a	2.5E−6		
Land use	PDF*m^{2}year	0.0099		
Mineral extraction/minerals	MJ surplus	0.0017	0.0017	0.0017
Non-carcinogens	kg C_2H_3Cl			0.72
Non-renewable energy	MJ primary			197.35
Ozone layer	DALY		0	
Ozone layer depletion	kg CFC-11			0
Photochemical oxidation	kg C_2H_4	0.0013		
Radiation	DALY		0	
Ionising radiation	Bq C-14			0
Respiratory inorganics	DALY		5.4E−6	
Respiratory inorganics	kg $PM_{2.5}$			0.0078
Respiratory organics	DALY		6.2E−9	
Respiratory organics	kg ethylene			0.0022
Solid waste	kg	1,300		
Water use	kL water	0.216		

PDF potentially disappeared fraction, *DALY* disability-adjusted life years, *PM* particulate matters, *m^{2}org.arable* m^2 equivalent organic arable land, *TEG* triethylene glycol, *P-lim* in P-limited water, *Ha a* = hectare. annum

Table 8 Using ReCiPe 2008 impact methods for Australian iron ore (per tonne) (Data source: Norgate and Haque 2010)

Impact	Midpoint unit	Midpoint values	End-point unit	End-point values
Climate change/human health			DALY	1.5E−05
Climate change/climate change ecosystem	kg (CO_2 to air)	12.9	species·year	1.1E−07
Ozone depletion	kg (CFC-11 to air)	3.2E−17	DALY	0.0E+00
Terrestrial acidification	kg (SO_2 to air)	0.018	species·year	2.7E−11
Freshwater eutrophication	kg (P to freshwater)	0	species·year	0.0E+00
Marine eutrophication	kg (N to freshwater)	0.0007	species·year	0.0E+00
Human toxicity	kg (14DCB to urban air)	0.799	DALY	5.6E−07
Photochemical oxidant formation	kg (NMVOC to air)	0.022	DALY	8.6E−10

(continued)

Table 8 (continued)

Impact	Midpoint unit	Midpoint values	End-point unit	End-point values
Particulate matter formation	kg (PM10 to air)	0.0059	DALY	1.5E–06
Terrestrial eco-toxicity	kg (14DCB to industrial soil)	0.0002	species·year	2.0E–11
Freshwater eco-toxicity	kg (14DCB to freshwater)	0.042	species·year	3.3E–11
Marine eco-toxicity	kg (14DCB to marine water)	0.507	species·year	4.3E–12
Ionising radiation	kg (U-235 to air)	0	DALY	0.0E+00
Agricultural land occupation	m^2.a (agricultural land)	0	species·year	0.0E+00
Urban land occupation	m^2.a (urban land)	0.012	species·year	2.0E–10
Natural land transformation	m^2 (natural land)	0.0001	species·year	1.8E–10
Water depletion	m^3 (water)	0.0011	$ (increased cost)	0.0E+00
Metal depletion	kg (Fe)	0.0011	$ (increased cost)	7.9E–05
Fossil depletion	kg (oil)	1.0E–10	$ (increased cost)	7.3E–10

CFC-11 chlorofluorocarbon, *NMVOC* non-methane volatile organic carbon compound, *14DCB* 1,4 dichlorobenzene, *m^2.a* square meter.annum

Table 9 Using EDIP 2003 and CML 2001 impact methods for Australian iron ore (per tonne) (Data source: Norgate and Haque 2010)

EDIP 2003 1.02 version, Danish LCA methodology			CML 2001 baseline with Australian toxicity factor		
Impact category	Unit	Value	Impact category	Unit	Value
Global warming 100a	kg CO_2 eq	12.68	Abiotic depletion	kg Sb eq	0.10
Ozone depletion	kg CFC11 eq	0	Global warming (GWP100)	kg CO_2 eq.	12.65
Ozone formation (vegetation)	m^2.ppm.h	36.99	Ozone layer depletion (ODP)	kg CFC-11 eq	0
Ozone formation (human)	person.ppm.h	0.0025	Human toxicity	kg 1,4-DB eq	0.79
Acidification	m^2	0.3627	Fresh water aquatic eco-toxicity	kg 1,4-DB eq	0.17
Terrestrial eutrophication	m^2	0.4573	Marine aquatic eco-toxicity	kg 1,4-DB eq	8982
Aquatic eutrophication EP(N)	kg N	0.0017	Terrestrial eco-toxicity	kg 1,4-DB eq	0.009
Aquatic eutrophication EP(P)	kg P	0	Photochemical oxidation	kg C_2H_4	0.0013
Human toxicity air	m^3	200,945	Acidification	kg SO_2 eq	0.042
Human toxicity water	m^3	25.37	Eutrophication	kg PO_4 eq	0.0025
Human toxicity soil	m^3	0.54			
Eco-toxicity water chronic	m^3	1,968.53			
Eco-toxicity water acute	m^3	243.95			
Eco-toxicity soil chronic	m^3	0.75			
Hazardous waste	kg	4.0E–11			
Slags/ashes	kg	7.0E–13			
Bulk waste	kg	1,300			
Radioactive waste	kg	0			
Resources (all)	kg	3.5E–07			

similar using different methods, but it is difficult to draw any meaningful conclusion from an LCA study when different impact methods are used. The possibility of duplication of the same impact category expressed in different units remains, and the utility of the large number of impact methods is questionable. This dilemma stresses the need for developing some uniformity for an impact category or a group of impact categories and their units in LCA studies whether using a midpoint- or end-point-based method. From the LCA point of view, a midpoint is also known as 'definition of category indicators close to environmental interventions', whereas an end point refers to 'definition of category indicators close to areas of protection'.

Conclusions

In this chapter, we have conducted a critical review of LCA literature in general and literature related to mineral and metal sectors in particular with respect to abiotic resource depletion, land use impacts and spatial and temporal issues within the LCA. It was found that there has been enormous debate happening within the LCA community over these contentious issues, yet the methodology remains in a constant flux. Though LCA is being widely applied and extensively used within minerals and metal industries for production and product improvement, the power of LCA would increase significantly if it were validated by fixing these gaps in current methodology and knowledge. It is imperative that the scientific community addresses the issues discussed in this chapter to foster the process of acceptability and applicability in the minerals and metal sector. An example case has been presented for Australian iron ore using SimaPro based on published inventory data to demonstrate that uniformity is required in LCA studies.

The issues discussed in this chapter are of interest to LCA in general and important to the mining and metals industries in particular. Additionally, these issues have not been adequately addressed in LCAs in their current form. These issues also bring about inconsistency in LCA practice and hence the results, their interpretation and utilisation.

References

Althaus H-J, Classen M (2005) Life cycle inventories of metals and methodological aspects of inventorying material resources in ecoinvent. Int J Life Cycle Assess 10(1):43–49

Atherton J (2007) Declaration by the metals industry on recycling principles. Int J Life Cycle Assess 12(1):59–60

Bare JC, Pennington DW, Udo de Haes HA (1999) Life cycle impact assessment sophistication-international workshop. Int J Life Cycle Assess 5:299–306

Bauer C, Zapp P (2005) Generic characterisation factors for land use and water consumption. In: Dubreuil A (ed) Life cycle assessment of metals – issues and research directions. SETAC USA, Pensacola, pp 147–152

Bösch, M.E., Hellweg, S., Huijbregts, M.A.J., Frischknecht, R., 2007. Applying cumulative exergy demand (CExD) indicators to the ecoinvent database. Int. J.LCA12,181e190

Bjorklund AE (2002) Survey of approaches to improve reliability in LCA. Int J Life Cycle Assess 7(2):62–72

Brent AC, Hietkamp S (2006) The impact of mineral resource depletion. Int J Life Cycle Assess 11(5): 361–362

Brentrup F, Küsters J, Lammel J, Kuhlmann H (2002a) Impact assessment of abiotic resource consumption – conceptual considerations. Int J Life Cycle Assess 7:301–307

Brentrup F, Küsters J, Lammel J, Kuhlmann H (2002b) Life cycle impact assessment of land use based on the Hemeroby concept. Int J Life Cycle Assess 7(6): 339–348

Ciroth A, Becker H (2006) Validation-The missing link in life cycle assessment towards pragmatic LCAs. Int J Life Cycle Assess 11(5):295–297

Commonwealth of Australia (2007) Leading practice sustainable development program for the mining industry – biodiversity management. Department of Industry, Tourism and Resources, Australian Government Press Canberra, 88 p

Diamond ML, Gandhi N, Adams WJ, Atherton J, Bhavsar SP, Bulle C, Campbell PGC, Dubreuil A, Fairbrother A, Farley K, Green A, Guinee J, Hauschild MZ, Huijbregts MAJ, Humbert S, Jensen KS, Jolliet O, Margni M, McGeer JC, Peijnenburg WJGM, Rosenbaum R, van de Meent D, Vijver MG (2010) The clearwater consensus: the estimation of metal hazard in fresh water. Int J Life Cycle Assess 15(2):143–147

Dubreuil A (ed) (2005) Life cycle assessment of metals: issues and research directions. Society of Environmental Toxicology and Chemistry (SETAC), Pensacola, 307 p

Dubreuil A (2008) The dynamics of mineral reserves – The impact of demand and technology on the supply of minerals and metals. OECD-UNEP resource efficiency conference, Module 4 sustainable mining and recycling, 23–25 April 2008, Paris, France. http://www.oecd.org/dataoecd/13/14/40798864.pdf. Accessed 3 Jan 2012

Dubreuil A, Young SB, Atherton J, Gloria TP (2010a) Metals recycling maps and allocation procedures in life cycle assessment. Int J Life Cycle Assess 15(6):621–634

Dubreuil A, Bushi L, Das S, Tharumarajah A, Xianzheng G (2010b) A comparative life cycle assessment of magnesium front end auto parts. Society of Automotive Engineers (SAE) International publication number 2010-01-0275, 19 p

European Commission (2010) ILCD handbook, general guidance for life cycle assessment – detailed guidance. Publications Office of the European Union, Luxembourg, 414 p

Finnveden G (1996) Resources and related impact categories, part II. In: Udo de Haes HA et al (eds) Towards a methodology for life cycle impact assessment. Report of the SETAC-Europe working group on Life Cycle Impact Assessment (WIA), SETAC-Europe, Brussels, Belgium

Finnveden G (2005) The resource debate needs to continue. Int J Life Cycle Assess 10(5):372

Frischknecht R, Jungbluth N, Althaus H-J, Doka G, Dones R, Heck T, Hellweg S, Hischier R, Nemecek T, Rebitzer G, Spielmann M (2005) The ecoinvent database: Overview and methodological framework. Int J Life Cycle Assess 10(1):3–9

Giudice F, La Rosa G, Risitano A (eds) (2006) Product design for environment – a life cycle approach. Taylor & Francis Publishers, Boca Raton, 87 p

Goedkoop, M., Spriensma, R., 2001. The Eco-Indicator 99dA Damage Oriented Method for Life Cycle Impact Assessment. Pré Consultants, Amersfoort, The Netherlands

Guinée JB (ed) (2002) Handbook of life cycle assessment–operational guide to the ISO standards. Kluwer Academic Publishers, Dordrecht

Guinée, B.J., Heijungs, R., 1995. A proposal for the definition of resource equivalency factors for use in product life-cycle assessment. Environ. Toxicol. Chem. 14, 917e925

Hamilton M, Brown M, Nolan G (2008) Comparing bio-diversity impacts of timber and other building materials. Forests and Wood Products Australia Ltd., Melbourne, 62 p

Heijungs R, Sleeswijk AW (1999) The structure of impact assessment: mutually independent dimensions as a function of modifiers. Letters to the editor: comment and reply. Int J Life Cycle Assess 4(1):2–3

Hellweg S, Frischknecht R (2004) Evaluation of long-term impacts in LCA. Int J Life Cycle Assess 9(5):339–340

Hellweg S, Hofstetter TB, Hungerbühler K (2003) Discounting and the environment: should current impacts be weighted differently than impacts harming future generations? Int J Life Cycle Assess 8(1):8–18

Hellweg S, Hofstetter TB, Hungerbühler K (2005) Time-dependent life-cycle assessment of slag landfills with the help of scenario analysis: The example of Cd and Cu. J Clean Prod 13(3):301–320

Huijbregts M (1998) Application of uncertainty and variability in LCA. Part I: A general framework for the analysis of uncertainty and variability in life cycle assessment. Int J Life Cycle Assess 3(5):273–280

Huppes G (1996) LCA yesterday, today and tomorrow. Centre of Environmental Science, Leiden University, Leiden, pp 29–49

ISO 14021 (1999) Environmental labels and declarations – Self-declared environmental claims (Type II environmental labelling). International Organization for Standardization, ISO, Geneva

ISO 14040 (2006) Environmental management – Life cycle assessment – Principles and framework. International Organization for Standardization, ISO, Geneva

ISO 14044 (2006) Environmental management – Life cycle assessment – Requirements and guidelines. International Organization for Standardization, ISO, Geneva

ISO/TS 14048 (2002) Environmental management – Life cycle assessment – Data documentation format. International Organization for Standardization, ISO, Geneva

ISO/TR 14049 (2000) Environmental management – Life cycle assessment – Examples of application of ISO 14041 to goal and scope definition and inventory analysis. International Organization for Standardization, ISO, Geneva

ISO/TR 14047 (2003) Environmental management – Life cycle impact assessment – Examples of application of ISO 14042. International Organization for Standardization, ISO, Geneva

Jolliet O, Müller-Wenk R, Bare J, Brent A, Goedkoop M, Heijungs R, Itsubo N, Peña C, Pennington D, Potting J, Rebitzer G, Stewart M, Udo de Haes H, Weidema B (2004) The LCIA midpoint-damage framework of the UNEP/SETAC life cycle initiative. Int J Life Cycle Assess 9(6):394–404

Köllner T (2000) Species-pool effect potentials (SPEP) as a yardstick to evaluate land-use impacts on biodiversity. J Clean Prod 8(4):293–311

Lambert IB, Miezitis Y, McKay AD (2009) Australia's national classification system for identified mineral resources and its relationships with other systems. AusIMM Bull 6:52–56

Lindeijer E, Müller-Wenk R, Steen B (2002) Impact assessment of resources and land use. Chapter 2. In: de Haes HA Udo, Finnveden G, Goedkoop M, Hauschild M, Hertwich E, Hofstetter P, Jolliet O, Klöpffer W, Krewitt W, Lindeijer E, Müller-Wenk R, Olsen I, Pennington D, Potting J, Steen B (eds) Life-cycle impact assessment: striving towards best practice. SETAC Press, Pensacola

Lindfors LG, Christiansen K, Hoffman L, Virtanen Y, Juntilla V, Hanssen OJ, Rønning A, Ekvall T, Finnveden G (1995) Nordic guidelines on life-cycle assessment. Nord 1995:20. Nordic Council of Ministers, Copenhagen

Meadows D, Randers J, Meadows D (2005) Limits to growth – the 30 year update. Earthscan, London

Mila i Canals L, Clift R, Basson L, Hansen Y, Brandao M (2006) Expert workshop on land use impacts in life cycle assessment. Int J Life Cycle Assess 11(5):363–368

Mila i Canals L, Bauer C, Depestele J, Dubreuil A, Knuchel RF, Gaillard G, Michelsen O, Muller-Wenk R, Rydgren B (2007) Key elements in a framework for land use impact assessment within LCA. Int J Life Cycle Assess 12(1):5–15

Moncur MC, Jambor JL, Ptacek CJ, Blowes DW (2009) Mine drainage from the weathering of sulfide minerals and magnetite. Appl Geol 24(12):2362–2373

Müller-Wenk R (1998) Land use – the main threat to species. How to include land use in LCA. IWÖ–Diskussionsbeitrag No. 64, IWÖ, Universität St. Gallen, Switzerland

Norgate TE, Haque N (2009) Using life cycle analysis to assess the toxicity impacts of waste streams from metal production processes. In: The proceedings of SDIMI 2009, sustainable development indicators for the minerals industry, 6–8 July 2009, Gold Coast, Australia

Norgate T, Haque N (2010) Energy and greenhouse gas impacts of mining and mineral processing operations. J Clean Prod 18(3):266–274

Norgate T, Jahanshahi S, Rankin WJ (2007) Assessing the environmental impact of metal production processes. J Clean Prod 15(8–9):838–848

Oberholzer H-R, Weisskopf P, Gaillard G, Weiss Fr, Freiermuth R (2006) Life cycle impact assessment method for the impact of agricultural activities on soil quality. Agroscope FAL Reckenholz (in German with English summary), 58 pp and annexes

Owens W (1996) LCA impact assessment categories. Technical feasibility and accuracy. Int J Life Cycle Assess 1(3):151–158

Owens W (1997) Life cycle assessment. Constraints on moving from inventory to impact assessment. J Ind Ecol 1(1):37–49

Pennington DW, Chomkhamsri K, Pant R, Wolf M-A, Bidoglio G, Kogler K, Misiga P, Sponar M, Lorz B, Sonnemann G, Masoni P, Wang H, Ling L, Castanho C, Soon CS, Fieschi M, Filareto A, Hauschild M (2010) ILCD handbook public consultation workshop – international reference life cycle data system (ILCD). Int J Life Cycle Assess 15(3):231–237

Pleijel K, Pleijel H, Karlsson GP (1999) Quantification of additional ozone load from local source emissions of NOx and VOC. Poster presented at the 9th annual meeting of SETAC-Europe, 25–29 May 1999, Liepzig

Potting JMB (2000) Spatial differentiation in life cycle assessment. PhD thesis, University of Utrecht, Utrecht

Potting JMB, Hauschild M (1997) Spatial differentiation in life cycle assessment via the site dependent characterisation of environmental impact from emissions. Int J Life Cycle Assess 2(4):209–216

Potting JMB, Hauschild M (2006) Spatial differentiation in life cycle assessment: a decade of method development to increase the environmental realism of LCIA. Int J Life Cycle Assess 11(Special issue (1)):11–13

Price WA (2009) Prediction manual for drainage chemistry for sulphidic geologic materials. CANMET Mining and Mineral Sciences Laboratories, Natural Resources Canada, Ottawa, 579 pp. http://www.abandoned-mines.org/pdfs/MENDPredictionManual-Jan05.pdf. Accessed March 2010

Schenck RC (2001) Land use and biodiversity indicators for life cycle impact assessment. Int J Life Cycle Assess 6(2):114–117

Schulze C, Matthies M (1999) Introducing spatial and temporal variations in the LCIA of household laundry. In: 9th annual meeting of SETAC-Europe, 25–29 May, Germany. Quality of life and environment in cultured landscapes. Abstracts, 86 p

Steen, B.A., 1999. A Systematic Approach to Environmental Priority Strategies in Product Development (EPS), Version 2000dModels and Data. Chalmers University of Technology, Göteborg, Sweden. Centre for Environmental Assessment of Products and Material Systems (CPM) Report 1999(5).

Steen BA (2006) Abiotic resource depletion: different perceptions of the problem with mineral deposits. Int J Life Cycle Assess 11(Spec Issue (1)):49–54

Stewart M, Weidema B (2005) A consistent framework for assessing the impacts from resource use. A focus on resource functionality. Int J Life Cycle Assess 10(4):240–247

Strauss K, Brent AC, Hietkamp S (2006) Characterisation and normalisation factors for life cycle impact assessment of mined abiotic resources categories in South Africa. Int J Life Cycle Assess 11(3):162–171

Tilton JE (2003) On borrowed time? Assessing the threat of mineral depletion. Resources for the Future, Washington, DC

Tilton JE, Lagos G (2007) Assessing the long-run availability of copper. Res Policy 32(1–2):19–23

Tilton JE, Skinner BJ (1987) The meaning of resources. In: Resources and world development. Wiley, Ney York, pp 13–27

Turner GMA (2008) A comparative of the limits to growth with 30 years of reality. Glob Environ Chang 18(3):397–411

Udo de Haes HA (ed) (1996) Towards a methodology for lifecycle impact assessment. Report of the SETAC Europe first Working Group on Life-Cycle Impact Assessment. SETAC-Europe, Brussels, Belgium

Weidema B (2000) Can resource depletion be omitted from environmental impact assessment? Poster presented at SETAC world congress, 21–25 May 2000, Brighton, UK

Weidema BP, Lindeijer E (2001) Physical impacts of land use in product life cycle assessment. Final report of the EUROENVIRON-LCAGAPS sub-project on land use. Technical University of Denmark, Denmark

Weidema B, Finnveden G, Stewart M (2005) Impacts from resource use – a common position paper. Int J Life Cycle Assess 10(6):382

Yellishetty M, Ranjith PG, Tharumarajah A, Bhosale S (2009) Life cycle assessment in the minerals and metals sector: a critical review of selected issues and challenges. Int J Life Cycle Assess 14(3):257–267

Yellishetty M, Mudd G, Ranjith PG (2011) The steel industry, abiotic resource depletion and life cycle assessment: A real or perceived issue? Journal of Cleaner Production, 19, 78–90

Young SB, Brady K, Fava J, Saur S (2001) Eco-efficiency and materials. International Council on Metals and the Environment (ICME), Ottawa, pp 86. ISBN 1-895720-34-6

Index

R. Sinding-Larsen and F.-W. Wellmer (eds.), *Non-Renewable Resource Issues: Geoscientific and Societal Challenges*, International Year of Planet Earth,
DOI 10.1007/978-90-481-8679-2, © Springer Science+Business Media B.V. 2012

MIX
Papier aus verantwortungsvollen Quellen
Paper from responsible sources
FSC® C105338

If you have any concerns about our products,
you can contact us on
ProductSafety@springernature.com

In case Publisher is established outside the EU,
the EU authorized representative is:
Springer Nature Customer Service Center GmbH
Europaplatz 3, 69115 Heidelberg, Germany

Printed by Libri Plureos GmbH
in Hamburg, Germany